PROGRESS IN DAIRY SCIENCE

Progress in Dairy Science

Edited by

C.J.C. Phillips

Department of Clinical Veterinary Medicine
University of Cambridge, UK

CAB INTERNATIONAL

CAB INTERNATIONAL
Wallingford
Oxon OX10 8DE
UK

Tel: +44 (0)1491 832111
Fax: +44 (0)1491 833508
E-mail: cabi@cabi.org
Telex: 847964 (COMAGG G)

A catalogue record for this book is available from the British Library.

ISBN 0 85198 974 8

Typeset in 10/12 pt Plantin by Colset Pte Ltd, Singapore
Printed and bound in the UK by Biddles Ltd, Guildford

Contents

Contributors

R.J. Baer, *Dairy Science Department, Minnesota-South Dakota Dairy Foods Research Center, South Dakota State University, Brookings, South Dakota 57007-0647, USA.*

P.B.M. Berentsen, *Department of Farm Management, Wageningen Agricultural University, Hollandseweg 1, 6706 KN Wageningen, The Netherlands.*

H. Bovenhuis, *Department of Animal Breeding, Wageningen Institute of Animal Sciences, PO Box 338, 6700 AH Wageningen, The Netherlands.*

B. Bravo-Ureta, *Department of Agricultural and Resource Economics, University of Connecticut, Storrs, Connecticut 06269, USA.*

P.C. Chiy, *Department of Clinical Veterinary Medicine, University of Cambridge, Madingley Road, Cambridge CB3 0ES, UK.*

Gemeda, Takele, *Institute of Agricultural Research (IAR), PO Box 2003, Addis Ababa, Ethiopia.*

G.W.J. Giesen, *Department of Farm Management, Wageningen Agricultural University, Hollandseweg 1, 6706 KN Wageningen, The Netherlands.*

P.R. Greenough, *Department of Veterinary Anaesthesiology, Radiology and Surgery, Western College of Veterinary Medicine, University of Saskatchewan, Saskatoon, Saskatchewan S7N 0W0, Canada.*

J.E. Hillerton, *Institute of Animal Health, Compton, Berkshire RG16 0NN, UK.*

J.S. Hogan, *Ohio Agricultural Research and Development Center, Department of Dairy Science, Ohio State University, 1680 Madison Avenue, Wooster, Ohio 44691–4096, USA.*

J.T. Huber, *Department of Animal Sciences, University of Arizona, Tucson, Arizona 85719, USA.*

C.H. Knight, *Hannah Research Institute, Ayr KA6 5HL, UK.*

A. Kuipers, *Research Station for Cattle, Sheep and Horse Husbandry (PR), Runderweg 6, 8219 PK Lelystad, The Netherlands.*

P.J. l'Huillier, *Laboratoire de Génétique Biochimique et de Cytogénétique, INRA-CRJ, 78352 Jouy-en-Josas Cedex, France.* Permanent Address: *AgResearch, Ruakura Agricultural Centre, Hamilton, New Zealand.*

T.B. Mepham, *Centre for Applied Bioethics, Faculty of Agricultural and Food Sciences, University of Nottingham, Sutton Bonington Campus, Loughborough LE12 5RD, UK.*

J.D. Oldham, *Genetics and Behavioural Sciences Department, Scottish Agricultural College, Bush Estate, Penicuik, Midlothian EH26 0QE, UK.*

M. Peaker, *Hannah Research Institute, Ayr KA6 5HL, UK.*

C.J.C. Phillips, *Department of Clinical Veterinary Medicine, University of Cambridge, Madingley Road, Cambridge CB3 0ES, UK.*

W. Rossing, *Institute for Agricultural and Environmental Engineering (IMAG-DLO), Mansholtlaan 10–12, 6708 PA Wageningen, The Netherlands.*

K.L. Smith, *Ohio Agricultural Research and Development Center, Department of Dairy Science, Ohio State University, 1680 Madison Avenue, Wooster, Ohio 44691–4096, USA.*

E. Strandberg, *Department of Animal Breeding and Genetics, Swedish University of Agricultural Sciences, PO Box 7023, S-75007 Uppsala, Sweden.*

Tegegne, Azage, *International Livestock Research Institute (ILRI), PO Box 5689, Addis Ababa, Ethiopia.*

J.A.M. Van Arendonk, *Department of Animal Breeding, Wageningen Institute of Animal Sciences, PO Box 338, 6700 AH Wageningen, The Netherlands.*

J.L. Vilotte, *Laboratoire de Génétique Biochimique et de Cytogénétique, INRA-CRJ, 78352 Jouy-en-Josas Cedex, France.*

W.P. Weiss, *Ohio Agricultural Research and Development Center, Department of Dairy Science, Ohio State University, 1680 Madison Avenue, Wooster, Ohio 44691–4096, USA.*

R.F. Weller, *Institute of Grassland and Environmental Research, Trawsgoed, Aberystwyth, Dyfed SY23 41L, UK.*

C.J. Wilde, *Hannah Research Institute, Ayr KA6 5HL, UK.*

Wold, Alemu Gebre, *Institute of Agricultural Research (IAR), PO Box 2003, Addis Ababa, Ethiopia.*

J.E. Womack, *Department of Veterinary Pathobiology, College of Veterinary Medicine, Texas A&M University, College Station, Texas 77843–4467, USA.*

E. Zerbini, *International Livestock Research Institute (ILRI), PO Box 5689, Addis Ababa, Ethiopia.*

S.A. Zinn, *Department of Animal Science, University of Connecticut, Storrs, Connecticut 06269, USA.*

Introduction

In 1987 a conference was held in Bangor, North Wales, UK, to review recent developments in diary (and beef) science that were likely to have an impact on the dairy (and beef) industry in the foreseeable future (Phillips, 1989). Nearly a decade later, this book also attempts to review progress in dairy science, but since the first review the direction of research in dairy science has altered fundamentally. In the 1980s research was still concentrated on improving the efficiency of milk production by addressing the different ways of managing dairy production systems. Some would argue that such an integrated, 'systems' approach is still necessary. Undoubtedly, though, the main thrust of dairy research today is in improving the efficiency of dairy production by an understanding and control of the fundamental biological processes that regulate milk production systems. This research is focused on the control of milk production by the cow (e.g. Chapters 4 and 15), but also includes milk quality (Chapter 12) and nutrient utilization (Chapters 1 and 2). Already this knowledge is being utilized in some parts of the world to control and stimulate milk secretion by, for example, recombinant bovine somatotrophin (rbST).

The research in control of milk production is assuming new importance as the cattle production industry loses its integrated function, with dairy and beef production increasingly occupying separate specialized niches in world agriculture. This has partly arisen because of divergent, and increasingly rapid, genetic progress for milk production and cattle growth. Also milk still has to be produced relatively near to its point of consumption, mainly in industrialized nations, whereas beef can be inexpensively produced in the Third World and transported to industrialized nations for consumption. It is debatable how long this scenario will continue! Nevertheless it presents a challenge to milk producers in industrialized nations to maintain efficient milk production without the cow necessarily being pregnant for three-quarters of her productive life.

In many countries consumers are unwilling to accept artificial manipulation of milk production, for fear that it will reduce cow welfare or reduce the safety of milk for human consumption (Chapter 18). Although doubts have been raised concerning the latter in relation to rbST, it seems inevitable that ways will be found of manipulating milk production in a way that is safe for the consumer (e.g. Chapter 15). The key question is the effect of increased and prolonged early lactation catabolism on cow welfare. The incidence of mastitis, lameness, metabolic disorders and prolonged post-partum anoestrus is likely to increase, and a much more concerted effort is needed to address these problems (e.g. Chapters 9 and 10). With good management the best farmers can control these problems and attain an efficient milk production level. Many, however, are not adequately trained to safely use the new technologies becoming available and this urgently needs to be addressed.

With the varied response of governments to regulating the use of bio-technology and intensive farming methods in dairying, many consumers are demonstrating their concern by purchasing only dairy products produced to strict standards of animal welfare and with minimum use of potentially damaging chemical and biochemical inputs into the dairy production system (Chapter 16). Researchers must take heed and explore the welfare and environmental impact of new technologies much more thoroughly for consumers to allow their use in the dairy industry. They must also take a much more radical approach to developing husbandry methods that take account of the animals' needs and are therefore acceptable to the public consumers.

Mechanization of the dairy industry continues at a rapid rate. In the interests of labour economy in the industrialized world attention is focused on the development of fully automatic robotic milking units (Chapter 13), and the industry is poised for the revolution in dairy farm management that will accompany the widespread introduction of these units. Again the impact on cow welfare needs much more detailed exploration if the public is not to be sceptical of such technology. As with genetic manipulation (Chapters 5, 6 and 14), and immunization against lactation inhibitors (Chapter 15), greater milk production per cow is inevitable and the development of improved feeding methods are a high priority to limit potentially damaging prolonged post-partum catabolysis.

In less industrialized regions the need for improvement of efficiency is just as necessary as in industrialized regions, since population increases will put further pressure on land and animal resources. The multipurpose use of cattle and limited nutrition are dictated by the need for efficient use of animal, land and feed, rather than human resources. The high biological output achieved, relative to input, is often a salutary lesson to industrialized countries that are now striving for less wasteful and polluting use of nutrients in dairy systems. However, this high output can put the same strain on the cow's metabolism that the biotechnological advances are causing in single-purpose systems. We need a better understanding of the effect of enhanced cow productivity on cow reproduction (Chapter 8) and repeated reproductive failure is a good indicator of excessive stress on the cow's metabolism. I do not suggest that high output systems necessarily reduce the welfare of the cow by excessive stress. To do so would be the same as suggesting that the welfare of top athletes is compromised by the excessive exercise that they take. However, both the chance of the system failing and the required level of management expertise are increased. It is only with this expertise that adequate longevity can be assured for the cow, which is likely to be a requirement for both the financial viability of the unit and the acceptability to consumers (Chapter 7).

As with other industries, the pace of change in the dairy industry is accelerating and, while this brings new horizons for improvements in the efficiency of dairy systems, it also brings new responsibilities to safeguard animal welfare and protect the environmental and genetic resources that our dairy systems use. If we do not safeguard genetic diversity we are farming only for today, and not insuring for the time when new challenges, such as diseases or climate change (Chapter 11) must be met by adapted or adaptable animals. To achieve 'Progress in Dairy Science' requires not just new technologies, but ways of allowing them to be implemented on the farm with due regard for the safety of the animals, consumers and farmers who are involved in the human food industry.

Reference

Phillips, C.J.C. (ed.) (1989) New techniques in cattle production. *Proceedings of the Second International Symposium on New Techniques in Agriculture*, Bangor, UK, September 1987. Butterworth Scientific, London.

C.J.C. Phillips
January, 1995

Nutrition and Physiology

Protein Requirement Systems for Ruminants

J.D. Oldham

Genetics and Behavioural Sciences Department,
Scottish Agricultural College, Bush Estate, Penicuik,
Midlothian EH26 0QE, UK

Introduction

Feeding systems for animals generally have two main elements: (i) a system for predicting food intake (where all or part of the ration is likely to be offered *ad libitum*) and (ii) a scheme to assess the amounts of energy (metabolizable energy or net energy), amino acids, minerals and vitamins that are needed to sustain various body functions such as maintenance, growth, pregnancy, lactation or fibre production. It is usually assumed that, at the level of the animal's metabolism, the total needs of the animal can be assessed factorially, i.e. by simple summation of the amounts of energy, protein, etc. that are needed for component processes.

It is, perhaps, an implicit expectation, that when livestock are offered, in their diet, amounts of energy and nutrients in appropriate proportions that will sustain a particular rate of performance (e.g. a prescribed rate of weight gain) the animals receiving that diet will perform at that rate. In practice, of course, it is widely recognized that this expectation is not always realized and that the performance response of animals varies between individuals according to genetic, phenotypic and other environmental factors. The problem is exacerbated where some, or all, feeds are offered *ad libitum*. In more recent years attempts have begun to develop feeding systems in such a way that performance responses to available feeds, as well as requirements for particular nutrients and energy, are estimated.

This general shift in focus for nutritional schemes, from simple appraisal of requirements to prediction of responses, has been applied across all aspects of nutrition, including amino acid nutrition. It is therefore relevant to think about recent developments in amino acid or protein nutrition of ruminants, including dairy cows, against a general background of a shift in emphasis which, in the foreseeable future, should see developments in the application of response prediction schemes.

At the level of their metabolism ruminants use amino acids and not protein as primary substrates. It has become conventional to refer to the 'protein nutrition' of ruminant animals and this loose terminology has some attraction for ease of use but it is misleading. At the level of the rumen it might be more appropriate to refer to crude protein or nitrogen substrate, whilst at the level of the animal's metabolism it is appropriate to refer to individual amino acids.

Historical Background

There is a common genesis to the schemes for calculating the 'protein requirements' of ruminant animals which have been published in recent years (Madsen, 1985; NRC, 1985; Vérité and Peyraud, 1989; CSIRO, 1990; AFRC, 1992; Tamminga *et al.*, 1994), namely the quantification of aspects of ruminant digestive function and metabolism, qualitative knowledge of which can be traced back into the latter years of the 19th century. By the early years of this century Kellner (1909) was able to summarize some of the qualitative aspects of rumen function in ways that reflect concepts which are held today. Thus his comment that 'Some (rumen) bacteria have the power of forming proteins from certain non-protein substances, probably with the assistance of nitrogen-free extract substances . . .' is remarkably close to a statement that microbial nitrogen capture depends on supplies of energy from the fermentation of carbohydrate.

Still half a century ago, Phillipson, who was then a member of Sir Joseph Barcroft's team in Cambridge, pointed out that 'The ruminant lives not only on fatty acids and microorganisms, but also on any food, starch or protein that leaves the rumen unaffected by the action of microorganisms. The final answer must depend on the amount of these materials that become available to the animal' (Phillipson, 1946). The issue in question was, of course, the relationship between food input and nutrient delivery to the ruminant animal. Phillipson's continuing interest in this area led him to be at the forefront of the development of quantitative techniques for measuring post-ruminal nutrient flow, although his classical work, published in the mid-1960s had been prefaced by an impressive series of experiments done 15 years or so earlier by Sineschekov who, working in Russia, was not so widely recognized in the, then, western world (see Sineschekov, 1964).

Many of the qualitative aspects of ruminant digestion and, to a lesser extent, metabolism, were by the late 1950s substantially established, with a very effective summary in the monograph published in 1957 by Annison and Lewis.

The developments that led up to the publication of schemes for estimating 'protein requirements' were based largely on the appearance of a number of methods and techniques that allowed qualitative pathways of dietary nitrogen use in the ruminant animal to be quantified. Methods to measure nutrient supply into and out of the small intestine using cannulation and marker techniques, which were largely developed in the 1960s and 1970s, were important to provide information on the amounts of total nutrients that were made available to ruminants from their feed (Faichney, 1975; MacRae, 1975). Subdivision of those nutrients into ones that were of microbial origin and ones that were not, was made possible either by using intrinsic microbial-specific constituents or tracer methodology to identify microbial, separate from nonmicrobial, components of post-ruminal digesta. Development and refinement of these methods progressed especially through the late 1960s into the 1980s (see Clark *et al.*, 1992).

Prediction of nutrient supply to ruminant animals from their feed depends on methods that can identify nutrients of direct food origin in digesta. Separation of the non-microbial element of digesta into that which is of direct food origin and that which is of endogenous origin has proved to be a difficult technical task *in vivo* and estimation of these fractions has come to rely substantially on feed evaluation technology, such as the use of artificial fibre bag methods for measuring degradability of feedstuffs. The history of such methods is very long indeed (beginning with the use of perforated copper spheres which were used to estimate the digestion of meat in dogs as early as the 18th century). But the most recent development was largely achieved in the 1970s and 1980s, although there is continuing discussion, debate and refinement of those techniques today (Madsen and Hvelplund, 1994).

By the mid-1970s substantial amounts of information on quantitative aspects of digestion in ruminants were in the published literature. The subject animals frequently were chosen for experimental convenience and data from 'mature wether sheep' were much in evidence at the time. These technical developments, which produced a quantitative literature on ruminant digestion, laid the foundation for working groups in various parts of the world to appraise the state of knowledge and propose the first framework schemes which are now commonplace.

An additional incentive to these groups was an interest in developing efficient systems to enable non-protein-N (NPN) use for ruminants. Throughout the 20th century, but particularly since the end of the Second World War, there has been a great deal of interest in finding ways to exploit, efficiently, the ability of ruminant animals to convert NPN in food, via the moderation of the rumen microbes, into desirable animal products. Efficient use of urea in particular has been a long-running interest and the search for objective ways to predict circumstances in which NPN such as urea could be used effectively was a considerable spur to the development of 'new' systems for estimating ruminant 'protein' requirements.

Reviewing the genesis of these schemes is useful as a reminder to a number of issues that are germane to the current state of knowledge and ways in which that knowledge might develop. In particular, we should note:

1. That an important initial aim of the development of the new schemes was to improve the prediction of circumstances in which efficient use of NPN could be achieved.
2. Many of the factors within ruminant 'protein' systems have been adopted as average values taken from surveys of the literature in which available data vary across species (sheep and cattle) and plane of performance (from maintenance to, for example, high rates of milk production in dairy cows).
3. Many aspects of current schemes depend absolutely on the reliability of techniques for quantifying digestion.
4. Estimation of nutrient yields from food is only as good as the techniques that are used for feed evaluation and this might apply particularly to the estimation of the fraction of feed that is not degraded in the rumen.
5. Much more attention has been paid to the quantification of digestion processes in ruminants than has been paid to measures of the efficiency of metabolism and factors that might affect it.
6. The schemes have been devised to enable the prediction of requirements, and the design of suitable diets to meet these requirements. They were not designed to enable the prediction of dairy cow responses to the amounts and concentrations of N-substrates in the feedstuffs that are eaten.

'Metabolizable Protein' Schemes

The term 'metabolizable protein' (MP) was first coined in the United States some time ago (Burroughs *et al.*, 1975) and has been adopted in the UK as the unit of nutritional currency that describes the mixture of amino acids which is made available to the host ruminant for its metabolism and is of immediate dietary origin. Other terms for the same currency are used in schemes that have been devised in other countries, but I will use the term 'metabolizable protein' (MP) here – it is exactly analogous to protéines digestibles dans l'intestin (PDI) (Vérité and Peyraud, 1987), amino acids truly absorbed (AAT) (Madsen, 1985), darm verteerbar eiwit (DVE) (Tamminga *et al.*, 1994) and absorbed protein (AP) (NRC, 1985).

The pathways of digestion and subsequent metabolism whereby nitrogen constituents which originate in food, are converted to MP and, subsequently, result in animal protein products, have been described and discussed in a vast number of recent publications, See, for access to the literature, Oldham, 1984, 1994; Clark *et al.*, 1992; Webster, 1992; Seale and Reynolds, 1993; Armentano, 1994; Baldwin *et al.*, 1994; Reynolds *et al.*, 1994; Stern *et al.*, 1994. The main pathways are summarized in Fig. 1.1.

The elements of the scheme are well known. Food crude protein is partly degraded by the mixed population of rumen microbes to yield N-substrates for microbial growth and a small yield of energy, from fermentation. Another part escapes modification by the rumen microbes and, after the processes of

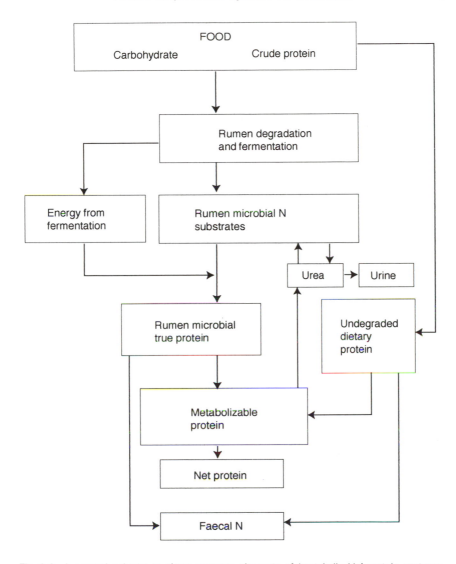

Fig. 1.1. A general scheme to show common elements of 'metabolizable' protein systems for dairy cows.

gastric digestion and absorption, can yield amino acids which contribute directly to metabolizable protein supply. In the UK this is referred to as digestible undegraded dietary protein (DUP). N-substrates are converted into microbial crude protein, subject to the availability of energy to support the synthetic processes. This arises largely, but not completely, from the fermentation of dietary carbohydrates within the rumen. Microbial crude protein, after passage beyond the rumen, contributes to metabolizable protein supply subject to the proportion that is true protein and the proportion of that which is truly absorbed from the small intestine. The proportion of microbial crude protein that is not true protein is predominantly in the form of nucleic acids.

Amino acids (MP) are metabolized either by incorporation into proteins, or deaminated and excreted as urea. There is the possibility of amino acid N from MP returning to the rumen, either in the form of urea which 'recycles' back into the rumen (by absorption across the rumen wall or in saliva) or, alternatively, by the entry into the rumen of endogenously synthesized proteins which can be degraded and used as N-substrates for microbial growth. There is also the possibility of other non-protein N (e.g. nucleic acids in sloughed cells) also returning to the rumen to provide N-substrates for microbial growth. It is usually implicit in MP systems that any recycling of N to the rumen that does occur is of urea-N but this need not be so.

In the scheme shown in Fig. 1.1, dietary N enters the system in food and is subsequently accounted for, either as waste product (faecal + urinary N), retained N (tissue protein or tissue non-protein – especially nucleic acids) or secreted in milk (or fibre).

Just about all schemes that have been published include functions or values for some or all of the following:

1. The partition of dietary crude protein between that which is degraded and made available for microbial capture in the rumen, and that which is not.
2. Rates of dietary crude protein degradation and digesta passage from the rumen.
3. The yield of rumen microbial crude protein as a function of energy which is made available from rumen fermentation of feed constituents.
4. The proportion of microbial crude protein that is true protein.
5. An allowance for recycling of urea N to the rumen.
6. The true absorption of microbial true protein in the small intestine.
7. The true absorption of undegraded dietary crude protein in the small intestine.
8. Endogenous losses of N from the body into urine and faeces.
9. Efficiencies of conversion of MP to animal protein products.
10. Rates of protein (or N) demand for maintenance, growth, pregnancy, milk or fibre production.

The values for some of these factors in some of the published schemes (but not all) are shown in Table 1.1 from which it is easy to see that there is a great deal of common ground in factors that have been chosen by different groups of people working in different countries. The reason for this is obvious in that the literature values that have been reviewed by different working groups are, essentially, the same, and it is therefore no surprise that similar conclusions are drawn. However, the apparent conviction that published 'chosen' values carry with them can be misleading. There is still considerable debate about a number of the factors listed above which we can consider in turn.

Dietary crude protein degradability

The partition of dietary crude protein into degraded and non-degraded fractions has most commonly been estimated using *in situ* rumen incubations of feedstuffs in artificial fibre bags. Some attempts have been made to standardize this methodology. Under the auspices of the European Association for Animal Production (EAAP) samples of soyabean meal, coconut meal, cottonseed meal, fishmeal and barley were evaluated in 21 different laboratories in Europe, using a standardized procedure. The results have recently been reported by Madsen and Hvelplund (1994). These show that there is considerable inter-laboratory variation in the application of this technique and that most of the variation is associated with the estimate of the 'rapidly degraded' or 'soluble' N fraction – sometimes referred to as the 'a' value. As distinctions are made in some schemes (e.g. AFRC, 1992) between rapidly and slowly degraded N fraction as regards the maximum efficiency with which those fractions can be incorporated into rumen microbial N, failure to estimate partition between rapidly and slowly degraded N has implications for the estimate of microbial MP which may be made using this technique. The search for an *in vitro* technique that can replace the *in situ* incubation methods has continued but has not yet yielded a convincing alternative (Roe *et al*; 1991; Cottrill, 1993; Stern *et al.*, 1994). There is, of course, a problem in developing a reliable alternative technique when the method that is used as reference is itself potentially flawed.

There has been considerable interest in using lower cost indirect methods, most particularly infrared reflectance spectrometry (NIR), to provide estimates of the degradability of food N fraction (Cottrill, 1993) but, again, this relies entirely on having an adequate calibration set of data against which to index NIR values.

Rates of rumen processes

The rate of dietary crude protein degradation is usually estimated from *in situ* incubations using an artificial fibre bag technique. The rate of digesta passage from the rumen (fractional rate of outflow of digesta from the rumen) is commonly used to calculated effective degradability using some representation of nitrogen degradability curves (various suggested mathematical treatments are available; Stern *et al.*, 1994). The AFRC (1992) scheme takes fractional rate of outflow to be a function of relative plane of feeding (multiple of maintenance intake). A much more complex treatment of kinetic processes in the rumen is built into the 'Cornell' (net carbohydrate and protein systems for evaluating cattle diets) scheme (Fox *et al.*, 1992; Russell *et al.*, 1992; Sniffen *et al.*, 1992). Whilst there have been some interesting developments in the mathematical treatment of kinetic data that deal with rates of degradation of

Table 1.1. Comparison of some of the elements of some published systems designed to allow calculation of 'metabolizable protein' requirements for ruminants.

	Country/Reference					
	Netherlands (Tamminga et al., 1994)	France (Verite and Peyraud, 1989)	Scandinavia (Madsen, 1985)	USA (NRC, 1985)	Australia (CSIRO, 1990)	UK (AFRC, 1992)
Protein degradability	In situ	In situ	In situ	In situ	In situ	In situ
Outflow rate (h^{-1})	0.045 - roughage 0.06 - concs	0.06	0.08	Variable	Variable	f(PF)
'Energy' unit for microbial yield calculation	FOM^1	FOM^2	Digested carbohydrates	TDN	DOM	FME
Microbial growth efficiency	150 g kg^{-1} FOM	145 g kg^{-1} FOM	125 g kg^{-1} DC	f(TDN intake)	f(Feed class)	f(PF)
MTP/MCP	0.75	0.8	0.7	0.8	0.8	0.7
Limiting efficiency of use of degradable N	1	0.9	Variable	0.9	0.8-1	0.8-1

N recycling	N/A	f(PF)	f(PL)	f(Diet CP %)	NSA	N/A
Absorption coefficient:						
MTP	0.85	0.80	0.85	0.8	0.7	0.85
UDP	Variable	Variable	0.82	0.8	Variable	Variable
Endogenous loss	f(BW)	f(BW)	f(BW)	f(BW)	f(BW)	f(BW)
Metabolic faecal N	f(UDM)	NSA	NSA	f(UDM)	f(DMI)	N/A
Efficiency of use of MP						
for: Maintenance	0.67	1	1	0.67	0.70	1
Growth	0.50	0.4-0.68	?	0.5	0.70	f(RV)
Lactation	f(DVE/NEL)	0.64	0.75-0.80	0.65	0.70	f(RV)
Analogous term for MP	DVE	PDI	AAT	AP	ADPLS	MP

[1] DOM corrected for crude fat, UDP, undegraded starch, fermentation end-products in feed.

[2] DOM corrected for crude fat, UDP and end-products in feed.

N/A no allowance; BW body weight; UDM undigested dry matter; PF plane of feeding; NSA no separate allowance; PL production level; RV relative value of AA mix; MTP microbial true protein; MCP microbial crude protein; UDP undegraded dietary protein; f( ) function of .....

feed constituents in the rumen and their passage out of the rumen, remarkably little of the fruits of those efforts seems to have been translated into practical means to improve 'protein' schemes. It is disappointing that, apart from the Cornell approach, the efforts that have gone into developing so-called 'mechanistic models' of rumen digestion processes (e.g. Baldwin *et al.*, 1987c; Danfaer, 1990) have yet to provide a reliable basis for the incorporation of dynamic variables into practical systems for designing rations.

Yield of rumen microbial crude protein

To measure rumen microbial protein synthesis and its contribution to post-ruminal nutrient flow is not easy. To account adequately for variations in net microbial protein supply beyond the rumen (according to dietary factors such as roughage:concentrate proportion, form of dietary carbohydrate, plane of feeding, etc.) is also proving very difficult (see AFRC, 1992; Clark *et al.*, 1992). The conventions adopted in different 'protein' schemes for estimating the (energetic) efficiency of microbial N capture varies from scheme to scheme. The estimate of energy contribution includes total digestible nutrients (TDN) intake (NRC, 1986), carbohydrate digestion in the rumen (Madsen, 1985), fermentable organic matter (FOM) (Vérité and Peyraud, 1989) and fermentable metabolizable energy (FME) (AFRC, 1992). No adequate model of rumen processes is yet available that appears to predict accurately rumen microbial protein synthesis – although there have been many valiant attempts.

Clark *et al.* (1992) provided a useful review of the available data and pointed out the wide variability in estimates of rumen microbial growth yields that are to be found in the literature. Whilst it does seem to be acknowledged that microbial growth efficiency (that is microbial N capture per unit of energy available from fermentation) is generally high in animals that are at high planes of feeding, such as dairy cows, the basis on which we currently estimate microbial MP yield and its variation still seems to be a significant area of uncertainty in all 'protein' schemes.

An interesting observation that does not seem to have been followed up to any extent is that some part of microbial crude protein might originate from endogenous proteins which enter the rumen (Marsden *et al.*, 1988). The estimate was that up to 15% of crude protein might be synthesized by this route. If that is the case, the contribution of rumen degraded N, of immediate dietary origin, to microbial MP yield, may be rather less than is conventionally estimated, with implications for amounts of undegraded dietary protein which may be needed to meet the animal's demand.

Microbial true protein

Until recently the proportion of microbial true protein that is true protein was generally thought to be reliably estimated at around 0.8. Different groups, however, have adopted different values for this factor (Table 1.1) and there seems to be increasing evidence (some of it arising out of further work by the EAAP Protein Working Party) that a value no higher than 0.7 would be appropriate. Clark *et al.* (1992) identified a very broad range of values indeed. If there were to be true variation in this value (perhaps according to the balance of species in the mixed rumen microflora and fauna) then, taken together with the potential large variation in microbial growth yield, there is obviously large scope for MP yield from microbial sources to vary significantly. This would have great importance for an understanding of the ways in which producing animals respond to different dietary regimes.

Recycling of urea to the rumen

It is sometimes assumed that N transfer between rumen ammonia and blood urea are in balance for nitrogen limiting circumstances (e.g. AFRC, 1992) and sometimes that a degree of nitrogen deficit across the rumen can be assumed within the calculation of MP supplies and needs (e.g. Madsen, 1985). This is an important issue as regards the design of diets to meet the requirements of rumen microbes for N-substrates. It is an implicit assumption of all of the protein schemes that to satisfy the substrate requirements of rumen microbes for their effective growth is important to the animal. The advantages are, first, to achieve an effective rumen fermentation (with consequences for food digestion and intake) and, second, to supply nutrients, in the form of microbial products, beyond the rumen. If urea recycling to the rumen can effectively counter-balance a shortfall in the supply of rumen degraded crude protein of immediate dietary origin then rations can be designed with more confidence, which will meet the needs of the animal without disrupting digestive processes (and possibly intake).

As I have pointed out elsewhere (Oldham, 1995), there is substantial evidence that a general effect of increasing dietary crude protein concentration in the diet for dairy cows is to increase digestibility and especially intake (Oldham, 1984). But there is also evidence that dairy cow intakes are not necessarily compromised when the allowance of rumen degraded protein per unit fermentable substrate is less than the prescribed value in at least some of the 'protein' schemes. These apparently contradictory observations point to real difficulties in predicting circumstances in which dietary (rumen degradable) crude protein intake may be a factor that limits digestion and intake. On this count, current approaches are not satisfactory.

The effects of rumen degradable protein, or NPN, on intake and digestion are not the only criteria for judgement. As pointed out above, an important aim of describing rumen microbial needs for N is to enable a rationale for calculating how much NPN can usefully be included in rations. The presumption that net urea recycling to the rumen can occur without detriment to rumen function is tantamount to a statement that recycled urea N and dietary NPN might replace each other (at similar efficiencies of use for microbial N capture) up to some limit, although the evidence that this is so is not plentiful.

True absorption coefficients

No new information has accrued to change the view that the true absorption coefficient for microbial true protein in the small intestine is different from the widely adopted value of 0.85. There is clearly much more variation in the values for true absorption coefficients of undegraded dietary crude protein, depending on the nature of the nitrogenous materials in the crude protein and their individual true digestibilities. Webster (1992) indicated that prediction equations which relate the digestibility of undegraded dietary crude protein to its acid detergent insoluble nitrogen (ADIN) content are reasonably robust across quite a wide range of feedstuffs. However, there are exceptions (perhaps particularly for distillery byproducts).

Endogenous loss

The majority of 'protein' schemes for estimating requirements include an allowance, directly or indirectly, for food related endogenous losses in faeces (metabolic faecal nitrogen; MFN). The AFRC (1992) structure is an exception. Owens (1987) reviewed the subject in some detail. While the argument that basal endogenous nitrogen (BEN) losses represents a good estimate of true endogenous losses of metabolic origin appears to be strong (AFRC, 1992) there is a rationale for suggesting that endogenous losses may be enhanced as a consequence of food transit through the gut. Some further consideration needs to be given to this area as the chosen values for BEN loss by AFRC (1992) are rather larger than estimates of endogenous losses in other systems that also include an allowance for MFN – but overall estimates of endogenous plus MFN losses are substantially lower in the AFRC (1992) scheme than in others.

Estimating MP Supply

The brief comments above deal with relationships between feed characteristics and MP supply. Since the first publications of 'new' protein schemes there have been many additional papers in the literature that report measurements of post-ruminal nutrient supply, or efforts to improve techniques to estimate post-ruminal nutrient supply. It is, however, disappointing that the state of knowledge in this area has not developed enough in recent years to allow much, if any, constructive modification to the kinds of estimating system which are built into the 'protein' scheme published in the 1980s and early 1990s and which are, largely, built on information from a few years earlier. Perhaps the growing application of techniques to measure nutrient uptake from the gut (Reynolds and Seal, 1994) will provide improved opportunities to test predictive relationships between estimated feed values and nutrient supply. An important technical difficulty, of course, is that such measurements do not allow for nutrient metabolism by the gut tissues themselves. However, confidence in the application of predictive systems to relate qualities of feeds to nutrient supply, and thence to nutrient use, will remain weak if the estimation of nutrient (MP) yield from consumed feed cannot be tested.

Use of MP

Dairy cows use the mixture of amino acids that is absorbed from the gut (MP) to maintain their functional integrity (maintenance), to synthesize milk proteins which are secreted, to synthesize tissue protein (either for their own or fetal growth or for the replenishment of potentially labile protein reserves) and to contribute carbon skeletons to glucose synthesis. To estimate requirements for MP for particular rates of production it is necessary to know the efficiency with which MP is used to satisfy particular demands. Many of the existing 'protein' schemes (Table 1.1) assume that needs for MP are met at a constant efficiency provided that performance is truly 'protein' (i.e. amino acid) limited. The AFRC (1992) approach is somewhat different and adopts a view for the estimation of efficiency that is conventional in non-ruminant animal systems, namely that there is a limiting efficiency with which an ideally balanced mixture of amino acids is used (estimated at 0.85) and that the actual efficiency observed will depend on the extent to which the mixture of amino acids actually supplied matches, or fails to match, the ideal amino acid mixture (or balance) (i.e. efficiency of use of MP depends on the balance of amino acids supplied as well as on a maximum limiting value). The INRA PDI system (Vérité and Peyraud, 1989) takes a somewhat different approach and uses an empirically derived function in which apparent efficiency of use of PDI (which is exactly analogous to MP) is used at a diminishing marginal rate as supply increases. Neither of these approaches is completely satisfactory.

Some progress is being made, now, to throw some light on this area of ruminant protein nutrition which, remarkably, has been rather overlooked up to now, perhaps mainly because of the difficulties of carrying out appropriate experimental tests of relevant ideas.

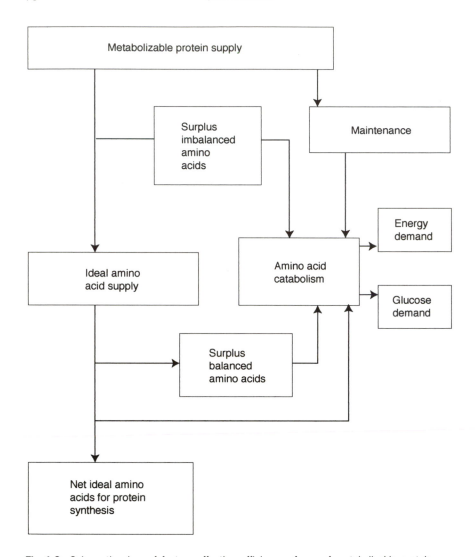

Fig. 1.2. Schematic view of factors affecting efficiency of use of metabolizable protein.

Figure 1.2 shows some of the factors that are likely to influence the efficiency with which MP will be used in dairy cows.

There have been some attempts to relate MP supply to milk protein yield in dairy cows under, supposedly, protein-limiting conditions. In the UK, AFRC (1992) refer to experiments that were designed to help in the development of the system. Recently, some additional tests of that system have begun to be reported (Mansbridge *et al.*, 1994; Newbold *et al.*, 1994) (Fig. 1.3). A characteristic of the relationship between MP supply and milk protein output in the experiments described in AFRC (1992), and as reported by Webster (1992), is that the marginal rate of response in milk protein output to increments in MP supply is substantially less than expected at around 0.2 as a partial efficiency, i.e. for each $100 \, g \, d^{-1}$ increment in calculated MP supply milk protein yield increases, on average, by only about $20 \, g \, d^{-1}$. The absolute performances of the cows in those experiments was, however, close to expectation for the amounts of MP consumed. In the more recent study (Newbold *et al.*, 1994) the situation was somewhat similar, although absolute performance was, on average, rather less than expected for the amounts of MP consumed, but the marginal rate of response in milk protein output to increments of MP supply was still low at around 0.4. The reasons for the low marginal rates of response to MP are intriguing and must be the result of one or more of the following factors:

1. Milk secretion was not protein limited.
2. The biological (or relative) value of MP was less than expected (AFRC, 1992).
3. MP supply was wrongly estimated (in all experiments it was predicted rather than measured).
4. MP was, in part, being catabolized as an energy and/or glucose source.
5. Part of MP was being partitioned towards other protein synthetic purposes rather than being secreted in milk.

The continuous, but low, marginal responses to MP increments shown in Fig. 1.3 suggest that it is wrong to discount the low marginal efficiencies of use on the grounds of MP not being limiting for performance. If that were the case then there would be no expectation of any response to MP beyond, say, the first step. Whether or not MP supply was correctly estimated is arguable, but the main effect of an error in estimating MP supply would be to displace the position of the response rather than to negate it. It therefore seems reasonable to think that the observed low marginal rates of response were the result of imbalanced amino acid supply (i.e. specific amino acid effect), or the use of protein for energy and/or glucose generation, or the partition of protein between alternative routes of synthetic use.

Amino acid responses
The availability of forms of specific amino acids (especially methionine and lysine) which appear, reliably, to protect those amino acids from degradation within the rumen but yet make them available for absorption beyond the rumen (Papas *et al.*, 1984) has helped, recently, to add useful information to our understanding of the responsiveness of dairy cows to changes in particular amino acid supply. Rulquin and Vérité (1993) and Rulquin *et al.* (1993) have summarized much of the more recently available evidence. Their interpretations of the available data identify concentrations of methionine and lysine in MP (PDI) that are needed for an appropriate amino acid balance in the mixture of

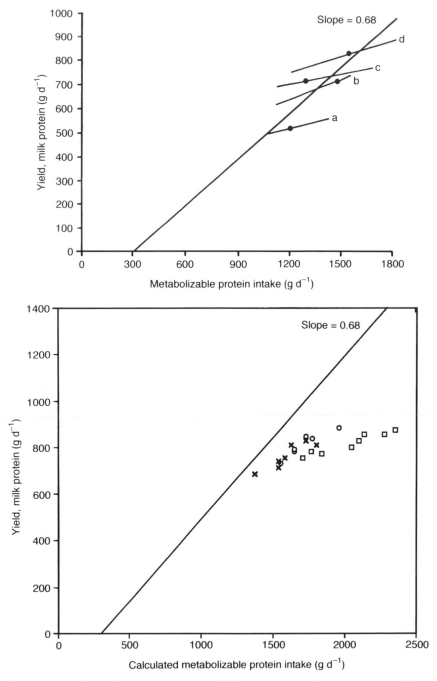

Fig. 1.3. (a and b) Performance responses of dairy cows to variation in calculated MP supply. Data are from experiments conducted to test AFRC (1992) proposals (see AFRC, 1992 and Newbold *et al.*, 1994 for details).

amino acids which is absorbed from the gut. Attempts are now being made to incorporate this kind of information into the PDI 'protein' system. Of course, in theory, it would be appropriate to identify the ideal balance of amino acids that is required for milk production (and also growth, maintenance, etc.) in the same way that has been done for non-ruminants (Fuller, 1991). The practical value of going beyond the identification of first or second most likely limiting amino acids is, however, questionable.

With diets that are based predominantly on maize and maize products there is clear evidence that lysine supply can be first limiting for milk production as shown elegantly in the work of Schwab *et al.* (1992). Whilst there is some evidence of milk protein yields being responsive to methionine supply (Rulquin and Vérité, 1993) the marginal rates of response to methionine are generally very low, whilst the marginal rates of response to lysine in Schwab's work were close to the theoretical (0.85). This may, in part, reflect the multifunctional role of methionine and its involvement as a methyl donor in processes that influence fat metabolism (see Oldham, 1981).

Partial protein catabolism for energy/glucose generation
If part of an increment of MP is catabolized to yield energy or glucose, so that the remainder can be synthesized into milk protein, then it would be expected that the marginal transfer efficiency of MP into milk protein would be low. As argued in detail elsewhere (Oldham, 1995), if supply of energy to synthesize milk protein is a limiting factor, then the partial catabolism of an increment of MP to provide the necessary energy could promote a response in milk protein yield by allowing an increase in milk protein concentration with no change in the secretion of other milk constituents. The argument here is that to change the secretion of those other constituents would require yet further energy and so concentration of protein would be expected to change, but not yields or concentrations of other constituents. On theoretical grounds (see Oldham, 1995) the efficiency of MP use under these circumstances would be expected to be around 0.25.

Similarly, if the ratio of protein:lactose in milk was as high as the metabolic system would allow (there appears to be an upper limit to protein:lactose ratio which is difficult to break) then further secretion of milk protein could only proceed provided that there was also glucose available to allow lactose synthesis to proceed in concert with protein synthesis. If an increment of MP was used to provide both the amino acids for milk protein and carbon skeletons to allow the extra lactose to be secreted, the efficiency of MP use is expected to be around 0.33 (Oldham, 1995). In this (theoretical) case, response in milk protein yield will be seen as an increase in milk volume with no change in concentration.

On these theoretical grounds therefore, low marginal rates of efficiency of MP use of around 0.2–0.4 are at least compatible with the idea that parts of increments of MP would be used either for energy or glucose generation (or, possibly, a little for each purpose).

Whilst it has become readily accepted that interrelationships between nitrogen use and energy supply are expected at the level of rumen microbial function (witness the various relationships between microbial N capture and fermentation activity in the various schemes listed in Table 1.1), few attempts have been made to incorporate protein:energy interrelationships into schemes to describe protein requirements and responses. Such interrelationships are, of course, implicated in the structure of complex models of dairy cow metabolism (e.g. Baldwin *et al.*, 1987b). But perhaps the concept of protein:energy interrelationships noted here could fairly readily be incorporated into practical systems for response prediction given a set of rules that could be used to identify circumstances in which protein, energy and/or glucose supply were close to being colimiting for production. The ideas about protein:energy and protein: glucogenic nutrient ratio that are implied here present some possible rationales to account for the nature of performance responses either as a change in milk protein concentration or yield (or both).

For the Dutch DVE (i.e. MP) protein evaluation system, the efficiency of use of MP for milk protein production is predicted (Subnel *et al.*, 1994) as a function of protein:energy ratio (DVE/NEL in the system's terminology) and level of (Fat and Protein Corrected; FPCM) milk yield according to:

$$\text{Efficiency} = \frac{117.6 - 3.044 \text{ DVE (g)} - 0.23 \text{ FPCM (kg)}}{\text{NEL (MJ)}}$$

This implies very low marginal efficiencies for increments of DVE (i.e. MP) of about 0.1 for cows at moderate (e.g. 30–35 kg day^{-1}) yield level. To be valid the data on which this approach has been built would have had to have been from cows whose performance was truly protein limiting.

Protein partition

The use of body fat reserves by dairy cows to support lactation when nutritional input is inadequate is a well-described phenomenon. It is also widely recognized that cows have reserves of protein in their bodies that are potentially labile and which may be drawn on to buffer against a short-term nutritional deficit. The size of the potentially labile protein reserve may be about 20–25% of the fully replete protein mass of the body. This may approach 20 kg protein in a mature cow (Botts *et al.*, 1978).

The direct evidence that cows actually do lose substantial amounts of body protein at any time in lactation is meagre. Gibb *et al.* (1992) found, from comparative slaughter studies, a maximum loss of about 8 kg body protein, which was subsequently replenished by the end of lactation. There are, however, many reports in the literature of substantial negative nitrogen balance estimates in the early phase of lactation. Given the usual presumptions about direction of errors in nitrogen balance measurements, these negative values should mean that the estimates are minimal as regards the amount of N being lost from the body.

We have shown recently (Pine *et al.*, 1994) that in lactating rats maternal protein reserves can indeed buffer lactational performance against nutritional adversity for a reasonable period of lactation. In dairy cows, Whitelaw *et al.* (1986) showed that, in response to increments of casein given by abomasal infusion, the milk protein secretory response represented only part of the overall response with the remainder being seen as an altered tissue N balance. Whilst the marginal response to abomasal casein in terms of the milk protein response alone was low (relative to the expected 'theoretical' value), the combined response as milk protein plus tissue protein retention accounted reasonably well for the expected response, i.e. the increments of MP were not being used inefficiently for protein synthesis, but they were being partitioned only partly towards milk protein synthesis.

Quite recently there have been some interesting reports of substantial increases in milk protein secretion following low energy but high protein concentrate supplements being offered to cows in the dry period (Van Saun *et al.*, 1993; Moorby *et al.*, 1994). The magnitude of these responses (which seem to last well into lactation) has been quite large compared with the total amounts of additional metabolizable protein given in the dry period. This might suggest that, even if part of the response is due to protein accretion in potentially labile reserves during the dry period, and their subsequent use in the milking period, that mechanism might not account for all of the response. The possibility that tissue protein reserves play a part in determining response to feeding is, however, a realistic one.

Response Prediction

Inevitably, as seen in the previous section, the discussion of 'protein' nutrition of dairy cows moves from a consideration of frameworks to estimate requirements to thoughts about the ways in which dairy cows respond to different feeding practices. Attempts to create so-called 'mechanistic models' of digestion and metabolism in dairy cows has been one approach aimed, in the longer term, at accounting for various kinds of response (e.g. Baldwin *et al.*, 1987a). An alternative framework was described by Oldham and Emmans (1988) and Fig. 1.4 illustrates the principles underlying that approach. It begins with a description of the animal's genotype and its current state (body composition, health, reproductive status). The left-hand side of the diagram represents the notion that dairy cows, like other animals, are trying to gain resources from their food that allow them to 'perform' (maintain themselves, lactate, etc.) at rates which are determined by their genotype and current state. In order to gain those resources the animal has to eat at a certain rate and provided that the qualities of the food do not inhibit food consumption then food intake is determined by the animal's 'needs'. The right hand side of the diagram suggests that the cow has only certain capacities to cope with its food – which

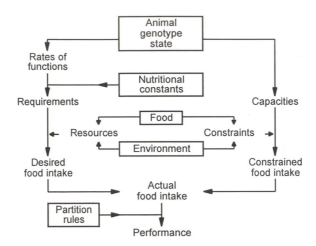

Fig. 1.4. A logic diagram to link descriptions of animals (cows) and food for predictions of food intake and performance.

may, for example, be a limited capacity to ingest food as determined by physical aspects of the food, presence of toxins, etc. If the qualities of the available food exceed the animal's capacity to ingest that food, then food intake is less than the animal 'targets' (which are the amounts it needs to eat in order to satisfy its demand for resources as determined by its genotype and its current state).

Once consumed, the nutrients that are made available from food are used, or partitioned, for particular functions according to some set of rules ('partition rules').

In the context of the earlier discussion, this framework is helpful in developing current concepts about demands for and responses to protein in dairy cows, in that the logic recognizes requirements for particular rates of production and partition rules for nutrients according to the animal's current state. It goes beyond the current MP schemes by asserting that the framework of requirements is determined by genotype and current state and that the animal is trying to eat to meet those requirements. Failure to achieve a satisfactory intake of resources (nutrients and energy) then requires that the animal partitions a scarce resource in an effective way. To make those ways predictable, we require rules that describe, sensibly, the partition of scarce resources. Such rules do not exist at present for dairy cows but research effort could usefully be expended to gain them.

A pivotal concept within this whole approach is that the animal is trying to eat in order to gain sufficient resources so that it can perform at rates which are determined by its phenotype. This may be an uncomfortable concept for some people. However, there is very compelling evidence from studies with growing pigs that, when offered a choice between two feeds, a combination of which allows maximum growth, they will grow at their maximum rate

(Kyriazakis and Emmans, 1991). Pigs of different breeds will make selections between foods of different protein:energy ratios according to the composition of body weight gain which is a characteristic of genotype (Kyriazakis *et al.*, 1993). We have also found that growing sheep, when offered a choice between two diets (differing in protein:energy ratio) – a combination of which allows them to grow at a maximum rate, will choose a combination that allows them to grow at a maximum rate but also to avoid an excessive intake of N (Kyriazakis and Oldham, 1993). These observations apply to feeds of high digestibility.

Such evidence points to new directions for understanding the 'protein' nutrition of ruminant animals. Choices between feeds of different N content, CP 'quality', or MP:ME ratio, for example, could be used to test ideas about nutritional demand in dairy cows. For ruminant animals though, it is obviously important to include in such studies choices amongst feedstuffs (forages, etc.) which depend on rumen fermentation for their effective utilization. Diet selection studies in sheep are beginning to show that selection between feeds of different crude protein:metabolizable energy ratio seems to be, at least in part, consistent with the animal's need to consume amounts of rumen degradable protein and fermentable carbohydrate in appropriate ratios. This allows effective microbial growth efficiency with minimal wastage of degraded nitrogen and with minimum risk to the physicochemical rumen environment (Kyriazakis and Oldham, 1993; Cooper *et al.*, 1994). It is quite possible that feeding behaviour, especially when capable of expression in the form of selection between feeds, will enable effective synchronization of nitrogen and energy yielding nutrients in the rumen which, after earlier studies (Oldham *et al.*, 1977) is enjoying a renaissance of interest (Sinclair *et al.*, 1993).

Conclusions

The 'protein' nutrition of dairy cows, despite the loose terminology, is a subject that has attracted widespread interest for a long period of time. Much of that interest has been focused on the development of systems for predicting the amounts of metabolizable protein and ruminally degraded crude protein which cows need in order to achieve certain rates of performance. Some of the adequacies and inadequacies of those schemes have been commented on here. Though by no means perfect in structure, the existing schemes are a useful practical step forward in designing effective rations for dairy cows. The next useful step will be to make the responses of dairy cows to the protein in their available feed more predictable. To do this effectively a view needs to be taken of the nature of the cow and *its* 'goal'. It would be naive to contemplate the prediction of animal responses to feed by describing only the essential properties of the feed, and not of the animal. Two major issues for the prediction of responses are: (i) how much of the available feed(s) will this animal eat; and (ii) how will it partition the nutrients and energy that become available from the feed which has been eaten? I suggest that an animal-centred approach to these issues is essential for useful progress to be made towards the goal of response prediction. Understanding of metabolism and physiological regulation will be helpful to guide the introduction of 'partition-rules' – for example, to cope with the problem of partitioning amino acids between protein synthetic and energy (or glucose) generating pathways when metabolizable protein:metabolizable energy ratios are imbalanced. But descriptions *only* at the level of metabolism are unlikely to yield practically useful models of response prediction. Whole-animal (holistic) descriptions of animals and their phenotypically directed 'goals' are relevant, practical integrations of a myriad of internal processes, many of which are not currently amenable to description or prediction. If our interest is in the performance of the whole animal then the over-riding framework for description of the animal should be set at the whole-animal level.

References

AFRC (1992) Nutritive Requirements of Ruminant Animals: Protein. Technical Committee on Responses to Nutrients, Report No. 9. *Nutrition Abstracts and Reviews, Series B: Livestock Feeds and Feeding* 62, 787–835.

Annison, E.F. and Lewis, D. (1957) *Metabolism in the Rumen.* Methuen Monograph, London.

Armentano, L.E. (1994) Impact of metabolism by extra gastrointestinal tissues on secretory rate of milk protein. *Journal of Dairy Science* 77, 2809–2820.

Baldwin, R.L., France, J., Beever, D.E., Gill, M. and Thornley, J.H.M. (1987a) Metabolism of the lactating cow. III. Properties of mechanistic models suitable for evaluation of energetic relationships and factors involved in the partition of nutrients. *Journal of Dairy Research* 54, 133–145.

Baldwin, R.L., France, J. and Gill, M. (1987b) Metabolism of the lactating cow. I. Animal elements of a mechanistic model. *Journal of Dairy Research* 54, 77–105.

Baldwin, R.L., Thornley, J.H.M. and Beever, D.E. (1987c) Metabolism of the lactating cow. II. Digestive elements of a mechanistic model. *Journal of Dairy Research* 54, 107–131.

Baldwin, R.L., Emery, R.S. and McNamara, J.P. (1994) Metabolic relationships in the supply of nutrients for milk protein synthesis: integrative modelling. *Journal of Dairy Science* 77, 2821–2836.

Botts, R.I., Hemken, R.W. and Bull, L.S. (1978) Protein reserves in the lactating dairy cow. *Journal of Dairy Science* 62, 433–440.

Burroughs, W., Nelson, D.K. and Mertens, D.R. (1975) Protein physiology and its application in the lactating cow: the metabolizable protein feeding standard. *Journal of Animal Science* 41, 933–944.

Clark, J.H., Klusmeyer, T.H. and Cameron, M.R. (1992) Microbial protein synthesis and flows of nitrogen fractions to the duodenum of dairy cows. *Journal of Dairy Science* 75, 2304–2323.

Cooper, S.D.B., Kyriazakis, I. and Oldham, J.D. (1994) The effects of physical form of feed, carbohydrate source and inclusion of sodium bicarbonate, on the diet selections of sheep. *Journal of Animal Science* (submitted).

Cottrill, B.R. (1993) Characterization of nitrogen in ruminant feeds. In: Garnsworthy, P.C. and Cole, D.J.A. (eds) *Recent Advances in Animal Nutrition*. University of Nottingham Press, Nottingham, pp. 39–53.

CSIRO, (1990) *Feeding Standards for Australian Livestock: Ruminants.* Report of Standing Committee on Agriculture, Ruminants sub-committee. CSIRO, Australia.

Danfaer, A. (1990) A dynamic model of nutrient digestion and metabolism in lactating dairy cows. Bertning fra Statens Husdrybrugsforsog No. 471. Foulum, Denmark.

Faichney, G.J. (1975) The use of markers to partition digestion within the gastro-intestinal tract of ruminants. In: McDonald, I.W. and Warner, A.C.I. (eds) *Digestion and Metabolism of the Ruminant*. Proceedings of the IVth International Symposium on Ruminant Physiology, Sydney, Australia, pp. 261–276.

Fox, D.G., Sniffen, C.J., O'Connor, J.D., Russell, J.B. and Van Soest, P.J. (1992) A net carbohydrate and protein system for evaluating cattle diets: III Cattle requirements and diet adequacy. *Journal of Animal Science* 70, 3578–3596.

Fuller, M.F. (1991) Present knowledge of amino acid requirements for maintenance and production: non-ruminants. In: Eggum, B.O., Boisen, S., Borsting, C. Danfaer, A. and Hvelplund, T. (eds) *Proceedings of the EAAP 6th International Symposium on Protein Metabolism and Nutrition. Herning, Denmark.* Institute of Animal Science, Foulum, Denmark, pp. 116–126.

Gibb, M.J., Ivings, W.E., Dhanoa, M.S. and Sutton, J.D. (1992) Changes in body components of autumn calving Holstein-Friesian cows over the first 29 weeks of lactation. *Animal Production* 54, 339–360.

Kellner, O. (1909) *The Scientific Feeding of Animals* (Translation by W. Goodwin). Duckworth & Co., Covent Garden.

Kyriazakis, I. and Emmans, G.C. (1991) Diet selection in pigs: Choices made by growing pigs following a period of underfeeding with protein. *Animal Production* 52, 337–346.

Kyriazakis, I. and Oldham, J.D. (1993) Diet selection in sheep: The ability of growing lambs to select a diet that meets their crude protein requirements. *British Journal of Nutrition* 69, 617–629.

Kyriazakis, I., Leus, K., Emmans, G.C., Haley, C.S. and Oldham, J.D. (1993) The effect of breed (Large White × Landrace vs purebred Meishan) on the diets selected by pigs given a choice between two foods that differ in their crude protein contents. *Animal Production* 56, 121–128.

MacRae, J.C. (1975) The use of re-entrant cannulae to partition digestion function within the gasto-intestinal tract of ruminants. In: McDonald, I.W. and Warner, A.C.I. (eds) *Digestion and Metabolism of the Ruminant*. Proceedings of the IVth International Symposium on Ruminant Physiology, Sydney, Australia, pp. 261–276.

Madsen, J. (1985) The basis for the proposed Nordic protein evaluation system. The AAT-PBV system. *Acta Agriculturae Scandinavica Suppl.* 25, 9–20.

Madsen, J. and Hvelplund, T. (1994) Prediction of *in situ* protein degradability in the rumen. Results of a European ring test. *Livestock Production Science* 39(2), 201–212.

Mansbridge, R.J., Cottrill, B.R., Newbold, J.R., Blake, J.S. and Spechter, H.H. (1994) The effects of increasing concentration of digestible undegraded protein (DUP) in the diet of dairy cows on silage intake, milk yield and milk composition. EAAP Meeting, Edinburgh, pp. 116. (Abstr.)

Marsden, M., Bruce, C.I., Bartram, C.G. and Buttery, P.J. (1988) Initial studies on leucine metabolism in the rumen of sheep. *British Journal of Nutrition* 60, 161–171.

Moorby, J.M., Dewhurst, R.J. and Madsen, S. (1994) Effects of supplementary protein in the dry period on milk production in the subsequent lactation. *Proceedings of the Nutrition Society* 53, 215A.

Newbold, J.R. (1994) Practical application of the MP system. In: Garnsworthy, P.C. and Cole, D.J.A. (eds) *Recent Advances in Animal Nutrition*. University of Nottingham Press, Nottingham.

Newbold, J.R., Cottrill, B.R., Mansbridge, R.M. and Blake, J.S. (1994) Effective metabolizable protein on intake of grass silage and milk protein yield in dairy cows. *Animal Production* 58, 455.

NRC (1985) *Ruminant Nitrogen Usage*. National Academy Press, Washington, DC.

Oldham, J.D. (1981) Amino acid requirements for lactation in high-yielding dairy cows. In: Haresign, W. and Lewis, D. (eds) *Advances in Animal Nutrition*. Butterworths, London, pp. 33–65.

Oldham, J.D. (1984) Protein–energy interrelationships in dairy cows. *Journal of Dairy Science* 67, 1090–1114.

Oldham, J.D. (1995) *Protein Requirements and Responses – a United Kingdom Perspective*. Occasional Publication, BSAS, Hillsborough, N. Ireland. (In Press).

Oldham, J.D. and Emmans, G.C. (1989) Prediction of responses to required nutrients in dairy cows. *Journal of Dairy Science* 72, 3212–3229.

Owens, F.N. (1987) Maintenance protein requirements. In: Jarrige, R. and Alderman, G. (eds) *Feed Evaluation and Protein Requirement Systems for Ruminants*. EU, Luxembourg, pp. 187–212.

Papas, A.M., Sniffen, C.J. and Muscato, T.V. (1984) Effectiveness of rumen protected methionine for delivering methionine post-ruminally in dairy cows. *Journal of Dairy Science* 67, 545–552.

Phillipson, A.T. (1946) The physiology of digestion in the ruminant. *Veterinary Record* 58, 81.

Pine, A.P., Jessop, N.S. and Oldham, J.D. (1994) Maternal protein reserves and their influence on lactational performance in rats. *British Journal of Nutrition* 71, 13–27.

Reynolds, C.K., Harmon, D.L. and Cecava, M.J. (1994) Absorption and delivery of nutrients for milk protein synthesis by portal drained viscera. *Journal of Dairy Science* 77, 2787–2808.

Roe, M.E., Chase, L.E. and Sniffen, C.J. (1991) Comparisons of *in vitro* techniques to the *in situ* technique for estimation of ruminal degradation of protein. *Journal of Dairy Science* 74, 1634–1640.

Rulquin, H. and Vérité, R. (1993) Amino acid nutrition of dairy cows: productive effects and animal requirements. In: Garnsworthy, P.C. and Cole, D.J.A. (eds) *Recent Advances in Animal Nutrition*. Nottingham University Press, Nottingham, pp. 55–77.

Rulquin, H., Pisulewski, P.M., Vérité, R. and Guenard, J. (1993) Milk production and composition as a function of post-ruminal lysine and methionine supply: a nutrient-response approach. *Livestock Production Science* 37, 69–90.

Russell, J.B., O'Connor, J.D., Fox, D.G., Van Soest, P.J. and Sniffen, C.J. (1992) A net carbohydrate and protein system for evaluating cattle diets: 1. Ruminal fermentation. *Journal of Animal Science* 70, 3551–3561.

Schwab, C.J., Bozak, C.K., Whitehouse, N.L. and Olsen, V.M. (1992) Amino acid limitation and flow to the duodenum at 4 stages of lactation. 2. Extent of lysine limitation. *Journal of Dairy Science* 75, 3503–3518.

Seale, C.J. and Reynolds, C.K. (1993) Nutritional implications of gastrointestinal and liver metabolism in ruminants. *Nutrition Research Reviews* 6, 185–208.

Sinclair, L.A., Garnsworthy, P.C., Newbold, J.R. and Buttery, P.J. (1993) Effects of synchronising the rate of dietary energy and nitrogen release on rumen fermentation and microbial protein synthesis in sheep. *Journal of Agricultural Science, Cambridge* 120, 251–263.

Sineschekov, A.D. (1964) *The Nutritional Physiology of Farm Animals.* Translation from Russian made by National Lending Library for Science and Technology: Boston Spa, Yorkshire.

Sniffen, C.J., O'Connor, J.D., Van Soest, P.J., Fox, D.G. and Russell, J.B. (1992) A net carbohydrate and protein system for evaluating cattle diets: II. Carbohydrate and protein availability. *Journal of Animal Science* 70, 3562–3577.

Stern, M.D., Varga, G.A., Clark, J.H., Firkins, J.L., Huber, J.T. and Palmquist, D.L. (1994) Evaluation of chemical and physical properties of feeds that affect protein metabolism in the rumen. *Journal of Dairy Science* 77, 2762–2786.

Subnel, A.P.J., Meijer, R.G.M., Van Straalen, W.M. and Tamminga, S. (1994) Efficiency of milk protein production in the DVE protein evaluation system. *Livestock Production Science* 40, 215–224.

Tamminga, S., Van Straalen, W.M., Subnel, A.P.J., Meijer, R.G.M., Steg, A., Wever, C.J.G. and Blok, M.C. (1994) The Dutch protein evaluation system: The DVE/OEB system. *Livestock Production Science* 40, 139–155.

Van Saun, R.J., Idleman, S.C. and Sniffen, C.J. (1993) Effect of undegradable protein amount fed prepartum on postpartum production in first lactation Holstein cows. *Journal of Dairy Science* 76, 236–244.

Vérité, R. and Peyraud, J.L. (1989) Protein: The PDI system. In: Jarrige, R. (ed.) *Ruminant Nutrition, Recommended Allowances and Feed Tables.* INRA, Paris, pp. 33–47.

Webster, A.J.F. (1992) The metabolizable protein system for ruminants. In: Garnsworthy, P.C., Haresign, W. and Cole, D.J.A. (eds) *Recent Advances in Animal Nutrition.* Butterworths, London, pp. 93–110.

Whitelaw, R.G., Milne, J.S., Orskov, E.R. and Smith, J.S. (1986) The nitrogen and energy metabolism of lactating cows given abomasal infusions of casein. *British Journal of Nutrition* 55, 537–556.

Sodium Nutrition of Dairy Cows | 2

P.C. Chiy and C.J.C. Phillips
*Department of Clinical Veterinary Medicine, University of
Cambridge, Madingley Road, Cambridge CB3 0ES, UK*

Introduction

Sodium is used routinely as a supplement for dairy cows even though in
temperate regions their forage usually contains more than the recommended
minimum concentration of 1.3–1.8 g Na kg^{-1} DM (MAFF, 1983; NRC, 1989).
It is well known that cows have a strong appetite for sodium. This appetite
for sodium is not surprising, because they have a constant high output of
sodium in milk, and internally sodium functions as the main regulator of the
ionic status of body fluids and is used in essential physiological functions such
as neurotransmission. Indeed the appetite for sodium is so acute, that perception of sodium status by the animal is well developed from the tongue to the
milieu interieur. The sodium appetite and the routine use of sodium supplements by farmers suggests that the recommended minimum sodium concentration in the diet only prescribes for the level of intake necessary to avoid a clinical
deficiency, and that there are beneficial effects of increasing sodium intake.

In free access mineral supplements the animal restricts its own intake
because the salt content is very high, potentially causing localized toxic effects
in the rumen and possible dehydration. However, given the range in intakes
from salt licks (Table 2.1), it must be questioned whether cows can self-regulate
sodium intakes from licks. However, the craving of ruminants, especially
dairy cows, for salt suggested to us that enrichment of their *forage* with sodium
could provide a more uniform sodium intake and possibly increase intake and
production.

Table 2.1. Daily consumption from salt licks by cows.

	Average Na uptake (g d^{-1})	Range
Winter period	13	4-42
Grazing period	7	2-35

Source: Hartmans (1971).

Table 2.2. Variation in Na, K, Mg and Ca contents in grazed pastures.

	Mean composition (g kg^{-1} DM)	Between-sample CV[1] (%)	Between-farm CV[1] (%)
Na	3.3	43.9	55.5
K	30.4	23.7	15.7
Mg	1.7	20.6	9.2
Ca	4.9	16.9	28.4

[1] Coefficient of variation.
Source: Thomson and Warren (1979).

Pasture and Milk Production Responses

Sodium availability to plants is more variable than for other nutrients, mainly because of differences in atmospheric deposition and soil geochemistry. As a result, and because some plant species have evolved as sodium-tolerant or sodium-excluding, the sodium content of plant species is more variable than other micronutrients (Table 2.2).

Responses in herbage yield to sodium tend to be restricted to situations where potassium is limiting or herbage is mainly composed of natrophilic species in free-draining soils. Pre-occupation of the fertilizer companies with herbage yield rather than compositional responses has resulted in a market dominated by just three plant nutrients – nitrogen, phosphorus and potassium, with particular emphasis on the first. The oligopoly that has arisen in recent years in the fertilizer industry prospered through the widespread recommendation of levels of N, P and K that can at best be described as providing insurance for plant requirements. In the era of agricultural intensification in industralized nations during the 1970s and early 1980s this was the best policy for farmers who could sell all their output at considerable profit. The widespread use of potassium fertilizers satisfied the potassium needs of the plant but unfortunately reduced the sodium content to an average in pasture of 1.5 g kg^{-1} DM, not much above the minimum recommended level for dairy cows (Hemingway, 1995). With subsequent restrictions on output and competition from Eastern

Europe for the N, P and K market, the fertilizer manufacturers have been forced to seek alternative nutrients that can give added value to their products. Sulphur has received widespread attention with the reduction in sulphur emissions from industry in many western nations. However, the folly of testing only on responses in herbage yield is exemplified by recent evidence that sulphur fertilizer application to pasture can reduce milk yield and fat content (Phillips *et al.*, 1995).

By contrast, sodium fertilizer application increases milk yield and fat content (Chiy and Phillips, 1991; Chiy *et al.*, 1993a), even if herbage production is not increased (Chiy and Phillips, 1995a). The increase in milk fat content has also been reported for sodium chloride added to a concentrate supplement fed to grazing cows (Chiy and Phillips, 1991). It is likely that this is caused by an elevation of rumen pH by sodium absorbed from the rumen and subsequently recycled in the gastrointestinal tract via salivary buffer salts (Chiy *et al.*, 1993b). The increased rumen pH is particularly needed by cows grazing young leafy pasture which has a low buffering capacity (Jasaitis *et al.*, 1987). Adding sodium to conserved feeds does not usually increase milk production (Rehearte *et al.*, 1984). The increase in milk yield is specific to the use of sodium as a fertilizer and does not occur when sodium is added to a concentrate supplement. It therefore directly reflects the effects of sodium on herbage composition, and the response of the cow to this herbage.

Although sodium is only essential for a few halophytic plants, such as *Atriplex vesicaria*, it is able to replace potassium in many other non-specific functions such as maintenance of cell turgor and depolarization of cell membranes. Potassium also functions in the activation of enzymes and is actively absorbed by plants for this purpose, as opposed to sodium which is passively absorbed (Spanswick and Williams, 1964). Indeed, potassium is up to four times as efficient in enzyme activation as sodium (Nitsos and Evans, 1969). When the soil medium is flooded with sodium ions, many of which are passively absorbed and replace potassium in the plant, the plant contains more univalent cations in total which serve as an osmotic force to draw water into the plant. This increases cell turgor. However, less starch is synthesized because sodium is less effective than potassium in activating starch synthetase (Hawker *et al.*, 1974), but there is an increase in water-soluble carbohydrate concentration in the plant (Chiy and Phillips, 1993a).

Probably the major reason for the effect of sodium fertilizer on herbage consumption by the dairy cow is the change in mineral content. The sodium content is increased with a diminishing response as sodium application increases (Chiy and Phillips, 1993a). Associated with the increase in sodium content is a linear reduction in potassium content and an increase in magnesium and calcium contents (Fig. 2.1). If there are large increments in herbage sodium contents due to sodium fertilizer, there is also an increase in herbage phosphorus content (Fig. 2.1).

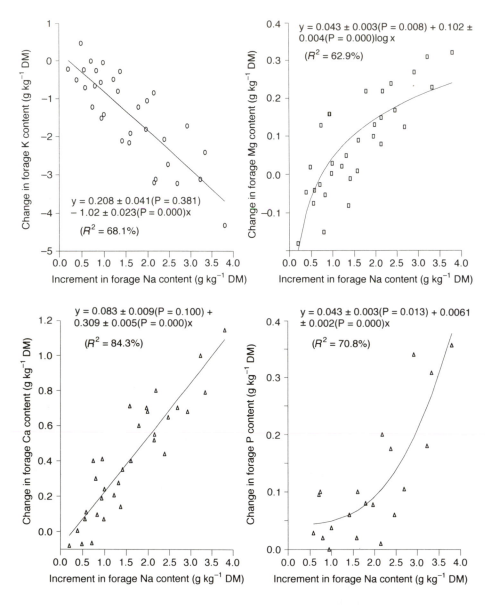

Fig. 2.1. Relationships between changes in herbage Na content due to Na fertilizer and response in inorganic nutrient contents (from the data of Abdullatif, 1991; Chiy and Phillips, 1991, 1993a, 1993b; Chiy, 1992 and Chiy *et al.*, 1993b).

Table 2.3. The effects of application of sodium fertilizer to pasture on the grazing behaviour of calves.

	Treatment			
	Sodium	Control	SED	*P*-value
Grazing (min day^{-1})	534	424	14.8	0.06
Biting rate (bites min^{-1})	77.2	75.7	0.59	<0.001
Walking rate[1] (m min^{-1})	2.20	1.98	0.06	<0.001
Rumination (min day^{-1})	505	421	9.9	<0.001
Drinking (min day^{-1})	1.1	0.7	0.09	0.002

[1] while grazing.
Source: Youssef *et al.* (1993).

Grazing Behaviour

The change in herbage composition, in particular the increase in sodium content, leads to changes in cow grazing behaviour that, in conjunction with improved herbage digestibility, appear responsible for the increases in milk yield that have been observed in five separate experiments at Bangor, North Wales, and with less productive pasture in Estonia (Chiy and Phillips, 1991, 1993c; Chiy *et al.*, 1993a; Phillips *et al.*, 1995; Arney *et al.*, 1995). Cows prefer to graze pasture fertilized with sodium than similar pastures with no sodium fertilizer (Chiy and Phillips, 1991). They graze for longer, bite faster and consume the sward down to a shorter height. They are essentially more active grazers. When grazed on a mosaic of pasture with and without sodium fertilizer, they walk faster while grazing on the pasture with sodium, indicating that there is less need for selection. Recent evidence suggests that they may even increase their intake of young expanding leaves, where the effect of the sodium fertilizer in increasing sodium content is greatest, and reduce their intake of older leaves (Chiy and Phillips, unpublished data). Paradoxically, when grazed on a mosaic of pasture with or without sodium fertilizer the cattle still graze for longer on pasture with sodium, even if the areas of fertilized and unfertilized pasture are the same. This remarkable feat of recognition demonstrates how acute the sodium appetite is in ruminants (Table 2.3) (Youssef *et al.*, 1993).

The changes in the grazing behaviour of cattle on pasture with sodium are accompanied by changes in drinking and rumination. The additional sodium load and the need to maintain hypotonicity and electrolyte balance in the body's cells necessitates the consumption of greater quantities of water (*c.* 30–50 ml water g^{-1} NaCl; Chiy and Phillips, 1995b). This increases the turnover rate of rumen fluid and associated small particles, which might be expected to increase DM intake and reduce digestibility. However, modest enhancement of sodium in herbage has been found to *increase* DM digestibility, both *in vivo* (Moseley, 1980; Chiy *et al.*, 1993b), *in vitro* (Chiy and Phillips, 1991, 1993a) and *in sacco* (Chiy *et al.*, 1993b). This may relate to enhanced activity of cellulolytic bacteria in sodium-rich media (Ueseka *et al.*, 1967; Wiedemeir *et al.*, 1987). Unlike most plants, rumen microorganisms tend to have a requirement for both Na and K, although the two are to some extent interchangeable. Enhanced cellulolysis could also result from the increased rumination that occurs when cattle graze sodium-enhanced pasture (Chiy and Phillips, 1991; Chiy *et al.*, 1993a; Youssef *et al.*, 1995). The reason for this increased rumination is unclear, since there is usually no change in the fibre content of herbage with sodium fertilizer, but the need to recycle sodium via saliva could increase salivation rates and facilitate rumination.

Calf Behaviour

Satiation of the sodium appetite by adding salt to the concentrate feed for calves has recently been demonstrated to lead to beneficial effects on calf behaviour (Youssef *et al.*, 1995). Young calves consume their concentrates slowly throughout the day so there is little danger of temporary toxic effects occurring at the site of digestion. The addition of 5 g Na (as NaCl) to each kg of concentrate was found to increase the concentrate intake of calves before weaning (Table 2.4).

As previously observed with adult cattle, the calves with extra sodium ruminated for longer. This may help to satiate the motivation for oral activity in the young calf. Most calves perform excessive oral behaviour, and traditionally it has been believed that this is because the calves are genetically programmed to search for their mothers' teat to receive sustenance. In penned calves, thwarting of the suckling stimulus by providing milk in a bucket and the restriction of socialization leads to the development of oral vices such as kissing, ear sucking and excessive licking of the surroundings. Once acquired these vices are difficult to eliminate and often develop into adult vices, such as teat sucking in dairy cows and prepuce sucking in steers (Phillips, 1993). From the results presented in Table 2.4 it appears that sodium deficiency is also contributing to the development of oral vices in the calves.

The sodium appetite may be partially responsible for the excessive licking activity that housed older cattle indulge in, as the crowded conditions often make the cattle sweat, which provides a ready source of sodium that can be licked off the coat during grooming.

The absolute nature of the sodium appetite must be further questioned since calves offered sodium enriched concentrates before weaning subsequently show a preference for feeds with a greater sodium content than calves with no

Table 2.4. The effects of adding sodium chloride to the concentrate ration for calves before weaning on their intake and behaviour.

	NaCl			
	+	−	SED	Probability
Concentrate intake (kg day^{-1})	1.19	0.97	0.029	<0.001
Water intake (l day^{-1})	2.98	1.93	0.09	<0.001
Rumination time (min per 12 h)	121	99	9.75	0.03
Oral vices (incidents per 12 h)				
Calf-kissing	1.9	2.4	0.35	0.13
Ear sucking	1.5	2.8	0.40	<0.001
Licking buckets	2.3	3.9	0.63	<0.001
Licking pen	4.1	9.1	0.80	<0.001

Source: Youssef *et al.* (1995).

sodium supplement before weaning (Fig. 2.2). This phenomenon could be exploited in saline regions to encourage ruminants to accept food with a high sodium content, e.g. *Atriplex vesicaria*. It is also known that the salt flavour cannot be considered in isolation. It can be masked, for example, by a sweet taste (Chiy and Phillips, 1995c).

Role of Sodium in Metabolic Disorders

The susceptibility of dairy cows to metabolic disorders has been increased in recent years by genetic improvements that have led to increased milk yields and increased risk of inadequate mineral intakes, especially in early lactation. In the future, the likelihood of further increases in milk yield with the advent of frequent milking by robotic systems, rbST and localized autocrine milk regulation (see Chapter 15) will further impose demands on the animal to absorb adequate nutrients to prevent metabolic diseases such as hypocalcaemia and hypomagnesaemia.

Efficient nutrient utilization by livestock is a function of adequate intake of a given nutrient, its unimpaired release from the ingested feed and optimal availability to, absorption by, and retention in animals. Optimization of these functions to fulfil livestock physiological requirements is not often achieved under normal feeding conditions. This frequently causes economic losses, mainly through clinical disease in the case of calcium (parturient paresis) and magnesium (grass tetany) and reduced performance in the case of phosphorus and sodium. Additional difficulties of improper nutrient balance occurs in diet formulation for two reasons. First, the addition of one nutrient almost invariably affects the concentration and utilization of others, particularly those

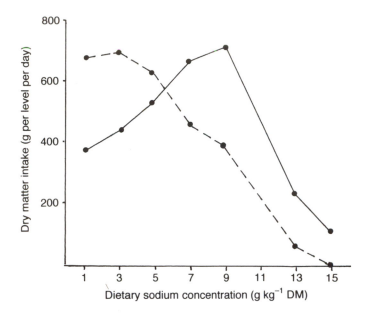

Fig. 2.2. Effects of early exposure of calves to high sodium feeds on their later preference for sodium content of feeds: —— exposed calves; - - - - calves not exposed (from Phillips and Youssef, 1994).

with similar chemical properties. The antagonism between sodium and potassium and their specific involvement at the cellular level in water metabolism, nutrient uptake and transmission of nerve impulses is a well-known example. Inadequate magnesium uptake is known to be partly caused by fertilizer potassium, which elevates herbage potassium content and prevents magnesium absorption in the rumen. Fertilizer sodium, by restricting herbage potassium uptake, offers a means to increase magnesium uptake in cattle and could potentially increase the uptake of other minerals (Table 2.5). Second, the significance of mineral ratios in causing disease or affecting production overrides the importance of the mineral *per se* in many situations. A classic example is the much greater significance of the Ca:P ratio in the utilization of either element and their relation to hypomagnesaemia than the absolute concentration of either element.

High levels of Na ingestion, such as often occur in saline environments or when loose forms of Na are included in conserved feeds, can be detrimental to animal performance. However, in more normal situations sodium supplementation can be beneficial to livestock productivity. Utilization of organic nutrients and cations by ewes is improved by increasing dietary Na up to about 6 g Na kg^{-1} DM. There is little additional benefit in nutrient intake, absorption and retention in increasing dietary Na in DM above 5.7 g kg^{-1} and

Table 2.5. Effect of sodium fertilizer on the uptake and apparent retention of minerals.

| | kg Na fertilizer ha^{-1} year^{-1} | | | | | |
	0	32	62	96	SED	P
Uptake (g day^{-1})						
Ca	10	14	16	22	0.71	0.016
Mg	2.7	3.5	4.3	5.2	0.62	0.004
P	6.0	7.2	7.3	8.6	0.35	0.003
Apparent retention (g kg^{-1} ingested mineral)						
Na	463	440	485	416	32.4	0.044
K	543	574	541	463	7.38	0.043
Mg	−324	9	0	−155	65.3	0.008
Ca	−280	−103	−104	−128	55.3	0.007
P	−8	138	134	107	39.7	0.006
S	−310	−260	−253	−397	32.9	0.005
N	533	532	511	457	21.3	0.046

Adapted from Chiy and Phillips (1993b) and Chiy *et al.* (1994).

levels greater than 6.5 g Na kg^{-1} DM are detrimental to livestock. Sodium, and to a lesser extent magnesium and calcium, are more effectively solubilized with the increase in sodium fertilizer (Chiy and Phillips, 1993b). Potassium tends to be less well solubilized, while release of phosphorus is unaffected by fertilizer sodium. Taking into account endogenous faecal losses, the true availability of calcium but not magnesium or phosphorus is increased by sodium fertilizer. Sodium fertilizer, however, increases the uptake of calcium, magnesium and phosphorus. Although potassium is the most soluble mineral in the rumen, it is less soluble with sodium fertilizer because of competition from the other minerals, especially sodium. As a result, magnesium and calcium uptake are approximately doubled at the highest sodium fertilizer level, and phosphorus uptake is increased by almost 50%. These changes in mineral balance should enable mineral requirements to be adequately met from grazed herbage.

Hypomagnesaemia

Hypomagnesaemia is one of the main metabolic disorders of cattle and sheep at pasture. Its incidence is sporadic and intensity variable. It is associated with a low Na:K ratio in forage and with high potassium pastures that are low in magnesium, calcium and sodium (Russel and Duncan, 1956). Potassium interferes with magnesium absorption, rather than substantially increasing its excretion (House and van Campen, 1971; Newton *et al.*, 1972). Feeding diets that are high in potassium reduces absorption of magnesium in the rumen

(Tomas and Potter, 1976; Wylie et al., 1985) and plasma magnesium, particularly when diets low in sodium are fed (Suttle and Field, 1967). Also, magnesium absorption from the recticulo-rumen is reduced by a low Na:K ratio in the rumen, which depends partly on the saliva Na:K (Martens and Rayssiguier, 1980; Martens et al., 1987). Several researchers have successfully reduced the incidence of hypomagnesaemia in dairy cows by feeding sodium supplements (Smith and Aines, 1959; Paterson and Crichton, 1960) or by using a sodium fertilizer (Smith et al., 1982).

The interaction between sodium, potassium and magnesium at the cellular level

The transfer of Na^+ and K^+ cations across cell membranes by the Na pump is dependent on an enzyme, ATPase, which is activated by both sodium and potassium and also requires magnesium for its function. Using the energy from ATP, this moves K^+ into, and Na^+ out of, the cell, both against their respective concentration gradients. The permeability of the cell membrane to K^+ and Cl^-, but not Na^+, allows K^+ to diffuse along its concentration gradient from the cell to the extracellular space and Cl^- in the reverse direction. This results in an increase in positive charges outside the cell and in negative charges inside, thus setting up a membrane potential. The resultant energy of this potential is used in many metabolic steps, including active transport of solutes, such as sodium and glucose co-transport into the cell.

Resting cell membrane potential ranges from -70 to $-90\,mV$, but sensory, nerve and muscle cells can alter this potential. When a strong external signal exceeds this threshold value, the cell membrane momentarily becomes permeable to Na^+, which diffuses into the cell over a short period, reversing the charge concentration – the cell becoming positive and the extracellular space negative. An action potential is thus induced, depolarizing the cell membrane. The resting potential is restored through the action of the Na/K pump, activated by Mg.

The interaction between sodium, potassium and magnesium in animals and plants

Under conditions where sodium replaces potassium in plants where it has been depleted, an increase in the magnesium content of legumes was found (Whitehead and Jones, 1972). In a survey of the factors predisposing cows to hypomagnesaemia in 64 farms in southern Scotland, Butler (1963) found a negative correlation ($r = -0.459; P < 0.05$) between sodium content of pasture and the incidence of hypomagnesaemia. As Na content declined, the incidence of hypomagnesaemia increased. They found that mean sodium content was higher in normal than in tetany-prone pastures (1.93 vs. 1.21 g kg^{-1} DM). Tetany-prone pastures were probably also inadequate in magnesium.

The primary effect of high potassium intakes is a reduction in magnesium absorption leading to an increase in faecal magnesium excretion (Suttle and Field, 1967; Tomas and Potter, 1976; Wylie *et al.*, 1985; Greene *et al.*, 1986), but the mechanism by which sodium can reduce the incidence of hypomagnesaemia has not been extensively studied. Increasing the *in vitro* concentration of sodium in the intestinal lumen of rats has been demonstrated to lead to an increased magnesium absorption (Ross, 1960). Uptake of magnesium prior to the duodenum is improved by a high Na:K ratio in rumen liquor (Erise, 1979; McGregor and Armstrong, 1979), possibly as a result of a change in electric potential across the gut wall (Care *et al.*, 1967) or electrophysiological changes at the rumen site of the membrane which determines uptake of magnesium into the epithelium (Martens *et al.*, 1987). Increasing herbage sodium content from 4 to 7 g Na kg^{-1} DM by applying 160 kg Na ha^{-1} increased the apparent availability and retention of sodium, potassium, magnesium and calcium (Moseley, 1980). In addition, salt supplementation of the low sodium diet in the latter study tended to improve the apparent availability of minerals. Not all of the extra minerals are retained, the urinary loss of minerals increased with salt addition to the diet within the range 3.9 to 18.2 g Na kg^{-1} DM. At higher rumen potassium concentrations, the transmural potential difference increases and more Mg^{2+} ions remain in the more negative rumen in comparison to diets high in sodium (Care *et al.*, 1967). Feeding high sodium grass pellets results in nearly equimolar concentrations of sodium and potassium in the rumen and increases magnesium absorption (Table 2.6) by reducing the transmural potential difference between the rumen (negative) and the blood (positive). Conversely, when a low sodium diet is fed, more potassium is retained in the rumen to maintain pH, which increases the transmural potential difference and magnesium resorption is reduced.

Grass containing more than 30 g K kg^{-1} DM has been associated with hypomagnesaemia in dairy cattle (Wolton, 1963) but potassium content is not as useful as the ratio K:(Ca + Mg) (mEq kg^{-1} DM) as an indicator of the risk of grass tetany (Kemp and t'Hart, 1957; Butler, 1963; Grunes *et al.*, 1970). The threshold value above which hypomagnesaemia is likely to occur is 2.2:1 (expressed in mEq kg^{-1} DM).

Both laboratory and field studies have therefore demonstrated the beneficial effect of sodium in reducing the risk of hypomagnesaemic tetany, mediated through the antagonistic effects of sodium and potassium.

Sodium Fertilizers

The research previously described suggests that there are benefits to dairy farmers of applying sodium as a fertilizer to pastures for grazing. This could

Table 2.6. Effect of adding sodium to a sodium-deficient diet on sodium and potassium concentration in the rumen and on magnesium resorption in sheep.

	No supplementary Na	With supplementary Na
Dietary mineral content (g kg^{-1} DM)		
Na	0.25	2.1
K	34.3	34.3
Rumen mineral content (m mol)		
Na	31.0	58.4
K	100.2	54.7
Mg resorption (g d^{-1})	0.7	0.95
Potential difference between rumen and blood (mV)	49.3	39.7

Source: Kubel (1982).

replace other fertilizers to provide a more nutritionally balanced feed for the cows and could in particular reduce the quantity of nitrogen that is lost from the system by leaching. Increased nitrogen recovery by 10–15% with sodium fertilizer has been recorded (Chiy and Phillips, 1993a), probably as a result of increased herbage growth. Sodium itself is not very readily leached (Bolton and Penny, 1978). The main possibilities for reductions in fertilizer leaching are by tailoring the rate and composition of fertilizer applied to each field to the crop requirements (Phillips and Chiy, 1995). In determining the rate of sodium to be applied, account must be taken of marine aerosol deposition, return from animal excreta and soil type (Phillips and Chiy, 1995).

Sodium is available mainly as sodium nitrate, $NaNO_3$, and sodium chloride, NaCl, although some fertilizers include sodium as Na_2O. Compound fertilizers are manufactured which combine sodium nitrate with ammonium nitrate and naturally occurring mixtures of sodium and potassium chlorides are also available. As sodium can last up to seven years in the soil (Bolton and Penny, 1978), applications can be infrequent. However, the sensitivity of cattle to herbage sodium content make it preferable to apply sodium at regular intervals, in particular by the application of small amounts of sodium in conjunction with the regular nitrogen applications during the growing season.

Conclusions

Sodium is not essential for the growth of most cattle feeds in temperate regions. However, an increase in sodium content of pasture will increase herbage intake and milk production of dairy cows. In addition, the reduction of herbage potassium content and increase in magnesium absorption from the rumen could reduce the risk of hypomagnesaemia. Supplying additional sodium for grazing cows in either increased herbage content or adding sodium to concentrate supplement will increase milk fat production by buffering the low rumen pH of cows grazing fresh herbage. Adding sodium to the limited number of fertilizer nutrients that are currently supplied to dairy cow pastures could therefore provide a more balanced feed, increase the efficiency of milk production and reduce the risk of metabolic disorders.

References

Abdullatif Al-Tulihan (1991) Salt effects on pasture composition and grazing preference of cattle. MSc thesis, University of Wales, Bangor.

Arney, D., Chiy, P.C. and Phillips, C.J.C. (1995) Lactational responses of dairy cows to sodium and potassium fertilization of pasture. *Annales de Zootechnie* 44 (suppl. 1), 371.

Bolton, J. and Penny, A. (1978) The longevity of sodium, potassium and magnesium fertilizer residual effects on the yield and composition of ryegrass grown on a sandy soil. *Journal of Agricultural Science, Cambridge* 91, 696–699.

Butler, E.J. (1963) The mineral element content of spring pasture in relation to the occurrence of grass tetany and hypomagnesaemia in dairy cows. *Journal of Agricultural Science, Cambridge* 60, 329–340.

Care, A.D., Vowles, L.E., Mann, S.O. and Ross, D.B. (1967) Factors affecting magnesium absorption in relation to the aetiology of acute hypomagnesaemia. *Journal of Agricultural Science, Cambridge* 68, 195–204.

Chiy, P.C. (1992) Sodium supply for pasture and dairy cow production. PhD thesis, University of Wales, Bangor.

Chiy, P.C. and Phillips, C.J.C. (1991) The effects of sodium chloride application to pasture, or its direct supplementation, on dairy cow production and grazing preference. *Grass and Forage Science* 46, 325–331.

Chiy, P.C. and Phillips, C.J.C. (1993a) Sodium fertilizer application to pasture. 1. Direct and residual effects on pasture production and composition. *Grass and Forage Science* 48, 189–202.

Chiy, P.C. and Phillips, C.J.C. (1993b) Sodium fertilizer application to pasture. 4. Effects on mineral uptake and the sodium and potassium status of steers. *Grass and Forage Science* 48, 260–270.

Chiy, P.C. and Phillips, C.J.C. (1993c) Sodium fertilizer for pasture and dairy cow production. *Occasional Publication of the Dairy Research Unit*, University of Wales, Bangor.

Chiy, P.C. and Phillips, C.J.C. (1995a) Sodium in forage crops. In: Phillips, C.J.C. and Chiy, P.C. (eds) *Sodium in Agriculture*. Chalcombe Publications, Canterbury, pp. 43–69.

Chiy, P.C. and Phillips, C.J.C. (1995b) Sodium in ruminant production, reproduction and health. In: Phillips, C.J.C. and Chiy, P.C. (eds) *Sodium in Agriculture*. Chalcombe Publications, Canterbury, pp. 107–144.

Chiy, P.C. and Phillips, C.J.C (1995c) Effect of flavours on dairy cow feeding behaviour. *Proceedings of the 9th International Congress of the International Society for Applied Ethology*, 3–5 August 1995, Exeter, pp. 163–164.

Chiy, P.C., Phillips, C.J.C. and Bello, M.R. (1993a) Sodium fertilizer application to pasture. 2. Effects on dairy cow production and behaviour. *Grass and Forage Science* 48, 203–212.

Chiy, P.C., Phillips, C.J.C. and Omed, H.M. (1993b) Sodium fertilizer application to pasture. 3. Rumen dynamics. *Grass and Forage Science* 48, 249–259.

Chiy, P.C., Phillips, C.J.C. and Ajele, C.L. (1994) Sodium fertilizer application to pasture. 5. Effects on herbage digestibility and mineral availability in sheep. *Grass and Forage Science* 49, 25–43.

Erise, B.M (1979) Magnesium absorption by the ruminant. PhD thesis, University of Reading.

Greene, L.W., Shelling, G.T. and Byeres, F.M. (1986) Effects of dietary monensin and potassium on absorption of magnesium and macroelements in sheep. *Journal of Animal Science*, 63, 1960–1967.

Grunes, D.L., Stout, P.R. and Brownell, J.R. (1970) Grass tetany of ruminants. *Advances in Agronomy*, 22, 331–374.

Hartmans, J. (1971) Effect of calcium on resorption and excretion of major and some minor elements in cattle. *Proceedings of the 8th Colloquium of the International Potash Insitute*, 207–211.

Hawker, J.S., Marchner, H. and Downton, W.J.S. (1974) Effect of sodium and potassium on starch synthesis in leaves. *Australian Journal of Plant Physiology* 1, 491–501.

Hemingway, R.G. (1995) Requirements for sodium by livestock and dietary allowances. In: Phillips, C.J.C. and Chiy, P.C. (eds) *Sodium in Agriculture*. Chalcombe Publications, Canterbury, pp. 43–69.

House, W.A. and Van Campen, D. (1971) Magnesium metabolism of sheep fed different levels of potassium and citric acid. *Journal of Nutrition* 101, 1483–1492.

Jasaitis, D.K., Wohlt, J.E. and Evans, J.L. (1987) Influence of fed ion content on buffering capacity of ruminant feedstuff *in vitro*. *Journal of Dairy Science* 70, 1391–1403.

Kemp, A. and t'Hart, M.L. (1957) Grass tetany in grazing milking cows. *Netherlands Journal of Agricultural Science* 5, 4–17.

Kubel, O.W. (1982) Der Einluss des Na-Mangels bei Schafen auf die passage und resorption von Mg, Na und K im gesanten verduungstrakt. Hannover, 109 pp.

McGregor, R.C. and Armstrong, D.G. (1979) The effect of increasing potassium intake on the absorption of Mg by sheep. *Proceedings of the Nutritional Society*, 38, 66A.

Martens, H. and Rayssiguier, Y. (1980) Magnesium metabolism and hypomagnesaemia. In: Ruckebusch, Y. and Thivend, P. (ed.) *Digestive Physiology and Metabolism in Ruminants*. AVI Publishing Co., Inc. Westport, CT, pp. 447–466.

Martens, H., Kubel, O.W., Gabel, G. and Honig, H. (1987) Effects of low sodium intake on magnesium metabolism of sheep. *Journal of Agricultural Science, Cambridge* 108, 237–243.

Ministry of Agriculture, Fisheries and Food, Department of Agriculture for Scotland, Department of Agriculture for Northern Ireland (MAFF) (1983) *Mineral, Trace Element and Vitamin Allowances for Ruminant Livestock*. HMSO, London.

Moseley, G. (1980) Effects of variation in herbage sodium levels and salt supplementation on the nutritive value of perennial ryegrass for sheep. *Grass and Forage Science* 35, 105–113.

National Research Council (NRC) (1989) *Nutritive Requirements of Dairy Cows*. National Academy of Sciences, Washington, D.C.

Newton, G.L., Fontenot, J.P., Tucker, R.E. and Polan, C.E. (1972) Effects of high dietary potassium intake on the metabolism of magnesium by sheep. *Journal of Animal Science* 35, 440–445.

Nitsos, R.E. and Evans, H.I. (1969) Effects of univalent cations on the activity of particulate starch synthethase. *Plant Physiology* 44, 1260–1266.

Paterson, R. and Crichton, C. (1960) Grass staggers in large scale dairying on grass. *Journal of British Grassland Society* 15, 100–105.

Phillips, C.J.C. (1993) *Cattle Behaviour*. Farming Press, Ipswich, pp. 193–196.

Phillips, C.J.C. and Chiy, P.C. (1995) Sodium fertilizers. In: Phillips, C.J.C. and Chiy, P.C. (eds) *Sodium in Agriculture*. Chalcombe Publications, Canterbury, pp. 70–81.

Phillips. C.J.C. and Youssef, M. (1994) Conditioning calves to accept high sodium feeds. *Animal Production* 58, 454 (abstr.)

Phillips, C.J.C., Avezinius, J. and Chiy, P.C. (1995) Effects of sodium and sulphur fertilizers on dairy cow production. *Animal Science* 60, 516(abstr.).

Rehearte, D.H., Kesler, E.M. and Stringer, W.C. (1984) Forage growth and performance of grazing dairy cows fed concentrates with or without sodium bicarbonate. *Journal of Dairy Science* 67, 2914–2921.

Ross, D.B. (1960) Influence of sodium on the transport of magnesium across the intestinal walls of the rat *in vitro*. *Nature* 189, 840–841.

Russel, J.B. and Duncan, D.L. (1956) In: *Minerals in Pasture: Deficiencies and Excesses in Relation to Animal Health*. Commonwealth Agricultural Bureaux, Farnham Royal, England.

Smith, G.S., Young, P.G. and O'Connor, M.B. (1982) Some effects of topdressing pasture with sodium chloride on plant and animal nutrition. *Proceedings of the New Zealand Grassland Society* 44, 179–183.

Smith, S.E. and Aines, P.D. (1959) Salt requirements of dairy cows. *Cornell University Agricultural Experimental Station Bulletin* 938, pp. 26.

Spanswick, R.M. and Williams, E.J. (1964) Electrical potentials and Na, K, and Cl concentrations in the vacuole and cytoplasm of *Nitella translucens*. *Journal of Experimental Botany* 55, 193–200.

Suttle, N.F. and Field, A.C. (1967) Studies on Mg in ruminant nutrition. *British Journal of Nutrition*, 21, 819–831.

Thomson, J.K. and Warren, R.W. (1979) Variations in composition of pasture herbage. *Grass and Forage Science* 34, 83–88.

Tomas, F.M. and Potter, B.J. (1976) The effect and site of action of K upon Mg absorption in sheep. *Australian Journal of Agricultural Research* 27, 873–880.

Uesaka, S., Kawashima, R., Zembayashi, M., Won Kim, S. and Toyama, T. (1967) Studies on the importance of trace elements in farm animal feeding. XXXII. Effects of trace elements on cellulose digestion by rumen bacteria (II). *Bulletin of the Research Insititute of Food and Science, Kyoto University* 30, 18–27.

Whitehead, D.C. and Jones, L.H.P. (1972) The effect of replacing potassium by sodium on cation uptake and transport to the shoots in four legumes and Italian ryegrass. *Annals of Applied Biology* 71, 81–89.

Wiedmeier, R.D., Arambel, M.J., Lamb, R.C. and Marrcinkowski, D.P. (1987) Effect of mineral salts, carbachol, and pilocarpine on nutrient digestibility and ruminal characteristics in cattle. *Journal of Dairy Science* 70, 592–600.

Wolton, K.M. (1963) Sodium nutrition of plants. *National Agricultural Advisory Service Quarterly Review* 79, 101–109.

Wylie, M.J., Fontenot, J.P. and Greene, L.W. (1985) Absorption of magnesium and macro-minerals infused with K in different parts of the digestive tract. *Journal of Animal Science* 61, 1219–1228.

Youssef, M.Y.I., Chiy, P.C., Phillips, C.J.C. and Metwally, M. (1993) The effects of species complexity and sodium fertilizer application on the grazing behaviour of calves. *Animal Production* 56, 455 (Abstract).

Youssef, M.Y.I., Phillips, C.J.C. and Metwally, M. (1995) Effects of sodium on farm animal behaviour. In: Phillips, C.J.C. and Chiy, P.C. (eds) *Sodium in Agriculture.* Chalcombe Publications, Canterbury, pp. 70–81.

Nutrition and Mammary Host Defences Against Disease in Dairy Cattle

J.S. Hogan, W.P. Weiss and K.L. Smith

Ohio Agricultural Research and Development Center, Department of Dairy Science, Ohio State University, 1680 Madison Avenue, Wooster, Ohio 44691–4096, USA

Introduction

Adequate dietary supplementation of micronutrients that affect bovine host defences is essential to maximize the profitability of a dairy herd. Deficiencies in a number of vitamins and minerals increase the susceptibility of dairy cows to infectious diseases. Vitamin A, β-carotene, vitamin E, copper, selenium, and zinc are among the list of essential micronutrients that influence resistance to diseases. While overt deficiencies in these micronutrients are unusual, marginal deficiencies are common in many dairy herds.

Most essential micronutrients that influence host defences are involved in antioxidant systems that maintain the integrity of phagocytic cells and lymphoid tissues. Impairments of the antioxidant system can result in a higher incidence and more severe clinical signs of diseases that have an economic impact on dairy herds. Mastitis is the disease of greatest economic concern to the dairy industry (Eberhart *et al.*, 1987). Therefore, the recent progress in determining the dietary levels necessary to optimize resistance in dairy cows has centred on control of intramammary infections.

Vitamin E and Selenium

Nutritional muscular dystrophy occurs in young dairy animals suffering from severe vitamin E and selenium deficiencies (Miller, 1979). Severe vitamin E and selenium deficiencies are uncommon in adult cows but borderline deficiencies are quite common. Clinical manifestations of marginal vitamin E and selenium deficiencies include increased incidences of mammary diseases and reproductive disorders (Harrison *et al.*, 1984; Smith *et al.*, 1985). Clinical trials and surveys have shown that vitamin E and selenium deficiencies result in increased rate and severity of clinical mastitis, elevated bulk milk somatic cell counts, and increased duration of intramammary infections (Smith *et al.*, 1984, 1985; Erskine *et al.*, 1987, 1989; Weiss *et al.*, 1990a). The impact of dietary supplementation of vitamin E and selenium is especially critical during the dry period and first few weeks of lactation.

Vitamin E and selenium share common biological activities. Vitamin E and the selenium containing enzyme glutathione peroxidase (GSH-Px) are antioxidants that protect tissue from oxidative attack by free radicals produced during the respiratory burst of neutrophils and macrophages. The respiratory burst by phagocytes is characterized by marked changes in oxygen metabolism that result in increased production of superoxide and hydrogen peroxide (Baboir, 1984). Though phagocyte-generated oxygen metabolites are necessary in antimicrobial defence mechanisms, these free radicals can also damage the neutrophil or macrophage and surrounding tissues (Weiss and LoBugilo, 1982). Glutathione peroxidase protects cells by detoxifying peroxides in the cytosol (Baboir, 1984). Vitamin E inhibits autoxidation of polyunsaturated fatty acids by oxygen metabolites in membranes (Boxer, 1986; Bendich, 1990).

Though data are limited linking deficiencies of either vitamin E or selenium in dairy cows to decreased immune response, deficiencies in these nutrients markedly reduced immune response to antigens in calves and other animals (Smith, 1986). Vitamin E and/or selenium deficiencies may adversely alter both membrane integrity of immune cells and arachidonic acid metabolism. Products of arachidonic acid metabolism are mediators of the inflammatory response (Kuehl and Egan, 1980). Vitamin E may suppress prostaglandin concentrations in tissues (Likoff *et al.*, 1981). Glutathione peroxidase participates directly in prostaglandin synthesis (Atroshi *et al.*, 1986a). Prostaglandin concentrations in milk were positively correlated with elevated somatic cell counts (Atroshi *et al.*, 1986b).

The greatest body of information concerning the bovine host defence and vitamin E and selenium (Se) deals with neutrophil function (Grasso *et al.*, 1990; Hogan *et al.*, 1990, 1992; Weiss *et al.*, 1994b). Neutrophils are considered to be a primary defence mechanism to bacterial infections in mammals. Incidence and severity of clinical signs associated with many bacterial infections depend on responsiveness of neutrophils (Craven and Williams, 1985). Therefore, herd management practices that result in optimal vitamin E and Se status of dairy cows also optimize neutrophil responses and increase resistance to diseases.

Neutrophils from either vitamin E-deficient or Se-deficient cows have been shown to have impaired bactericidal activities (Boyne and Arthur, 1979, 1981; Grasso *et al.*, 1990; Hogan *et al.*, 1990). Vitamin E and GSH-Px are both cellular antioxidants that protect against the cytotoxic capabilities of oxygen metabolites. Vitamin E protects within the membrane, whereas GSH-PX activity is in the cytosol. Glutathione peroxidase converts hydrogen peroxide

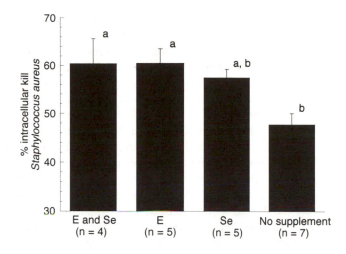

Fig. 3.1. Percentage of intracellular kill for *Staphylococcus aureus* in blood neutrophils from cows fed diets supplemented with vitamin E (E), Se, or both or neither. Figure redrawn from Hogan *et al.* (1993). [a,b]Means with different superscripts differ (*P* < 0.05).

to water and lipid hydroperoxides to the corresponding alcohol (Baker and Cohen, 1983). Vitamin E inhibits autoxidation of polyunsaturated fatty acids in neutrophil membranes (Baehner *et al.*, 1977; Baker and Cohen, 1983). Vitamin E is localized in cellular membranes in close proximity to the mixed function oxidase enzymes that initiate the production of free radicals (Chan *et al.*, 1989).

The effects of vitamin E and Se supplementation on intracellular kill of bacteria by neutrophils are not additive. Supplementation with both vitamin E and Se does not result in greater intracellular kill of bacteria by blood neutrophils than does supplementation with either one of the nutrients alone (Fig. 3.1). However, vitamin E and GSH-Px have sparing effects on the requirements for one another relative to intracellular kill of bacteria (Hogan *et al.*, 1990). The protection afforded cellular membranes by vitamin E may spare the requirement for GSH-Px by oxidizing free radicals at the membrane, thereby preventing leakage of free radicals into the cytosol and maintaining intracellular kill capacity of the cell. Conversely, GSH-Px activity in the cytosol may spare the requirement for vitamin E in the membranes. A similar sparing effect of vitamin E for protecting GSH-Px-deficient human neutrophils against oxidative damage has been documented (Boxer *et al.*, 1979).

Tissue concentrations of Se are correlated with GSH-Px activity and are directly related to dietary intake (Scholz and Hutchinson, 1979). Plants grown on Se-deficient soils generally do not provide adequate dietary Se to meet nutritional requirements of dairy cows. Selenium-deficient soils are geographically widespread, and approximately two-thirds of dairy cows in the United States and Europe are in areas of known Se-deficient soils (Smith *et al.*, 1992). A major source of vitamin E for dairy cows is forages (Schingoethe *et al.*, 1978), but the concentration of α-tocopherol in forages decreases as plants mature. Substantial loss of vitamin E also occurs when feedstuffs are processed and stored (Lynch, 1992). Total confinement of dairy herds (zero grazing) may result in increased feeding of locally grown and stored forages that are low in both vitamin E and Se.

Dietary intake and blood concentrations of Se can both be used to assess properly the Se status of a herd. Few data exist to suggest that dietary Se greater than $3-6$ mg day^{-1} results in additional enhancement of host defences against mastitis (Weiss *et al.*, 1990b). However, factors exist that interfere with Se absorption in the intestinal tract, such as sulphates, nitrates, and high concentrations of dietary Ca. To ensure that herd Se status is adequate, blood samples from a representative group of cows in the herd should be analysed. Whole blood concentrations of Se should be at least $0.2\,\mu g\,ml^{-1}$ but not exceed $1\,\mu g\,ml^{-1}$.

Dry and lactating cows should consume 1000 IU day^{-1} of vitamin E based on the significant reductions in mastitis when cows were supplemented with adequate vitamin E to achieve this consumption level (Smith *et al.*, 1984, 1985; Hogan *et al.*, 1990, 1992, 1993; Weiss *et al.*, 1990a, 1992). For cows fed stored forages, vitamin E may need to be supplemented at 1000 IU day^{-1} for dry cows and at 500 IU day^{-1} for lactating cows, dependent on forage quality and dry matter intake. Plasma concentrations greater than $3.5-4\,\mu g\,ml^{-1}$ of α-tocopherol are considered to be adequate (Fig. 3.2) as evidenced by the relationship between intracellular kill of bacteria by neutrophils and plasma vitamin E concentrations (Weiss *et al.*, 1994b).

Vitamin A and β-Carotene

Vitamin A is necessary for growth and maintenance of epithelial cells. Severe deficiencies in vitamin A are characterized by night blindness, drying of epithelial and mucosal surfaces and heightened susceptibility to diseases (Sommer, 1990). Improvements in mammary health and reproduction can be made by supplementing vitamin A and its precursor β-carotene to dairy cows during the periparturient period (Chew *et al.*, 1982; Johnston and Chew, 1984; Dahlquist and Chew, 1985; Daniel *et al.*, 1986). Vitamin A and β-carotene influence a number of host defence mechanisms. Vitamin A and β-carotene may influence cellular development and host responses by several mechanisms. These models include a role in post-translational modification of glycoproteins and vitamin A-cellular binding proteins (Zile and Cullum, 1983). In addition, vitamin A and β-carotene have unique antioxidant properties (Burton and Ingold, 1984; Zamora *et al.*, 1991).

Vitamin A is essential for proliferation and maturation of tissues including

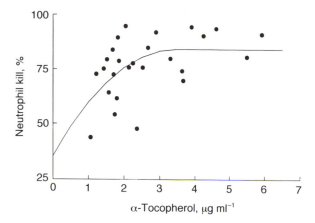

Fig. 3.2. Relationship between intracellular kill of bacteria by neutrophils and plasma α-tocopherol at calving. Figure from Hogan *et al.* (1993).

lymphoid organs. Animals fed vitamin A-deficient diets have impaired development of thymus, spleen, and lymph nodes. Associated with atrophy of lymphoid tissue, reduced trafficking and localization of lymphocytes within tissues occurs in vitamin A-depleted animals (McDermott *et al.*, 1982; Takagi and Nakano, 1983). Deficiencies in vitamin A can result in reduced blastogenesis by T cells, but function of B lymphocytes remains unaltered (Tjoelker *et al.*, 1988b). However, enhanced humoral responses in animals due to vitamin A supplementation have been reported (Cohen and Cohen, 1973). Vitamin A supplementation has a limited effect on phagocytosis and intracellular kill by phagocytes (Tjoelker *et al.*, 1988a).

β-carotene has immuno-enhancing properties independent of it's role as a vitamin A precursor. β-Carotene supplementation of cows enhanced mitogen stimulated proliferation of lymphocytes similar to the responses seen to vitamin A supplementation (Tjoelker *et al.*, 1988b). In addition, β-carotene is involved in proliferation of specific T lymphocyte subsets. Dietary supplementation of β-carotene increased blood concentrations of helper T lymphocytes by 30% while cytotoxic and suppressor T lymphocyte populations remained constant (Alexander *et al.*, 1985). Bactericidal activity of neutrophils has been found to be responsive to β-carotene at drying off and during the periparturient period in dairy cows (Chew, 1993). Although lymphocytes contain substantial amounts of β-carotene, bovine neutrophils contain very little (Weiss *et al.*, 1994a).

Dietary intake of 70,000–100,000 IU day^{-1} of vitamin A appears optimal for cellular and humoral immunity (Chew, 1987). Substantial amounts of vitamin A can be stored in the liver and mobilized during periods of reduced feed intake. Liver vitamin A values below $1.0 \, \mu g \, ml^{-1}$ indicate a critical deficiency in cattle (Miller, 1979). Requirements for β-carotene are not as well established. Cows can absorb significant quantities of β-carotene from their diets. Fresh forages contain large amounts of β-carotene and a response to supplemental β-carotene by cows fed fresh forage is unlikely. Improvement in host defences against disease following β-carotene supplementation of diets is dependent on the stage of lactation. Concentrations of β-carotene in plasma decrease at calving and host defences can be improved by β-carotene supplementation. Although routine supplementation with β-carotene is not currently recommended, supplementing $300\text{–}600 \, mg \, day^{-1}$ of β-carotene during the periparturient period may be beneficial to cows fed stored forages. The optimal blood concentration of β-carotene in dairy cows is at least $2 \, \mu g \, ml^{-1}$ (Chew, 1993).

Zinc

A wide variety of zinc-responsive pathological conditions have been observed in animals (Miller, 1970). The most apparent zinc-associated disease in dairy cows is acrodermatitis enteropathica. This is a lethal disorder affecting Friesian calves which is transmitted as an autosomal recessive trait. Death is caused by immune deficiency. Prolonged zinc depletion in cattle results in a pathological response comparable to acrodermatitis enteropathica (Sugarman, 1983). Marginal deficiencies in zinc may result in similar, but less severe, immunological suppression. Zinc metallo-enzymes are involved in metabolic processes including DNA synthesis, protein, carbohydrate and nucleic acid metabolism, cytokine production, and oxidation of free radicals (Underwood, 1977, 1981).

Zinc deficiencies decrease host defence capacity via reduction in total number of lymphocytes. Specifically, animals fed zinc deficient diets have decreased proliferation of T lymphocytes (Wirth *et al.*, 1984). This reduction in T cell proliferation in zinc deficient animals is attributed to interference of DNA and RNA synthesis of leucocytes and atrophy of lymphoid tissue (Kruse-Jarres, 1989). Zinc is a component of DNA polymerase essential for cell proliferation. In contrast, prolonged zinc deficiencies have little effect on B-lymphocyte function and humoral responses. Maturation of B lymphocytes does not appear to be influenced by zinc availability and phagocytic cells generally are not altered by zinc deficiencies (Cook-Mills and Fraker, 1993). Although lymphopenia is characteristic of zinc deficiencies, neutrophil and monocyte counts are not dependent on zinc status (Underwood, 1981).

Zinc plays an important role in the prevention of diseases by stabilizing epithelial cells (Moynahan, 1981). Zinc deficiencies produce marked reduction in incorporation of several amino acids into skin proteins. Clinical manifestation of severe to moderate zinc deficiencies in dairy cows includes skin parakeratosis with lesions commonly occurring on teat skin (Miller, 1979). Potentially important in the prevention of mastitis is the role zinc plays in desquamification of epidermal cells. Teat canal keratin, comprised of desquamified epidermal cells and milk constituents, is the primary barrier to intramammary infections. Zinc deficiencies may alter keratin composition and render the mammary glands more susceptible to infections. Zinc deficiencies have been related to increases in somatic cell counts in milk (Kincaid *et al.*, 1984; Aguilar *et al.*, 1988).

Dietary zinc intake of 40 ppm is recommended for lactating dairy cows. When zinc deficient cows are given supplemental zinc, there is a very rapid and dramatic recovery in appetite and thriftiness (Miller, 1970). Plasma zinc normally decreases with reduced feed intake during the periparturient period. Plasma zinc concentrations of less than 0.4 ppm indicate a severe deficiency.

Copper

Signs of copper deficiency include anaemia, diarrhoea, weight loss, and non-specific unthriftiness (Miller, 1979). As is the case with other micronutrients, the immune system is impaired before clinical signs of copper deficiencies are detected (Koller *et al.*, 1987). Primary and secondary lymphoid tissues are altered when copper intake is suboptimal in rodents. Copper deficient rodents have reduced thymic weight and enlarged spleens characterized by deterioration of subcellular organelles (Prohaska *et al.*, 1983). Animals with affected lymphoid tissue often experience dose dependent reduction in splenic and blood T lymphocytes as copper content of diets is reduced (Mulhern and Koller, 1988). Lymphopenia is common in some species of copper deplete animals. The leucocyte population that appears to be most sensitive to copper deficiencies are the T lymphocytes (Lukasewycz *et al.*, 1989; Lukasewycz and Prohaska, 1990). Although maturation of B lymphocytes appears to be unaffected by copper status, T lymphocyte mediated antibody class switching mechanism is altered by copper deficiency (Prohaska and Lukasewycz, 1981; Lukasewycz and Prohaska, 1990).

Marginal copper deficiencies in dairy cows can result in increased incidence of intramammary infections and decreased response to acute mammary inflammation (Harmon *et al.*, 1994a,b,c). The increased susceptibility of the mammary gland to mastitis appears related to an impairment in phagocyte function (Boyne and Arthur, 1981; Jones and Suttle, 1981). Intracellular killing of microbes by phagocytes is decreased in copper deplete animals (Babu and Failla, 1990). Copper is a component of the enzyme superoxide dismutase (SOD) that inactivates superoxide anion by converting it to hydrogen peroxide in the cytosol of neutrophils. Activity of Cu, Zn-SOD in neutrophils is depressed even with marginal copper deficiencies (Jones and Suttle, 1981). Killing ability of neutrophils decreases proportionally to Cu,Zn-SOD activity in cells.

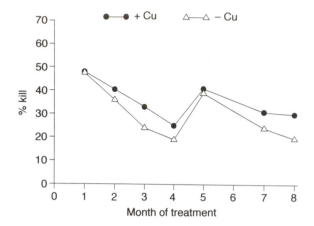

Fig. 3.3. Mean percentage killing of *Staphylococcus aureus* by polymorphonuclear neutrophils *in vitro* in steers fed diets supplemented with copper ($^{+}$Cu) or molybdenum ($^{-}$Cu) over eight months. Reprinted from Xin *et al.* (1991).

Intracellular killing of microbes by neutrophils is affected by Cu,Zn-SOD deficiencies by interfering with the initiation of the respiratory burst necessary for intracellular killing and the accumulation of superoxide anions that cause oxidative damage to the phagocytes (Babu and Failla, 1990). However, phagocytosis and survival of neutrophils and macrophages are not affected by copper status.

The dietary copper requirements of dairy cows are affected by a number of factors, but the most critical is the molybdenum content of feed (Fig. 3.3). Copper metabolism is interfered by molybdenum (Miller, 1979). Dietary intake of 10–20 ppm of copper usually is adequate to maintain immune functions in dairy cows; however copper status of cows should be tested to assure adequate copper is adsorbed and metabolized for maintaining host defences. Blood and tissue concentrations of copper respond rapidly to changes in copper intake. Liver copper is the most accurate estimate of copper status in dairy cows and a copper concentration of 150 ppm in liver is considered adequate. Plasma copper can be used to indicate a deficiency but does not reflect higher copper stores in the liver (Miller, 1979). Plasma concentrations of 0.75–1.2 ppm reflect adequate copper status.

Practical Applications

Relationships between nutrition and disease are evident. One can easily rationalize that feeding diets deficient in essential micronutrients will eventually result in decreased resistance to infections. In addition, the physiological role of many micronutrients appear to be dependent upon actions of other micronutrients. Direct relationships between vitamin A and zinc, copper and zinc, and vitamin E and selenium have been documented. Micronutrients such as chromium and vitamin D also are likely to be involved in host defences. Practical methods of supplying the proper diets are needed to optimize mammary defences of dry and lactating cows against infections.

Management practices that assure adequate supplementation of micronutrients are critical to avoid the economic consequences of clinical deficiencies during the periods of greatest risk. The stage of lactation that dairy cows are at highest risk to infectious diseases is at calving. The ability of cows to mount host defence responses against microbial pathogens is diminished at this time, corresponding with normal decreases in tissue concentrations of many immune responsive nutrients. Therefore, proper nutrition during the dry period is essential to assure adequate tissue levels of these micronutrients during the periparturient period as feed intake decreases. Parenteral administration of these nutrients may be necessary in some cases to maintain tissue levels. Providing the proper concentrations of micronutrients to dairy cows is an essential component of herd health programmes.

References

Aguilar, A.A., Kujawa, M. and Olson, J.D. (1988) Zinc-methionine supplementation in lactating dairy cows. In: *Proceedings of the 27th Annual National Mastitis Council Meeting*, Arlington, Virginia, USA, p. 119.

Alexander, M., Newmark, H. and Miller, R.G. (1985) Oral beta-carotene can increase cells in human blood. *Immunology Letters* 9, 221–225.

Atroshi, F., Tyopponen, J., Sankari, S., Kangasniemi, R. and Parantainen, J. (1986a) Possible roles of vitamin E and glutathione metabolism in bovine mastitis. *International Journal of Vitamin Nutrition Research* 54, 37–43.

Atroshi, F., Parantainen, J., Sankari, S. and Osterman, T. (1986b) Prostaglandins and glutathione peroxidase in bovine mastitis. *Research in Veterinary Science* 40, 361–366.

Baboir, B.M. (1984) The respiratory burst of phagocytes. *Journal of Clinical Investigation* 73, 599–601.

Babu, U. and Failla, M.L. (1990) Copper status and function of neutrophils are reversibly depressed in marginally and severely copper-deficient rats. *Journal of Nutrition* 120, 1700–1709.

Baehner, R.L., Boxer, L.A., Allen, J.M. and Davis, J. (1977) Autoxidation as a basis for altered function by polymorphonuclear leucocytes. *Blood* 50, 327–335.

Baker, S.S. and Cohen, H.J. (1983) Altered oxidative metabolism in selenium-deficient rat granulocytes. *Journal of Immunology* 130, 2856–2860

Bendich, A. (1990) Antioxidant micronutrients and immune responses. *Annals of the New York Academy of Science* 587, 168–180.

Boxer, L.A. (1986) Regulation of phagocyte function by α-tocopherol. *Proceedings of the Nutrition Society* 45, 333–344.

Boxer, L.A., Oliver, J.M., Speilberg, S.P., Allen, J.M. and Schulman, J.D. (1979) Protection of granulocytes by vitamin E in glutathione synthetase deficiency. *New England Journal of Medicine* 301, 901–905.

Boyne, R. and Arthur, J.R. (1979) Alterations of neutrophil function in selenium-deficient cattle. *Journal of Comparative Pathology* 89, 151–158.

Boyne, R. and Arthur, J.R. (1981) Effects of selenium and copper deficiency on neutrophil function in cattle. *Journal of Comparative Pathology* 91, 271–276.

Burton, G.W. and Ingoid, K.U. (1984) β-Carotene: an unusual type of antioxidant *Science* 224, 569–573.

Chan, A.C., Tran, K., Pyke, D.D. and Powell, W.S. (1989) Effects of dietary vitamin E on the biosynthesis of 5-lipoxygenase products by rat polymorphonuclear leucocytes (PMNL). *Biochimica et Biophysica Acta* 1005, 265–269.

Chew, B.P. (1987) Vitamin A and β-carotene on host defence. *Journal of Dairy Science* 70, 2732–2743.

Chew, B.P. (1993) Role of carotenoids in the immune response. *Journal of Dairy Science* 76, 2804–2811.

Chew, B.P., Hollen, L.L., Hilers, J.K. and Herlugson, M.L. (1982) Relationship between vitamin A and β-carotene in blood plasma and milk and mastitis in Holsteins. *Journal of Dairy Science* 65, 2111–2118.

Cohen, B.E. and Cohen, I.K. (1973) Vitamin A: adjuvant and steroid antagonist in the immune response. *Journal of Immunology* 111, 1376–1380.

Cook-Mills, J.M. and Fraker, P.M. (1993) The role of metals in the production of toxic oxygen metabolites by mononuclear phagocytes. In: Cunningham-Rundles, S. (ed.) *Nutrition Modulation of the Immune Response* Marcel-Dekker, New York, pp. 127–140.

Craven, N. and Williams, M.R. (1985) Defences of the bovine mammary gland against infection and prospects for their enhancements. *Veterinary Immunology and Immunopathology* 2, 71–127.

Dahlquist, S.P. and Chew, B.P. (1985) Effects of vitamin A and β-carotene on mastitis in dairy cows during the early dry period. *Journal of Dairy Science* 68(Suppl. 1), 191.

Daniel, L.S., Chew, B.P., Tanaka, T.S. and Tjoelker, L.W. (1986) *In vitro* vitamin A and β-carotene influence on bovine blood lymphocyte transformation. *Journal of Dairy Science* 69(Suppl. 1), 119.

Eberhart, R.J., Harmon, R.J., Jasper, D.E., Natzke, R.P., Nickerson, S.C., Reneau, J.K., Row, E.H., Smith, K.L. and Spencer, S.B. (1987) *Current Concepts of Bovine Mastitis*. National Mastitis Council, Arlington, Virginia, USA.

Erskine, R.J., Eberhart, R.J., Hutchinson, L.H. and Scholz, R.W. (1987) Blood selenium concentrations and glutathione peroxidase activities in dairy herds with high and low somatic cell counts. *Journal of the American Veterinary Medical Association* 190, 1417–1421.

Erskine, R.J., Eberhart, R.J., Grasso, P.J. and Scholz, R.W. (1989) Induction of *Escherichia coli* mastitis in cows fed selenium-deficient or selenium-supplemented diets. *American Journal of Veterinary Research* 50, 2093–2100.

Grasso, P.J., Scholz, R.W., Erskine, R.J. and Eberhart, R.J. (1990) Phagocytosis, bactericidal activity, and oxidative metabolism of mammary neutrophils from dairy cows fed selenium-adequate and selenium-deficient diets. *American Journal of Veterinary Research* 51, 269–275.

Harmon, R.J., Clark, T.W., Trammell, D.S., Smith, B.A., Torre, P.M. and Hemken,

R.W. (1994a) Influence of copper status in heifers on response to intramammary challenge with *Escherichia coli* endotoxin. *Journal of Dairy Science* 77(Suppl. 1), 198.

Harmon, R.J., Clark, T.W., Trammell, D.S., Akers, K., Smith, B.A., Torre, P.M., Langlois, B.E. and Hemken, R.W. (1994b) Copper status and mastitis in heifers with or without prepartum copper supplementation. *Journal of Dairy Science* 77(Suppl. 1), 198.

Harmon, R.J., Clark, T.W., Smith, B.A., Trammell, D.S., Torre, P.M., Akers, K., Langlois, B.E. and Hemken, R.W. (1994c) Influence of copper status in heifers on response to intramammary challenge with *Staphylococcus aureus*. *Journal of Dairy Science* 77(Suppl. 1), 198.

Harrison, J H., Hancock, D.D. and Conrad, H.R. (1984) Vitamin E and selenium for reproduction in the dairy cow. *Journal of Dairy Science* 67, 123–132.

Hogan, J.S., Smith, K.L., Weiss, W.P., Todhunter, D.A. and Shockey, W.L. (1990) Relationships among vitamin E, selenium, and bovine blood neutrophils. *Journal of Dairy Science* 73, 2372–2378.

Hogan, J.S., Weiss, W.P., Todhunter, D.A., Smith, K.L. and Schoenberger, P.S. (1992) Bovine neutrophil responses to parenteral vitamin E. *Journal of Dairy Science* 75, 399–405.

Hogan, J.S., Weiss, W.P. and Smith, K.L. (1993) Role of vitamin E and selenium in the host defense responses to mastitis. *Journal of Dairy Science* 76, 2795–2803.

Johnston, L.A. and Chew, B.P. (1984) Peripartum changes of plasma and milk vitamin A and β-carotene among dairy cows with or without mastitis. *Journal of Dairy Science* 67, 1832–1840.

Jones, D.G. and Suttle, N.F. (1981). Some effects of copper deficiency on leucocyte function in sheep and cattle. *Research in Veterinary Science* 31, 151–156.

Kincaid, R.L., Hodgson, A.S., Riley, R.E. and Cronrath, J.D. (1984) Supplementation of diets for lactating cows with zinc as zinc oxide and zinc methionine. *Journal of Dairy Science* 67(Suppl. 1), 103.

Koller, L.D., S.A. Mulhern, N.C. Franker, M.G. Steven, and J.R. Williams (1987) Immune dysfunction in rats fed a diet deficient in copper. *American Journal of Clinical Nutrition* 45, 997–1006.

Kruse-Jarres, J.D. (1989) The significance of zinc for humoral and cellular immunity. *Journal of Trace Elements and Electrolytes in Health and Disease* 3, 1–8.

Kuehl, A.K. and Egan, R.W. (1980) Prostaglandins, arachidonic acid and inflammation. *Science* 210, 978–984.

Lickoff, R.O., Guptill, D.R., McKay, L.M., Mathias, C.C., Nockels, C.F. and Tengerdy, R.P. (1981) Vitamin E and aspirin depress prostaglandins in protection of chickens against *Escherichia coli* infection. *American Journal of Clinical Nutrition* 34, 245–251.

Lukasewycz, O.A. and Prohaska, J.R. (1990) The immune response in copper deficiency. *Annals of the New York Academy of Sciences* 587, 147–159.

Lukasewycz, O.A., Prohaska, J.R., Meyer, S.G., Schmidtke, J.D., Hatfield, S.M. and Marder, P. (1985) Alterations in lymphocyte subpopulations in copper-deficient mice. *Infection and Immunity* 48, 644–647.

Lynch, G. L. (1992) Natural occurrence and content of vitamin E in feedstuffs. In: *Vitamin E in Animal Nutrition and Management*. BASF Corporation, Parsippany, New Jersey, pp. 43–48.

McDermott, M.R., Mark, D.A., Befus, A.D., Baliga, B.S., Suskind, R.M. and

Bienenstock, J. (1982) Impaired intestinal localization of mesenteric lymphoblasts associated with vitamin A deficiency and protein-calorie malnutrition. *Immunology* 45, 1-5.

Miller, W.J. (1970) Zinc nutrition of cattle: a review. *Journal of Dairy Science* 53, 1123-1135.

Miller, W.J. (1979) *Dairy Cattle Feeding and Nutrition.* Academic Press, New York, USA.

Moynahan, E.J. (1981) Acrodermatitis enteropathica and the immunological role of zinc. In: Safai, B. and Good, R.A. (eds) *Immunodermatology.* Plenum Medical Book Company, New York, USA, pp. 437-447.

Mulhern, S.A. and Koller, L.D. (1988) Severe or marginal copper deficiency results in a graded reduction in immune status in mice. *Journal of Nutrition* 118, 1041-1047.

Prohaska, J.R. and Lukasewycz, O.A. (1981) Copper deficiency suppresses the immune response of mice. *Science* 213, 559-561.

Prohaska, J.R., Downing, S.W. and Lukasewycz, O.A. (1983) Chronic dietary copper deficiency alters biochemical and morphological properties of mouse lymphoid tissues. *Journal of Nutrition* 113, 1583-1590.

Schingoethe, D.J., Parsons, J.G., Ludens, F.C., Tucker, W.L. and Schave, H.J. (1978) Vitamin E status of dairy cows fed stored feeds continuously or pastured during summer. *Journal of Dairy Science* 61, 1582-1589.

Scholz, R.W. and Hutchinson, L.J. (1979) Distribution of glutathione peroxidase activity and selenium in the blood of dairy cows. *American Journal of Veterinary Research* 40, 245-249.

Smith, K.L. (1986) Vitamin E-enhancement of immune response and effects on mastitis in dairy cows. In: *Proceedings of the Roche Symposium.* Hoffman LaRoche, Nutley, New Jersey, USA, pp. 1-22.

Smith, K.L., Conrad, H.R., Amiet, B.A., Schoenberger, P.S. and Todhunter, D.A. (1985) Effect of vitamin E and selenium dietary supplementation on mastitis in first lactation dairy cows. *Journal of Dairy Science* 68(Suppl. 1), 190.

Smith, K.L., Hogan, J.S. and Weiss, B.P. (1992) Dietary vitamin E and selenium influence the resistance of cows to mastitis. In: *Vitamin E in Animal Nutrition and Management.* BASF, Parsippany, New Jersey, USA, pp. 151-157.

Smith, K.L., Harrison, J.H., Hancock, D.D., Todhunter, D.A. and Conrad, H.R. (1984) Effect of vitamin E and selenium supplementation on incidence of clinical mastitis and duration of clinical symptoms. *Journal of Dairy Science* 67, 1293-1300.

Sommer, A. (1990) Vitamin A status, resistance to infection, and childhood mortality. *Annals of the New York Academy of Sciences* 587, 17-23.

Sugarman, B. (1983) Zinc and infection. *Review of Infectious Diseases* 5, 137-147.

Takagi, H. and Nakano, K. (1983) The effect of vitamin A depletion on antigen-stimulated trapping of peripheral lymphocytes in local lymph nodes. *Immunology* 48, 123-128.

Tjoelker, L.W., Chew, B.P., Tanaka, T.S. and Daniel, L.R. (1988a) Bovine vitamin A and β-carotene intake and lactational status. 1. Responsiveness of peripheral blood polymorphonuclear leucocytes to vitamin A and β-carotene challenge *in vitro.* *Journal of Dairy Science* 71, 3112-3119.

Tjoelker, L.W., Chew, B.P., Tanaka, T.S. and Daniel, L.R. (1988b) Bovine vitamin

A and β-carotene intake and lactational status. 2. Responsiveness of mitogen-stimulated peripheral blood lymphocytes to vitamin A and β-carotene challenge *in vitro. Journal of Dairy Science* 71, 3120–3127.

Underwood, E.J. (1977) *Trace Elements in Human and Animal Nutrition.* Academic Press, New York, USA.

Underwood, E.J. (ed.) (1981) Zinc. In: *The Mineral Nutrition of Livestock.* Commonwealth Agricultural Bureaux, London, UK, pp. 135–146.

Weiss, S.J. and LoBuglio, A.F. (1982) Phagocytic-generated oxygen metabolites and cellular injury. *Laboratory Investigation* 47, 5–18.

Weiss, W.P., Hogan, J.S., Smith, K.L. and Hoblet, K.H. (1990a) Relationships among selenium, vitamin E, and mammary gland health in commercial dairy herds. *Journal of Dairy Science* 73, 381–390.

Weiss, W.P., Todhunter, D.A., Hogan, J.S. and Smith, K.L. (1990b) Effect of duration of supplementation of selenium and vitamin E on periparturient cows. *Journal of Dairy Science* 73, 3187–3194.

Weiss, W.P., Hogan, J.S., Smith, K.L., Todhunter, D.A. and Williams, S.N. (1992) Effects of supplementing periparturient cows with vitamin E on distribution of α-tocopherol in blood. *Journal of Dairy Science* 75, 3479–3485.

Weiss, W.P., Hogan, J.S., Smith, K.L. and Williams, S.N. (1994a) Effect of dietary fat and vitamin E in α-tocopherol and β-carotene in blood of peripartum dairy cows. *Journal of Dairy Science* 77, 1422–1429.

Weiss, W.P., Hogan, J.S. and Smith, K.L. (1994b) Relationship between neutrophil function and α-tocopherol concentrations in blood of periparturient dairy cows. *Agri-Practices* 15, 5–8.

Wirth, J.J., Fraker, P.J. and Kierszenbaum, F. (1984) Changes in the level of marker expression by mononuclear phagocytes in zinc deficient mice. *Journal of Nutrition* 114, 1826–1833.

Xin, Z., Waterman, D.F., Hemken, R.W. and Harmon, R.J. (1991) Effects of copper status on neutrophil function, superoxide dismutase, and copper distribution in steers. *Journal of Dairy Science* 74, 3078–3085.

Zamora, R., Hidalgo, F.J. and Tappel, A.L. (1991) Comparative antioxidant effectiveness of dietary β-carotene, vitamin E, selenium, and coenzyme Q10 in rat erythrocytes and plasma. *Journal of Nutrition* 121, 50–56.

Zile, M.H. and Cullum, M.E. (1983) The function of vitamin A: current concepts. *Proceedings of the Society for Experimental Biology and Medicine* 172, 139–152.

The Effect of Bovine Somatotrophin on Dairy Production, Cow Health and Economics

S.A. Zinn[1] and B. Bravo-Ureta[2]

Departments of [1]Animal Science and [2]Agricultural and Resource Economics, University of Connecticut, Storrs, Connecticut 06269, USA

Introduction

Major goals of the dairy industry are to increase the efficiency, quality and (or) profitability of milk production. To address these objectives, research in physiology, nutrition, engineering and other related fields are integrated and developed into new management systems and new technologies. To determine if these new systems influence efficiency, quality or the profitability of milk production, academic, industrial and government researchers and ultimately producers evaluate new technologies under laboratory and practical conditions. Over time, many of these technologies have been incorporated into current management practices (e.g. milking machines), while others have been rejected (e.g. feeding thyroprotein). Beginning in 1994, a new technology was made available for dairy producers in the USA. The Food and Drug Administration (FDA) approved the use of recombinant bovine somatotrophin (rbST) in lactating dairy cows. The commercial use of rbST in dairy cows is an excellent example of the development and integration of basic research derived from several fields into a new management practice that can influence the efficiency and profitability of milk production without adversely affecting the quality of the product. This chapter will address the development and use of bST in dairy cows, the influence of bST on production and composition of milk, animal health, consumer acceptance and potential factors influencing the rate of adoption and on-farm profitability of bST.

Pituitary-derived Somatotrophin and Lactation

Somatotrophin was first identified in the rat and was associated with increased growth and muscle mass and decreased fat (Evans and Long, 1921; Lee and Schaffer, 1934). Since these early studies, the physiological actions attributed to the hormone have increased dramatically; among these is the stimulatory effect of bST on milk production in the lactating cow. Readers are referred to Beerman and DeVol (1991) for additional physiological actions of ST.

The mechanism by which ST increases milk yield is not yet completely understood. Since no receptors for ST have been identified in mammary secretory tissue (Gertler *et al.*, 1984), the effects of the hormone are most likely indirect (Peel and Bauman, 1987). Peel *et al.* (1981) proposed that ST acts to coordinate the partitioning of nutrients among body tissues with preferential use by the mammary gland, which enables the animal to increase milk production. In addition, ST may stimulate mammary epithelial cells to synthesize more milk by increasing the metabolic activity of these cells (reviewed in Johnsson and Hart, 1986; Peel and Bauman, 1987; Bauman and Vernon, 1993).

Studies with Crude Pituitary Extracts

Short-term studies (less than a complete lactation) that focused on the effect of ST on lactation began in the 1930s. Asdell (1932) reported that injection of extracts from ovine anterior pituitary glands increased milk yield in lactating goats. Similarly, Asimov and Krouze (1937) injected more than 500 dairy cows with ox pituitary gland extracts. Milk yield was increased by 6–29%. Somatotrophin was later identified as the primary galactopoietic component of the pituitary gland extract (Young, 1947; Cotes *et al.*, 1949). Milk yield was increased when injections of bST were administered for as little as 2–14 days (e.g. Brumby, 1956; Bullis *et al.*, 1965) or for as long as 10–12 weeks (Brumby and Hancock, 1955; Hutton, 1957). In addition, bST increased feed efficiency (Brumby and Hancock, 1955). Taken together, these experiments indicated that ST would increase milk yield in an established lactation and the increases in milk could be maintained for the duration of the treatment. Moreover, the increase in milk yield is accompanied by an increase in feed efficiency. However, since there was significant contamination with thyrotropin secreting hormone and prolactin in these ST preparations, the galactopoietic activity of the preparations could not definitely be ascribed to ST.

Studies with Purified Pituitary-derived (pbST) Somatotrophin

In 1973, Machlin reported that injection of a purified pbST for durations of 4 days to 10 weeks increased milk production by between 12 and 35%. Increased milk production was associated with increased feed efficiency. Similarly, injection of pbST, but not fragments of the hormone, increased milk yield and feed efficiency in cows in late lactation (Bines *et al.*, 1980). Although these two studies utilized relatively pure sources of bST, treated cows had relatively low milk production (under 25 kg) and cows were in mid- to late-lactation. Peel *et al.* (1981) injected high-yielding (greater than 34 kg milk per day) Holstein cows in peak lactation (74 days in milk) and, similar to results with low producing cows, ST increased milk yield by 9.5%.

Subsequent experiments reported that injection of ST for 7–21 days increased milk yield by 12–30% in cows with little change in feed intake, resulting in increased feed efficiency (e.g. Peel *et al.*, 1982; Lough *et al.*, 1988). However, increasing nutrient supply (casein, glucose or fat) postruminally in cows treated with bST had little additional effect on milk production (Peel *et al.*, 1982; Lough *et al.*, 1988). Somatotrophin continued to stimulate milk production compared with controls after 22 weeks of treatment (Peel *et al.*, 1985) and feed intake increased to meet the demand of the increased milk yield, but the increase in feed efficiency was maintained (Peel *et al.*, 1985).

Recombinant Somatotrophin and Lactation

Given the limited supply of pituitary glands and the cost of extracting and purifying bST, the use of the hormone to increase milk production on a commercial basis was not viable. With the introduction of recombinant DNA technology, large quantities of highly purified rbST could be produced from bacteria (Seeburg *et al.*, 1983). This essentially removed the limited supply problem and the commercial application of utilizing bST to increase milk production in dairy cows was revived. In addition, several sustained-release formulations of rbST were developed. Instead of daily injections, the sustained release bST required injections once per 7, 12 or 28 days, depending on the formulation. With a larger supply of rbST available and a variety of formulations to evaluate, the number and the duration of experiments and the number of animals per treatment increased dramatically. In fact, it is estimated that in the last 10 years, more than 1500 scientific studies on bST have been published (Bauman *et al.*, 1993).

Recombinant Somatotrophin Versus Pituitary-derived Somatotrophin

Available formulations of rbST can differ from pbST by up to eight amino acids (Bauman and Vernon, 1993), which are added at the NH_2-terminus end of the protein (Juskevich and Guyer, 1990). Purified pbST and rbST have these differences in amino acid sequence, but they have similar immunological and biological properties (Wood *et al.*, 1989).

BST-induced increases in milk yield were similar between pbST and rbST following a 6 or 7 day treatment period (Bauman *et al.*, 1982; Heap *et al.*, 1989). Treatment of high producing Holsteins with either rbST or pbST for 188 days increased milk yield compared with controls. However, rbST ($27\,\mathrm{mg\,day}^{-1}$) increased milk yield by 36.2%, whereas milk yield was increased by 16.5% in cows injected with the same dosage of pbST (Bauman *et al.*, 1985). In agreement, cows treated for 21 days with pbST produced less milk than cows treated with several forms of rbST (Eppard *et al.*, 1992).

Differences in response to equal dosages of rbST and pbST are not due to differences in blood concentrations of ST nor to development of anti-bST antibodies (Eppard *et al.*, 1992). Heap *et al.* (1989) postulated that the differences in response may be due to presence of multiple forms of the pbST that are not present in the rbST. In addition, the reduced galactopoietic potency of pbST is due, at least in part, to deletions in amino acid residues in the NH_2-terminus portion of the pbST (Eppard *et al.*, 1992).

Recombinant Somatotrophin and Milk Production

Treatment with rbST has increased milk production in all dairy breeds examined (Bauman and Vernon, 1993), including *Bos indicus* cows (Phipps *et al.*, 1991) and buffaloes (*Bubalus bubalis*; Ludri *et al.*, 1989). Detectable increases in milk yield are apparent after 2–3 days of rbST treatment and increased milk yield is maintained for the duration of the treatment period (Crooker and Otterby, 1991). Over an entire lactation, it is estimated that 40% of the enhanced milk production is due to short-term (between injections) increases in milk yield and 60% is due to increased persistency of lactation (Gallo *et al.*, 1994).

The majority of studies cited to this point were conducted for one lactation or less. However, an important consideration for the commercial application of rbST is the milk yield response over multiple lactations. Several studies have now examined rbST treatments for two or more lactations. For example, rbST-induced increases in milk yield were maintained over four lactations (Huber *et al.*, 1991). Similarly, the response to rbST was maintained in cows treated for two successive lactations (Leonard *et al.*, 1990; Lotan *et al.*, 1993). However, the advantage in feed efficiency may be lost in the second lactation (McBride *et al.*, 1990). Although rbST continues to increase milk production in the second lactation, the magnitude of the response is decreased (Phipps *et al.*, 1990; Eppard *et al.*, 1991).

Factors that Influence the Magnitude of Response

Management factors are commonly identified as a major source of variation in the magnitude of the rbST response (Chilliard, 1989; Bauman, 1992). That is, cows that are better managed are known to have a greater response to exogenous rbST than poorly managed cows (Bauman, 1989). Producers that manage their operation to maximize milk production have the greatest potential to maximize the milk yield response to rbST. Poorly managed herds that fail to meet the production potential of the cows will probably fail to get a maximum response to the hormone.

Several management factors have been identified as partially responsible for the variation, including dosage of rbST, injection interval, parity, genetic potential and environmental conditions (Chilliard, 1989).

Dosage and Injection Interval

Similar to pbST (Eppard *et al.*, 1985), a curvilinear response occurs between dosage of the hormone and increases in milk production when rbST is adminis-tered as a daily or sustained-release formulation (Chilliard, 1989; Franson *et al.*, 1989). The optimal dose of rbST as a galactopoietic agent is in the range of 25–$50 \, \mathrm{mg \, day^{-1}}$. Given an adequate dosage, increased milk yield to rbST is maintained following administration of the hormone every 7, 14 or 28 days (McClary *et al.*, 1990; Zinn *et al.*, 1993). Milk yield response to several formula-tions of rbST, injected at different intervals and dosages, are summarized in Table 4.1.

Although milk yield is increased with sustained-release formulations of rbST, production within a single injection interval varies (Bauman *et al.*, 1989; Zinn *et al.*, 1993). That is, following each injection the milk yield will increase to a peak, approximately at the mid-point of the injection interval, and then decline until the next injection (Fig. 4.1).

Stage of Lactation and Parity

Studies that compare the effect of rbST at various periods of lactation indicate that the milk yield response is proportionately greater later in lactation than earlier (prior to day 60). For example, Thomas *et al.* (1991) reported that rbST-induced increments in milk production were greater later in lactation (days

S.A. Zinn and B. Bravo-Ureta

Table 4.1. Milk yield response to several dosages, injection frequencies and formulations of rbST in lactating cows from selected studies.

Dosage (mg day⁻¹)[1]	Treatment				Milk yield (kg day⁻¹)		% response	Reference
	Frequency	Duration (days)	Source[2] of rbST	N(B)[3]	Controls	bST		
11.4	28 d	84–252	1	60(H)	24.7[4]	29.0	17	Leonard *et al.* (1990)
22.9						28.9	17	
34.3						30.6	24	
11.4	28 d	252	1	48(H)	22.2	24.7	11	Oldenbroek *et al.* (1993)
22.8								
34.2								
11.4	28 d	196	1	48(H)	24.6	26.7	9	Laurent *et al.* (1992)
22.8	14 d					28.1	14	
22.9	28 d					27.5	12	
5.0	daily	230	2	40(H + J)	19.4	20.8	7	West *et al.* (1990)
10.0						23.1	19	
15.0						23.5	21	
20.0						24.1	24	
25.0	daily	125	2	36(H)	36.9	39.7	8	Lormore *et al.* (1990)

Reference								
Zinn *et al.* (1993)	0	29.0	28.7	44(H)	2	230	7 d	7.1
	12	31.8						14.3
	25	35.9						21.4
Gibson *et al.* (1992a)	31	34.8	26.5	35(H)	3	259	daily	10.3
	36	36.1						20.6
Eisenbeisz *et al.* (1990)	20	29.2	24.2	48(H)	3	203	daily	10.3
	31	31.7						20.6
	22	29.5						30.9
Downer *et al.* (1993)	9	29.2	26.8	264(H)	3	210	14 d	10.0
	8	29.0						25.0
	5	30.4						50.0
Zhao *et al.* (1992)	7	27.3	25.6	74(H)	3	280	daily	10.3
	9	27.8					14 d	25.0
Burton *et al.* (1990b)	14	29.3	25.7	26(H)	3	266	daily	10.3
	21	31.1						20.6
	16	29.9						41.2
Erdman *et al.* (1990)	11	30.4	27.5	40(H)	3	203	daily	10.3
	17	32.1						20.6
	19	32.7						30.9

Table 4.1. (Continued)

Dosage (mg day⁻¹)[1]	Treatment				Milk yield (kg day⁻¹)		% response	Reference
	Frequency	Duration (days)	Source[2] of rbST	N(B)[3]	Controls	bST		
10.3	daily	266	3	43(H)	28.2	33.5	19	McBride et al. (1990)
20.6						35.0	24	
20.6	daily	210	3	64(H)	35.7[5]	41	15	Tessman et al. (1991)
35.7	14 d	150	4	46(J)	13.3	17.4	31	Pell et al. (1992)
42.8	14 d	150	4	82(H)	21.1	28.3	34	Eppard et al. (1991)
128.6						30.5	45	
214.3						29.5	40	
17.8	14 d	350	4	254(H)	25.2	28.3	12	Hartnell et al. (1991)
35.7						29.2	16	
53.6						31.5	25	

[1] Daily dosage is either actual dosage for daily administration or calculated theoretical dosage per day of sustained release formulations.
[2] 1 = Lilly, 2 = Upjohn, 3 = American Cyanamid, 4 = Monsanto.
[3] N = Total number of animals in experiment; B = Breed; (H) = Holstein, (J) = Jersey.
[4] Yields averaged over three stages of lactation.
[5] Yields averaged for medium and high energy diets.

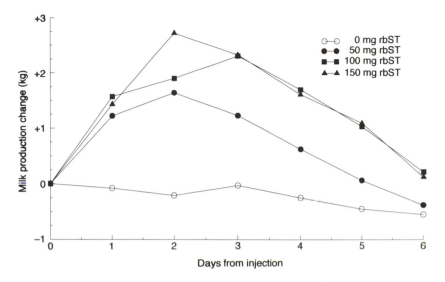

Fig. 4.1. Average daily response to an injection of one of four dosages of rbST. Response is the deviation from average milk production on the milkings just before an injection (from Zinn *et al.*, 1993).

101–140 or 141–189) than earlier (days 57–100). Although the percentage response may be greater later in lactation, actual increments in milk production may be similar (Peel *et al.*, 1983).

Increases in the milk yield response to rbST are often reported to be greater in multiparous than primiparous cows (e.g. Chilliard, 1989; Sullivan *et al.*, 1992). Muller (1992) suggested that the differences in response between parity groups may be due to the fact that first calf heifers continue to gain weight during their first lactation which reduces rbST-induced increases in milk yield.

Genetic Merit and Level of Milk Production

There is a correlation between concentrations of bST in the blood and genetic potential for milk production and(or) actual milk production (Hart *et al.*, 1979; Kazmer *et al.*, 1990, 1991). There may also be differences in rbST-induced milk production in cows of different genetic merit. Results from experiments that addressed this question are variable and often contradictory. For example, Bines *et al.* (1980) reported proportionately similar bST-induced increases in milk production between high (Friesians) and low (Herefords) yielding cows, but intake in these animals was restricted and may have limited the response in the Friesians. In contrast, in pasture-fed Friesians, response to rbST was greater in genetically inferior lines than superior lines (Michel *et al.*, 1990).

In agreement, cows with lower potential for milk yield tend to have larger responses to rbST than cows with greater potential (Leitch *et al.*, 1990). One study has reported that cows with greater genetic merit may have a greater potential to respond to rbST (Gibson *et al.*, 1992b) while other studies have reported that cows of varying genetic merit respond equally to rbST (Leitch *et al.*, 1990; Nytes *et al.*, 1990; Sullivan *et al.*, 1992). Potentially, with widespread commercial use of rbST, data from a sufficient number of animals will become available to accurately determine this relationship.

Environmental Conditions

Extreme climatic conditions, both cold and hot, can affect milk production (Collier, 1985). Several studies have addressed the potential interaction between climatic conditions and responses to rbST. In short-term studies, rbST increased milk yield in cows exposed to cold ($-5°C$ to $5°C$; Becker *et al.*, 1990) and hot, humid climates ($22-35°C$ and $30-100\%$ relative humidity; Mohammed and Johnson, 1985; West *et al.*, 1990; Johnson *et al.*, 1991). However, the magnitude of the response compared with normal temperatures may be reduced (Zoa-Mboe *et al.*, 1989). Just as milk yield is reduced in non-treated cows maintained under extreme climatic conditions, milk yield response to rbST is also reduced.

Somatotrophin and Feed Intake

Despite relatively large increases in milk production, in cows treated with rbST for less than 6 weeks, no changes in feed intake were observed, thereby resulting in increased feed efficiency (e.g. Peel *et al.*, 1981, 1982). The extra nutrients required for this increased milk yield were thought to come from mobilization of body tissues (reviewed in Bauman and Vernon, 1993). Based on these results it was estimated that concentrate feeding would have to be increased substantially to meet the additional milk production and maintain body tissue reserves (Chalupa and Galligan, 1989), implying that formulation of special diets would be necessary for cows administered rbST. With longer periods of treatment it was apparent that cows injected with rbST gradually increased feed intake to meet increased metabolic demand (see Chalupa and Galligan, 1989; Bauman, 1992). The increase in feed intake generally begins after 6–8 weeks of rbST treatment and is maintained for the remainder of lactation (Muller, 1992). Although the magnitude of the increase varies with dosage of rbST and composition of the diet, increases in dry matter intake in rbST-treated cows are similar to non-treated cows whose milk production is increased to the same extent (Chilliard, 1989).

Since treatment with rbST does not change the feed digestibility or partial efficiency of energy utilization for maintenance or milk synthesis (Tyrrell *et al.*, 1988; Crooker *et al.*, 1990), current nutritional standards for lactating cows remain applicable for rbST-treated cows (Crooker and Otterby, 1991). The interaction of specific components in the diet and the milk yield response to rbST have been reviewed elsewhere (Chalupa and Galligan, 1989; Chilliard, 1989; McGuffey and Wilkinson, 1991). Diets formulated to meet the requirements of cows producing 10,000 kg of milk will be the same whether that quantity of milk is produced with or without rbST. The most important aspect of nutrition and rbST treatment is that cows must be fed to maximize milk yield in a given management situation. Cows that are under-nourished, either in quantity or quality of the diet, will respond poorly to rbST.

Somatotrophin and Milk Composition

The composition of milk influences both the manufacturing properties and the price of milk (Barbano *et al.*, 1992). The effects of bST on gross milk composition, fatty acid composition, protein composition and several variables associated with the manufacturing properties of milk have been extensively reviewed (Zullo and Martin, 1990; Van Den Berg, 1991). Although there have been reports of small differences in fatty acid (Baer *et al.*, 1989) and protein (Barbano *et al.*, 1992) composition, Van Den Berg (1991) and others (Baer *et al.*, 1989; Barbano *et al.*, 1992) have concluded that the processability of milk from bST-treated animals is not altered. Given little change in milk composition, together with an increase in milk yield, yield of milk components is increased in the same proportion as the yield (Zullo and Martin, 1990; Van Den Berg, 1991; Barbano *et al.*, 1992).

Daily milk composition varies in response to sustained-release rbST in parallel to changes in daily milk production. For example, Vérité *et al.* (1989) reported that with a 28-day injection interval, milk fat percentage increased rapidly over the first few days following injection, reached a plateau in 5–10 days and returned to preinjection values three weeks postinjection. This cyclic nature of milk components between injections has been confirmed (Oldenbroek *et al.*, 1989; Gallo *et al.*, 1994). However, it is unlikely that these changes in milk components will influence the manufacturing qualities of the overall milk supply (Van Den Berg, 1991).

Little to no change in mineral content, vitamin content, melting properties, enzymes, pH and sensory evaluation has been reported in milk from rbST-treated cows (Zullo and Martin, 1990; Van Den Berg, 1991). In conclusion, evidence to date indicates that rbST has little or no effect on the composition, processing properties or taste of milk.

Somatotrophin and Cow Health

Given the positive effects of rbST on milk production and feed efficiency, the potential consequences of long-term use of rbST on the health of the cow deserve close attention. Health implications can be divided into three broad areas: general health, reproduction and udder health. Many experiments have now been completed that examine the effects of rbST on cow health and the data have, in some cases, been combined from different sites to give statistically meaningful animal numbers (see White *et al.*, 1994).

General Health

The effects of rbST treatment on the general health of cows has recently been reviewed (Moore and Hutchinson, 1992). Treatment with the hormone does not affect body temperature or respiration rate (Eppard *et al.*, 1987; Soderholm *et al.*, 1988; Phipps, 1989), but heart rate has been reported to increase by 5–15% (Soderholm *et al.*, 1988). In general, blood chemistry in rbST-treated cows is within normal ranges (Eppard *et al.*, 1987).

Reports that focus on lameness are contradictory. For example, Eppard *et al.* (1987) and Burton *et al.* (1990a) reported that there is no significant increase in leg or hoof problems in treated cows. In contrast, Phipps (1989) and Cole *et al.* (1992) have reported increased incidence of lameness in cows treated with rbST.

Treatment with rbST did not significantly affect the incidence of metabolic diseases, including milk fever or ketosis (Bauman, 1992; Moore and Hutchinson, 1992). Despite unsubstantiated reports of catastrophic metabolic effects, leading to 'cow burnout' (Kronfeld, 1993), there have been no reported associations between metabolic distress and rbST treatment (Peel and Bauman, 1987; Phipps, 1989; Crooker and Otterby, 1991; Moore and Hutchinson, 1992).

Reproduction

Since subsequent lactations are an essential component of an economically viable dairy cow, pregnancy, parturition and calving intervals influence farm income (Schmidt, 1989). Therefore, the influence of rbST on reproductive performance should be evaluated. Potential effects of rbST on reproductive performance have been reviewed (e.g. Crooker and Otterby, 1991; Moore and Hutchinson, 1992). Reproductive variables have been monitored in production studies with rbST, including days open (parturition to conception interval, PCI), days to first service (interval from parturition to first insemination) and services per conception (number of inseminations required to achieve conception). Data that focus on these variables from selected studies published since 1990 are listed in Table 4.2.

Although the magnitude of response is variable, treatment with rbST generally increases the PCI. Several experiments report no differences in PCI between control and treated cows (Table 4.2). In other studies, treatment with rbST increased PCI by 28–30% (Table 4.2). McGuffey *et al.* (1991) also reported that increases in PCI were observed in cows in which treatment was initiated early in lactation (28–45 days postpartum) but not when treatment started in mid (111–166 days postpartum) or late lactation (166–334 days postpartum). Although several reports indicate there are no differences in services per conception and days to first oestrus, there is a trend for increased services per conception and days to first service in cows treated with rbST (Table 4.2).

Decreases in reproductive performance in rbST-treated cows may be due more to increased milk yield and short-term negative energy balance than to direct effects of rbST (Phipps, 1989; Weller *et al.*, 1990). For example, Erb and Smith (1987) reported that for every 4.54 kg increase in milk yield in the first 120 days of lactation, there is a corresponding increase in PCI (0.3 days) and services per conception (0.005). Similarly, Hard *et al.* (1988; cited by Phipps, 1989) reported that days open was more related to level of production than rbST. In addition, rbST increases milk energy output before there is a concomitant increase in feed intake and therefore, in the period directly following the initiation of treatment, treated cows tend to be in more negative energy balance than controls (Hartnell *et al.*, 1991), and negative energy balance is known to reduce reproductive performance (Villa-Godoy *et al.*, 1990).

Mastitis and Udder Health

Mastitis is one of the most economically important diseases in a dairy herd. Therefore, effects of rbST on udder health and the incidence of mastitis are important. Many experiments have compared the number of clinical cases of mastitis in rbST and control cows. In addition, somatic cell counts (SCC) in milk have been evaluated as an indirect index of subclinical mastitis.

Similar to other health related variables, the influence of rbST on SCC and mastitis is variable. For example, several reports indicate that SCCs tend to increase in rbST-treated cows (e.g. Peel *et al.*, 1988; Weller *et al.*, 1990; Lissemore *et al.*, 1991) whereas, others have found no differences in SCC in treated and control cows (Eppard *et al.*, 1987; Zinn *et al.*, 1993). There are reports of either no difference or an increase in mastitis in rbST-treated cows (e.g. Phipps, 1989; Lissemore *et al.*, 1991; Pell *et al.*, 1992). However, increased incidence of mastitis in several experiments appears to be more related to random effects than specifically to rbST (Weller *et al.*, 1990; Pell *et al.*, 1992).

Recently two reports were published that specifically addressed the relationship between rbST treatment and mastitis in dairy cows (McClary *et al.*, 1994; White *et al.*, 1994). McClary *et al.* (1994) examined milk SCC in 352 rbST-treated cows. Although SCC increased with dosage, there was no

Table 4.2. Reproductive outcomes in cows treated with bST from selected studies.

Dosage frequency	N(B)[1] DIM	Days open	Days to first service	Services per conception	Reference	Comment
0	64H	94.3	NR	1.7	Tessman *et al.*, 1991	2
20.6 mg d^{-1}	91 DIM	94.8		1.8		
0	43H	155.2^a	85.3^a	3.1	Burton *et al.*, 1990a	
20.6 mg d^{-1}	28–35 DIM	199.2^b	118.7^b	3.4		
0	64H	120.5	NR	2.52	Downer *et al.*, 1993	3
56–700 mg per 14 d	98 DIM	127.2		2.33		
0	301H	100.0	71.3	2.03	Lotan *et al.*, 1993	
500 mg per 14 d	35–120 DIM	100.0	99.0	1.85		
0	46J	88.0	71.0	1.7	Pell *et al.*, 1992	
500 mg per 14 d	60 DIM	106.0	71.0	2.2		
0	90F	82.0	58.0	2.02	Weller *et al.*, 1990	4
500 mg per 14 d	60 DIM	101.5	52.5	2.48		

Dosage	N	DIM				Reference
0	60F		78.0	55.0	1.78	
500 mg per 14 d		60 DIM	82.5	52.5	2.04	Eisenbeisz *et al.*, 1990 [3]
0	48H		122.0	81.0	2.18	
10.3-30.9 mg d^{-1}		104 DIM	163.0	85.0	2.57	
0	36H		107.1	82.6	2.4	Lormore *et al.*, 1990
25 mg d^{-1}		25 DIM	139.7	88.8	8.5	
0	40H		153.0	107.0	2.0	
10.3 mg d^{-1}		98-104 DIM	165.0	105.0	3.0	Erdman *et al.*, 1990 [5]
30.9 mg d^{-1}			220.0	88.0	2.8	
0	74H		112.0	65.0[b]	2.5	
10.3 mg d^{-1}		28-35 DIM	136.0	77.0[a]	2.6	Zhao *et al.*, 1992
350 mg per 14 d			122.0	73.0[a,b]	2.6	

[1] N = Total number of animals in experiment; B = Breed; F = Friesian; H = Holstein; J = Jersey; DIM = days in milk at start of injections.
[2] NR = not reported.
[3] Results averaged across several dosages.
[4] Data from cows treated for a second lactation (60 cows) separated from cows treated in first lactation (90 cows).
[5] Data from cows pregnant prior to initiation of treatment excluded.
[a,b] Means with different superscripts within a column (within an experiment) differ.

association between treatment and the incidence or duration of clinical mastitis. White *et al.* (1994) analysed the incidence of mastitis in 914 cows from 15 different full lactation studies and 2697 cows from 70 different short-term studies treated with 0 or 500 mg rbST every two weeks. These authors concluded that under normal conditions there is a positive relationship between the incidence of mastitis and peak and total milk yield and that treatment with rbST did not alter this relationship. That is, cows that produce more milk normally have a greater tendency to develop mastitis and this relationship exists regardless of the use of rbST (White *et al.*, 1994). Therefore, the potential increase in mastitis in cows treated with rbST is due more to increased milk yield than to any direct effects of the hormone.

Economics of Somatotrophin

The economic implications of biotechnology, in general and specifically rbST, are far ranging and have received increasing attention from scientists, policy makers, consumers and producers (Lacy *et al.*, 1992). Research has primarily focused on developing models that attempt to predict the impact of rbST usage on the industry. For example, the Executive Branch (1994) of the USA federal government projected that with adoption rates increasing from 5% of the USA dairy herds in 1994 to 34% in five years, there would be a 4% increase in production per cow and a 3% decline in cow numbers by 1999. Due to these changes, there would be a 0.3 billion kg increase in total milk production and price would decline by 2% per year. Over the 1994–1999 period, net farm income would decline by US $1.3 billion (a net cost), consumer welfare would increase by US $2.3 billion (a net saving) and government costs would increase by US $297 million. The combined effect over the six years would be a net benefit to the USA of $0.7 billion. This model assumes specific rates of adoption, net farm income and profitability, and no change in consumption of dairy products.

This section will focus on issues related to several of these assumptions. Specifically, issues examined are the rate of adoption of rbST, the potential for on-dairy-farm profitability and consumer acceptance of products derived from rbST-treated cows.

Consumer acceptance

Two of the most critical factors related to the use of rbST in commercial dairy operations ultimately are the attitude and action of the milk consuming public. The critical question is, to what extent, if any, consumers will alter consumption of milk produced from cows treated with rbST. Any large-scale reductions in consumption would potentially offset any gains in production or efficiency due to rbST. Smith and Warland (1992) summarized data from eight different surveys that attempted to identify consumer attitudes towards milk derived from cows treated with rbST in the USA. Approximately 60% of the population would not change their level of milk consumption, 30% will likely reduce consumption and 10% will stop milk consumption entirely. Thus, data from these surveys indicate that use of rbST would have a negative impact on milk consumption. In contrast, Hoban (1994) reported that in a USA national survey of 1004 people, only 2–3% would reduce milk consumption with 86% indicating that their consumption will not change. The apparent contradiction in these studies may be due, in part, to the fact that the wording of the question may bias the response, especially if a respondent is unsure of the issue (Smith and Warland, 1992).

Given that rbST has been available for commercial use in the USA since February 1994, patterns in consumption since that time should be an indicator of consumer acceptance of dairy products from rbST-treated cows. By July 1994, based on approximately 89.5% of all fluid sales in the USA, fluid milk sales had increased by 0.6% (Dairy Market News, 1994). These data indicate that sales of milk are remaining relatively constant in the USA and that the use of rbST in commercial dairy production has had little impact so far on milk consumption.

Adoption of somatotrophin

The impact that rbST will have on the dairy industry will depend on the extent and speed with which this technology is adopted (Yonkers, 1992). Several studies have been undertaken to better understand, before the fact or *ex ante*, the determinants and likely rates of rbST adoption. Much of the data in these economic studies are generated from mathematical models and in some cases rates of rbST adoption are sometimes generated from small, maybe even unique, subsets of the producer population (e.g. Wisconsin farmers) and then extrapolated to all dairy producers. As a result, conclusions from different studies are often contradictory in nature. With the commercial availability of rbST in the USA, the true rate of adoption, at least within the USA, will be apparent within the next year or two.

Kalter *et al.* (1985), analysed data from two samples of New York dairy farmers (40 farmers in the first study; 133 farm managers in the second) and found that nearly one-half of those surveyed would adopt rbST within the first year (early adopters) of commercial availability. One-third would adopt rbST between the second and fifth year (middle adopters), while 13% said they would never adopt the technology. Results from two surveys conducted in the southern USA (1000 dairy farmers located in four southern states in one study; 142 Georgia farmers in the second) indicated that approximately 30–40% of the producers would adopt rbST within the first year, but approximately 40% indicated that they would never utilize rbST (Carley *et al.*, 1988; Kinnucan *et al.*, 1990).

Cost of rbST was an important consideration for adoption. Other factors that influence the rate of adoption are the age of the producer, level of education and herd size. That is, early adopters tended to be younger farmers with more education and larger herds (Carley *et al.*, 1988; Kinnucan *et al.*, 1990; Zepeda, 1990). Marion *et al.* (1988) estimated that the upper limit of adoption during the first five years of use in Wisconsin would be 45% of the herds. This equals the percentage of cows participating in the Dairy Herd Improvement Program, and it is probably the producers with better managed herds who are more likely to adopt this technology.

In general, producers that have expressed a willingness to adopt new technology will adopt rbST more quickly than producers that are opposed to or fear technological advancements in the industry. Although there appears to be geographical variation, younger farmers and better managers are more likely to be early adopters of rbST.

On-farm profitability

The most critical on-farm issue facing producers concerning the use of rbST, or any new technology, is its effect on overall farm profitability. In principle, many scientists agree that rbST can be effective and affordable regardless of herd size. However, the differential impact that any technology, not just rbST, might have on farms differing in size is often a major consideration from a public policy perspective. Although several variables influence the response to rbST (see above), as long as these variables are similar among large and small herds, there should be no per cow productivity differences across farm size.

Butler (1992) constructed a partial budget framework to examine the potential impact of rbST adoption on farm level net revenues. This analysis was performed for annual production rates ranging from 5450 to 9545 kg assuming rbST increases milk yield from 5 to 20%, daily costs of rbST per cow from 20 cents to 80 cents, and milk prices from US $10 to $13 per 45.5 kg. Resulting net returns had a wide distribution. The most sensitive variable in the model was the price of milk. Although rbST adoption might be profitable under a wide range of assumptions, the author indicated that the available data do not permit a thorough understanding of the additional managerial requirements that rbST use might impose. Therefore, producers need to evaluate their management situation prior to adopting rbST.

Based on Wisconsin producers, Marion *et al.* (1988) calculated a 50 cent reduction per 45.5 kg in milk price due to rbST. Although this might seem economically significant, the impact of rbST may be to accelerate the trend of rising milk supply by two years. Data collected in the USDA 1989 Farm Costs and Returns Survey was used to evaluate the potential impact of rbST adoption on farm costs and returns. This analysis indicated that, at 1989 prices, rbST adoption would increase average net cash balances from 15 to 22% (Executive Branch, 1994).

Existing knowledge concerning the farm level economics of rbST indicates that the diffusion of this technology will be quite rapid and that its adoption is likely to benefit better managed farms more than poorly managed ones. The net effect will be to accentuate, but not to alter dramatically, the trend towards fewer but better managed dairy farms that has been observed over the last several decades. The effects of this practice on dairy farming are not expected to be large and might very well be overshadowed by other developments. Similar to the rate of adoption, specific and actual farm data that focus on the profitability of rbST in the USA can now be generated and the mathematical models that were used to predict profitability can now be tested.

Acknowledgements

Storrs Agricultural Experiment Station Scientific Contribution Number 1586. The authors thank Mira Moreshead and Amy Mowrey for technical assistance and Tina Burnham and Jody Klinger for assistance with preparation of this manuscript.

References

Asdell, S.A. (1932) The effect of the injection of hypophyseal extract in advanced lactation. *American Journal of Physiology* 100, 137–140.

Asimov, G.J. and Krouze, N.K. (1937) The lactogenic preparations from the anterior pituitary and the increase of milk yield in cows. *Journal of Dairy Science* 20, 289–306.

Baer, R.J., Tieszen, K.M., Schingoethe, D.J., Casper, D.P., Eisenbeizs, W.A., Shaver, R.D. and Cleale, R.M. (1989) Composition and flavor of milk produced by cows injected with recombinant bovine somatotrophin. *Journal of Dairy Science* 72, 1424–1434.

Barbano, D.M., Lynch, J.M., Bauman, D.E., Hartnell, G.F., Hintz, R.L. and Memeth, M.A. (1992) Effect of a prolonged-release formulation of n-methionyl bovine somatotrophin (sometribove) on milk composition. *Journal of Dairy Science* 75, 1775–1793.

Bauman, D.E. (1989) Biology of bovine somatotrophin in dairy cattle. *Animal Science Mimeograph Series*, Cornel Cooperative Extension, No. 133, pp. 1–8.

Bauman, D.E. (1992) Bovine somatotrophin: Review of an emerging animal technology. *Journal of Dairy Science* 75, 3432–3451.

Bauman, D.E. and Vernon, R.G. (1993) Effects of exogenous bovine somatotrophin on lactation. *Annual Review of Nutrition* 13, 437–461.

Bauman, D.E., DeGeeter, M.J., Peel, C.J., Lanza, G.M., Gorewit, R.C. and Hammond, R.W. (1982) Effect of recombinantly derived bovine growth hormone (bGH) on lactational performance of high yielding dairy cows. *Journal of Dairy Science* 65(Suppl. 1), 121 (Abstract).

Bauman, D.E., Eppard, P.J., DeGeeter, M.J. and Lanza, G.M. (1985) Responses of high-producing dairy cows to long-term treatment with pituitary somatotrophin and recombinant somatotrophin. *Journal of Dairy Science* 68, 1352–1362.

Bauman, D.E., Hard, D.L., Crooker, B.A., Partridge, M.S., Garrick, K., Sandles, L.D., Erb, H.N. Franson, S.E., Hartnell, G.F. and Hintz, R.L. (1989) Long-term evaluation of a prolonged-release formulation of n-methionyl bovine somatotrophin in lactating dairy cows. *Journal of Dairy Science* 72, 642–651.

Bauman, D.E., McBride, B.W., Burton, J.L. and Sejrsen, K. (1993) Somatotrophin (BST): International Dairy Federation Technical Report. *Bulletin of the IDF* 293, 2–5.

Becker, B.A., Johnson, H.D., Li, R. and Collier, R.J. (1990) Effect of farm and simulated laboratory cold environmental conditions on the performance and physiological responses of lactating dairy cows supplemented with bovine somatotrophin (BST). *International Journal of Biometeorology* 34, 151–156.

Beermann, D.H. and DeVol, D.L. (1991) Effects of somatotrophin, somatotrophin releasing factor and somatostatin on growth. In: Pearson, A.M. and Dutson, T.R (eds), *Growth Regulation in Farm Animals: Advances in Meat Research Volume 7*. Elsevier Applied Science, New York, pp. 373–426.

Bines, J.A., Hart, I.C. and Morant, S.V. (1980) Endocrine control of energy metabolism in the cow: The effect on milk yield and levels of some blood constituents of injecting growth hormone and growth hormone fragments. *British Journal of Nutrition* 43, 179–188.

Brumby, P.J. (1956) Milk production in first-calf heifers following growth-hormone therapy. *New Zealand Journal of Science and Technology* 37A, 152–156.

Brumby, P.J. and Hancock, J. (1955) The galactopoietic role of growth hormone in dairy cattle. *New Zealand Journal of Science and Technology* 36, 417–436.

Bullis, D.D., Bush, L.J. and Barto, P.B. (1965) Effect of highly purified and commercial grade growth hormone preparations on milk production of dairy cows. *Journal of Dairy Science* 48, 338–341.

Burton, J.L., McBride, B.W., Burton, J.H. and Eggert, R.G. (1990a) Health and reproductive performance of dairy cows treated for up to two consecutive lactations with bovine somatotrophin. *Journal of Dairy Science* 73, 3258–3265.

Burton, J.H., MacLeod, G.K., McBride, B.W., Burton, J.L., Bateman, K., McMillan, I. and Eggert, R.G. (1990b) Overall efficacy of chronically administered recombinant bovine somatotrophin to lactating dairy cows. *Journal of Dairy Science* 73, 2157–2167.

Butler, L.J. (1992) Economic evaluation of BST for on farm use. In: Hallberg, M.C. (ed.) *Bovine Somatotrophin and Emerging Issues – An Assessment*. Westview Press, San Francisco, pp. 145–175.

Carley, D., Fletcher, S.M. and Alexander, D.C.S. (1988) The adoption of developing technologies by Georgia dairy farmers. *Georgia Agricultural Experiment Station*, Athens, GA.

Chalupa, W. and Galligan, D.T. (1989) Nutritional implications of somatotrophin for lactating cows. *Journal of Dairy Science* 72, 2510–2524.

Chilliard, Y. (1989) Long-term effects of recombinant bovine somatotrophin (rBST) on dairy cow performances: A review. In: Sejrsen, K., Vestergaard, M. and

Neimann-Sorensen, A. (eds) *Use of Somatotrophin in Livestock Production*. Elsevier Applied Science, New York, pp. 61–87.

Cole, W.J., Eppard, P.J., Boysen, B.G., Madsen, K.S., Sorbet, R.H., Miller, M.A., Hintz, R.L., White, T.C., Ribelin, W.E., Hammond, B.G., Collier, R.J. and Lanza, G.M. (1992) Response of dairy cows to high doses of a sustained-release bovine somatotrophin administered during two lactations. *Journal of Dairy Science* 75, 111–123.

Collier, R.J. (1985) Nutritional, metabolic and environmental aspects of lactation. In: Larson, B.L. (ed.) *Lactation*. The Iowa State University Press, Ames, Iowa, pp. 80–128.

Cotes, P.M., Crichton, J.A., Folley, S.J. and Young, F.G. (1949) Galactopoietic activity of purified anterior pituitary growth hormone. *Nature* 164, 992–993.

Crooker, B.A. and Otterby, D.E. (1991) Management of the dairy herd treated with bovine somatotrophin. *Veterinary Clinics of North America: Food Animal Practice* 7, 417–437.

Crooker, B.A., McGuire, M.A., Cohick, W.S., Harkins, M., Bauman, D.E. and Sejrsen, K. (1990) Effect of dose of bovine somatotrophin on nutrient utilization in growing dairy heifers. *Journal of Nutrition* 120, 1256–1263.

Dairy Market News (1994) July fluid milk highlights. 61, 9.

Downer, J.V., Patterson, D.L., Rock, D.W., Cleale, R.M., Firkins, J.L., Lynch, G.L., Clark, J.H., Chalupa, W.V., Brodie, B.O., Jenny, B.F. and De Gregorio, R. (1993) Dose titration of sustained-release recombinant bovine somatotrophin in lactating dairy cows. *Journal of Dairy Science* 76, 1125–1136.

Eisenbeisz, W.A., Schingoethe, D.J., Casper, D.P., Shaver, R.D. amd Cleale, R.M. (1990) Lactational evaluation of recombinant bovine somatotrophin with corn and barley diets. *Journal of Dairy Science* 73, 1269–1279.

Eppard, P.J., Bauman, D.E. and McCutcheon, S.N. (1985) Effect of dose of bovine growth hormone on lactation of dairy cows. *Journal of Dairy Science* 68, 1109–1115.

Eppard, P.J., Bauman, D.E., Curtis, C.R., Erb, H.N., Lanza, G.M. and DeGeeter, M.J. (1987) Effect of 188-day treatment with somatotrophin on health and reproductive performance of lactating dairy cows. *Journal of Dairy Science* 70, 582–591.

Eppard, P.J., Hudson, S., Cole, W.J., Hintz, R.L., Hartnell, G.F., Hunter, T.W., Metzger, L.E., Torkelson, A.R. Hammond, B.G., Collier, R.J. and Lanza, G.M. (1991) Response of dairy cows to high doses of a sustained-release bovine somatotrophin administered during two lactations. *Journal of Dairy Science* 74, 3807–3821.

Eppard, P.J., Bentle, L.A., Violand, B.N., Ganguli, S., Hintz, R.L., Kung, L., Jr. Krivi, G.G and Lanza, G.M. (1992) Comparison of the galactopoietic response to pituitary-derived and recombinant-derived variants of bovine growth hormone. *Journal of Endocrinology* 132, 47–56.

Erb, H.N. and Smith, R.D. (1987) The effects of periparturient events on breeding performance of dairy cows. *Veterinary Clinics of North America* (Food Animal Practice) 3, 501–508.

Erdman, R.A., Sharma, B.K., Shaver, R.D. and Cleale, R.M. (1990) Dose response to recombinant bovine somatotrophin from weeks 15 to 44 postpartum in lactating dairy cows, *Journal of Dairy Science* 73, 2907–2915.

Evans, H.M. and Long, J.A. (1921) Characteristic effects upon growth, oestrus and ovulation induced by the intraperitoneal administration of fresh anterior hypophyseal substance. *Proceedings of the National Academy of Sciences USA* 8, 38–39.

Executive Branch (1994) Use of bovine somatotrophin (BST) in the United States: Its potential effects. A study by The Executive Branch of the Federal Government.

Franson, S.E., Cole, W.J., Hoffman, R.G., Meserole, V.K., Sprick, D.M., Madsen, K.S., Hartnell, G.F., Bauman, D.E., Head, H.H., Huber, J.T. and Lamb, R.C. (1989) Response of cows throughout lactation to sometribove, recombinant methionyl bovine somatotrophin, in a prolonged release system – A dose titration study. *Journal of Dairy Science* 72, (Suppl. 1), 451 (Abstract).

Gallo, L., Cassandro, M., Carnier, P., Mantovani, R., Ramanzin, M., Bittante, G., Tealdo, E. and Casson, P. (1994) Modeling response to slow-releasing somatotrophin administered at 3- or 4-week intervals. *Journal of Dairy Science* 77, 759–769.

Gertler, A., Ashkenazi, A. and Madar, Z. (1984) Binding sites of human growth hormone and ovine and bovine prolactins in the mammary gland and liver of the lactating dairy cow. *Molecular and Cellular Endocrinology* 34, 51–57.

Gibson, J.P., McBride, B.W., Burton, J.H., Politis, I. and Zhao, X. (1992a) Effect on production traits of bovine somatotrophin for up to three consecutive lactations. *Journal of Dairy Science* 75, 837–846.

Gibson, J.P., Van Der Meulen, M., McBride, B.W. and Burton, J.H. (1992b) The effects of genetic and phenotypic production potential on response to recombinant bovine somatotrophin. *Journal of Dairy Science* 75, 878–884.

Hard, D.L., Cole, W.J., Franson, S.E., Samuels, W.A., Bauman, D.E., Erb, H.N., Huber, J.T. and Lamb, R.C. (1988) Effect of long term sometribove treatment in a prolonged release system on milk yield, animal health and reproductive performance pooled across four sites. *Journal of Dairy Science* 71(Suppl. 1), 210 (Abstract).

Hart, I.C., Bines, J.A. and Morant, S.V. (1979) Endocrine control of energy metabolism in the cow: Correlations of hormones and metabolites in high and low yielding cows for stages of lactation. *Journal of Dairy Science* 62, 270–277.

Hartnell, G.F., Franson, S.E., Bauman, D.E., Head, H.H., Huber, J.T., Lamb, R.C., Madsen, K.S., Cole, W.J. and Hintz, R.L. (1991) Evaluation of sometribove in a prolonged-release system in lactating dairy cows – production responses. *Journal of Dairy Science* 74, 2645–2663.

Heap, R.B., Fleet, I.R., Fullerton, F.M., Davis, A.J., Goode, J.A., Hart, I.C., Pendleton, J.W., Prosser, C.G., Silvester, L.M. and Mepham, T.B. (1989) A comparison of the mechanisms of action of bovine pituitary-derived and recombinant somatotrophin (ST) in inducing galactopoiesis in the cow during late lactation. In: Heap, R.B., Prosser, C.G. and Lamming, G.E. (eds), *Biotechnology in Growth Regulation*. Butterworths, Boston, pp. 73–84.

Hoban, T.J. (1994). Consumer awareness and acceptance of bovine somatotrophin. Grocery Manufacturers of America.

Huber, J.T., Sullivan, J.L., Willman, S., Arana, M., Denigan, M., DeCorte, C., Hoffman, R.G. and Hartnell, G.F. (1991) Response of Holstein cows to biweekly sometribove (SB) injections for 4 consecutive lactations. *Journal of Dairy Science* 74(Suppl. 1), 211 (Abstract).

Hutton, J.B. (1957) The effect of growth hormone on the yield and composition of cows' milk. *Journal of Endocrinology* 16, 115–125.

Johnson, H.D., Li, R., Manalu, W., Spencer-Johnson, K.J., Becker, B.A., Collier, R.J. and Baile, C.A. (1991) Effects of somatotrophin on milk yield and physiological responses during summer farm and hot laboratory conditions. *Journal of Dairy Science* 74, 1250–1262.

Johnsson, I.D. and Hart, I.C. (1986) Manipulation of milk yield with growth hormone. In: Haresign, W. and Cole, O.J.A. (eds) *Recent Advances in Animal Nutrition.* Butterworths, London, pp. 105–120.

Juskevich, J.C. and Guyer, C.G. (1990) Bovine growth hormone: Human food safety evaluation. *Science* 249, 875–884.

Kalter, R.J., Milligan, R., Lesser, W., Magrath, W., Tauer, L. and Bauman, D.E. (1985) Biotechnology and the dairy industry: Production costs and commercial potential and the economic impact of the bovine growth hormone. Cornell University Centre for the Biotechnology, Committee for Economic Development, Ithaca, New York, Research Bulletin 85–20.

Kazmer, G.W., Canfield, R.C. and Bean, B. (1991) Somatotrophin and prolactin profile characteristics in proven AI dairy sires. *Journal of Animal Science* 69, 1601–1606.

Kazmer, G.W., Canfield, R.C. and Bean, B. (1990) Plasma growth hormone and prolactin concentrations in young dairy sires before and after a 24 h fast. *Journal of Dairy Science* 73, 3112–3117.

Kinnucan, H., Upton, H., Molnar, J.J. and Pencergrass, R. (1990) Adoption and diffusion potentials for bovine somatotrophin in the southeast dairy industry. Bulletin 605, Alabama Experiment Station, Auburn University, Auburn, AL.

Kronfeld, D.S. (1993) Recombinant bovine growth hormone: Cow responses delay drug approval and impact public health. In: Liebhardt, W.C. (ed.) *The Dairy Debate.* Sustainable Agriculture Research and Education Program, Davis, California, pp. 65–112.

Lacy, W.B., Lacy, L.R. and Busch, L. (1992) Emerging trends, consequences and policy issues in agricultural biotechnology. In: Hallberg, M.C. (ed.) *Bovine Somatotrophin and Emerging Issues – An Assessment.* Westview Press, San Francisco, pp. 3–32.

Laurent, F., Vignon, B., Coomans, D., Wilkinson, J. and Bonnel, A. (1992) Influence of bovine somatotropin on the composition and manufacturing properties of milk. *Journal of Dairy Science* 75, 2226–2234.

Lee, M.O. and Schaffer, N.K. (1934) Anterior pituitary growth hormone and the composition of growth. *The Journal of Nutrition* 7, 337–363.

Leitch, H.W., Burnside, E.B. and McBride, B.W. (1990) Treatment of dairy cows with recombinant bovine somatotrophin: Genetic and phenotypic aspects. *Journal of Dairy Science* 73, 181–190.

Leonard, M., Gallo, M., Gallo, G. and Block, E. (1990) Effects of a 28-day sustained-release formulation of recombinant bovine somatotrophin (rbST) administered to cows over two consecutive lactations. *Canadian Journal of Animal Science* 70, 795–809.

Lissemore, K.D., Leslie, K.E., McBride, B. W., Burton, J. H., Willan, A.R. and Bateman, K.G. (1991) Observations on intramammary infection and somatic cell counts in cows treated with recombinant bovine somatotrophin. *Canadian Journal of Veterinary Research* 55, 196–198.

Lormore, M.J., Muller, L.D., Deaver, D.R. and Griel Jr., L.C. (1990) Early lactation responses of dairy cows administered bovine somatotrophin and fed diets high in energy and protein. *Journal of Dairy Science* 73, 3237–3247.

Lotan, E., Sturman, H., Weller, J.I. and Ezra, E. (1993) Effects of recombinant bovine somatotrophin under conditions of high production and heat stress. *Journal of Dairy Science* 76, 1394–1402.

Lough, D.S., Muller, L.D., Kensinger, R.S., Sweeny, T.F. and Griel , L.C.Jr. (1988)

Effect of added dietary fat and bovine somatotrophin on the performance and metabolism of lactating dairy cows. *Journal of Dairy Science* 71, 1161–1169.

Ludri, R.S., Upadhyay, R.C., Singh, M., Guneratne, J.R.M. and Basson, R.P. (1989) Milk production in lactating buffalo receiving recombinantly produced bovine somatotrophin. *Journal of Dairy Science* 72, 2283–2287.

McBride, B.W., Burton, J.L., Gibson, J.P., Burton, J.H., and Eggert, R.G. (1990) Use of recombinant bovine somatotrophin for up to two consecutive lactations on dairy production traits. *Journal of Dairy Science* 73, 3248–3257.

McClary, D., McGuffey, R.K. and Green, H.B. (1990) Bovine somatotrophin: part 2. *Agri-Practice* 11, 5–11.

McClary, D.G., Green, H.B., Basson, R.P. and Nickerson, S.C. (1994) The effects of a sustained-release recombinant bovine somatotrophin (Somidobove) on udder health for a full lactation. *Journal of Dairy Science* 77, 2261–2271.

McGuffey, R.K., Basson, R.P., Snyder, D.L, Block, E., Harrison, J.H., Rakes, A.H., Emery, R.S. and Muller, L.D. (1991) Effect of somidobove sustained release administration on the lactation performance of dairy cows. *Journal of Dairy Science* 74, 1263–1276.

McGuffey, R.K. and Wilkinson, J.I.D. (1991) Nutritional implications of bovine somatotrophin for the lactating dairy cow. *Journal of Dairy Science* 74(Suppl. 2), 63–71.

Machlin, L.J. (1973) Effect of growth hormone on milk production and feed utilization in dairy cows. *Journal of Dairy Science* 56, 575–580.

Marion, B.W., Wills, R.L. and Butler, L.J. (1988) The social and economic impact of biotechnology on Wisconsin agriculture. College of Agricultural and Life Sciences. Report prepared for the State Legislature.

Michel, A., McCutcheon, S.N., MacKenzie, D.D.S., Tait, R.M. and Wickham, B.W. (1990) Effects of exogenous bovine somatotrophin on milk yield and pasture intake in dairy cows of low or high genetic merit. *Animal Production* 51, 229–234.

Mohammed, M.E. and Johnson, H.D. (1985) Effect of growth hormone on milk yields and related physiological functions of Holstein cows exposed to heat stress. *Journal of Dairy Science* 68, 1123–1133.

Moore, D.A. and Hutchinson, L.J. (1992) BST and animal health. In: Hallberg, M.C. (ed.) *Bovine Somatotrophin and Emerging Issues – An Assessment.* Westview Press, San Francisco, pp. 99–141.

Muller, L.D. (1992) BST and Dairy Cow Performance. In: Hallberg, M.C. (ed.) *Bovine Somatotrophin and Emerging Issues – An Assessment.* Westview Press, San Francisco, pp. 53–71.

Nytes, A.J., Combs, D.K., Shook, G.E., Shaver, R.D. and Cleale, R.M. (1990) Response to recombinant bovine somatotrophin in dairy cows with different genetic merit for milk production. *Journal of Dairy Science* 73, 784–791.

Oldenbroek, J.D., Garssen, G.J., Forbes, A.B. and Jonker, L.J. (1989) The effect of treatment of dairy cows of different breeds with recombinantly derived bovine somatotrophin in a sustained-delivery vehicle. *Livestock Production Science* 21, 13–34.

Oldenbroek, J.K., Garssen, G.J., Jonker, L.J. and Wilkinson, J.I.D. (1993) Effects of treatment of dairy cows with recombinant bovine somatotrophin over three or four lactations. *Journal of Dairy Science* 76, 453–467.

Peel, C.J. and Bauman, D.E. (1987) Somatotrophin and lactation. *Journal of Dairy Science* 70, 474–486.

Peel, C.J., Bauman, D.E., Gorewit, R.C. and Sniffen, C.J. (1981) Effect of exogenous growth hormone on lactational performance in high yielding dairy cows. *Journal of Nutrition* 111, 1662–1671.

Peel, C.J., Fronk, T.J., Bauman, D.E. and Gorewit, R.C. (1982) Lactational response to exogenous growth hormone and abomasal infusion of a glucose-sodium caseinate in high-yielding dairy cows. *Journal of Nutrition* 112, 1770–1778.

Peel, C.J., Fronk, T.J., Bauman, D.E. and Gorewit, R.C. (1983) Effect of exogenous growth hormone in early and late lactation on lactational performance of dairy cows. *Journal of Dairy Science* 66, 776–782.

Peel, C.J., Sandles, L.D., Quelch, K.J. and Herington, A.C. (1985) The effects of long-term administration of bovine growth hormone on the lactational performance of identical-twin dairy cows. *Animal Production* 41, 135–142.

Peel, C.J., Eppard, P.J. and Hard, D.L. (1988) Evaluation of sometribove (methionyl bovine somatotrophin) in toxicology and clinical trials in Europe and the United States. In: Heap, R.B., Prosser, C.G. and Lamming, G.E. (eds) *Biotechnology in Growth Regulation*. Butterworths, Boston, pp. 107–116.

Pell, A.N., Tsang, D.S., Howlett, B.A., Huyler, M.T., Messerole, V.K., Samuels, W.A., Hartnell, G.F. and Hintz, R.L. (1992) Nutrition, feeding and calves – effects of a prolonged-release formulation of sometribove (n-methionyl bovine somatotrophin) on Jersey cows. *Journal of Dairy Science* 75, 3416–3431.

Phipps, R.H. (1989) A review of the influence of somatotrophin on health, reproduction and welfare in lactating dairy cows. In: Sejrsen, K., Vestergaard, M. and Neimann-Sorensen, A. (eds) *Use of Somatotrophin in Livestock Production*. Elsevier Applied Science, New York, pp. 88–119.

Phipps, R.H., Weller, R.F., Craven, N. and Peel, C.J. (1990) Use of prolonged-release bovine somatotrophin for milk production in British Friesian dairy cows. *Journal of Agricultural Science* 115, 95–104.

Phipps, R.H., Madakadze, C., Mutsvangwa, T., Hard, D.L., and Kerchove, G. (1991) Use of bovine somatotrophin in the tropics: The effect of sometribove on milk production of *Bos indicus*, dairy crossbred and *Bos taurus* cows in Zimbabwe. *Journal of Agricultural Science* 117, 257–263.

Schmidt, G.H. (1989) Effect of length of calving intervals on income over feed and variable costs. *Journal of Dairy Science* 72, 1605–1611.

Seeburg, P.H., Sias, S., Adelman, J., de Boes, H.A., Hayflich, J., Jhurani, P., Goeddel, D.V. and Heyneker, H.L. (1983) Efficient bacterial expression of bovine and porcine growth hormones. *DNA* 2, 37–45.

Smith, B.J. and Warland, D. (1992) Consumer response to milk from BST-supplemented cows. In: Hallberg, M.C. (ed.) *Bovine Somatotrophin and Emerging Issues – An Assessment*. Westview Press, San Francisco, pp. 53–71.

Soderholm, C.G., Otterby, D.E., Linn, J.G., Ehle, F.R., Wheaton, J.E., Hansen, W.P. and Annexstad, R.J. (1988) Effects of recombinant bovine somatotrophin on milk production, body composition and physiological parameters. *Journal of Dairy Science* 71, 355–365.

Sullivan, J.L., Huber, J.T., DeNise, S.K., Hoffman, R.G., Kung, L., Jr, Franson, S.E. and Madsen, K.S. (1992) Factors affecting response of cows to biweekly injections of sometribove. *Journal of Dairy Science* 75, 756–763.

Tessmann, N.J., Dhiman, T.R., Kleinmans, J., Radloff, H.D. and Satter, L.D. (1991) Recombinant bovine somatotrophin with lactating cows fed diets differing in energy density. *Journal of Dairy Science* 74, 2633–2644.

Thomas, J.W., Erdman, R.A., Galton, D.M., Lamb, R.C., Arambel, M.J., Olson, J.D., Madsen, K.S., Samuels, W.A., Peel, C.J. and Green, G.A. (1991) Responses by lactating cows in commercial dairy herds to recombinant bovine somatotrophin. *Journal of Dairy Science* 74, 945–964.

Tyrrell, H.F., Brown, A.C.G., Reynolds, P.J., Haaland, G.L., Bauman, D.E., Peel, C.J. and Steinhour, W.D. (1988). Effect of bovine somatotrophin on metabolism of lactating dairy cows: Energy and nitrogen utilization as determined by respiration calorimetry. *Journal of Nutrition* 118, 1024–1030.

Van Den Berg, G. (1991) A review of quality and processing suitability on milk from cows treated with bovine somatotrophin. *Journal of Dairy Science* 74(Suppl. 2), 2–11.

Vérité, R., Rulquin, H. and Faverdin, Ph. (1989) Effect of slow-released somatotrophin on dairy cow performances. In: Sejrsen, K., Vestergaard, M. and Neimann-Sorensen, A. (eds) *Use of Somatotrophin in Livestock Production*. Elsevier Applied Science, New York, pp. 269–273.

Villa-Godoy, A., Hughes, T.L., Emery, R.S., Enright, W.J., Ealy, A.D., Zinn, S.A. and Fogwell, R.L. (1990) Energy balance and body condition influence luteal function in Holstein heifers. *Domestic Animal Endocrinology* 7, 135–148.

Weller, R.F., Phipps, R.H., Craven, N. and Peel, C.J. (1990) Use of prolonged-release bovine somatotrophin for milk production in British Friesian dairy cows. *Journal of Agricultural Science* 115, 105–112.

West, J.W., Bondari, K. and Johnson Jr., J.C. (1990) Effects of bovine somatotrophin on milk yield and composition, body weight and condition score of Holstein and Jersey cows. *Journal of Dairy Science* 73, 1062–1068.

White, T.C., Madsen, D.S., Hintz, R.L., Sorbet, R.H., Collier, R.J., Hard, D.L., Hartnell, G.F., Samuels, W.A., de Kerchove, G., Adrianens, F., Craven, N., Bauman, D.E., Bertrand, G., Bruneau, P.H., Gravert, G.O., Head, H.H., Huber, J.T., Lamb, R.C., Palmer, C., Pell, A.N., Phipps, R., Weller, R., Piva, G., Rijpkema, Y., Skarda, J., Vedeau, F. and Wollny, C. (1994) Clinical mastitis in cows treated with sometribove (recombinant bovine somatotrophin) and its relationship to milk yield. *Journal of Dairy Science* 77, 2249–2260.

Wood, D.C., Salsgiver, W.J., Kasser, T.R., Lange, G.W., Rowold, E., Violand, B.N., Johnson, A., Leimgruber, R.M., Parr, G.R., Seigel, N.R., Kimack, N.M., Smith, C.E., Zobel, J.F., Ganguli, S.M., Garbow, J.R., Bild, G. and Krivi, G.G. (1989) Purification and characterization of pituitary bovine somatotrophin. *The Journal of Biological Chemistry* 264, 14741–14747.

Yonkers, R.D. (1992) Potential adoption and diffusion of BST among dairy farmers. In: Hallberg, M.C. (ed.) *Bovine Somatotrophin and Emerging Issues – An Assessment*. Westview Press, San Francisco, pp. 177–192.

Young, F.G. (1947) Experimental stimulation (galactopoiesis) of lactation. *British Medical Bulletin* 5, 155–160.

Zepeda, L. (1990) Predicting bovine somatotrophin use by California dairy farmers. *Western Journal of Agricultural Economics* 15, 55–62.

Zhao, X., Burton, J.H. and McBride, B.W. (1992) Lactation, health and reproduction of dairy cows receiving daily injectable or sustained-release somatotrophin. *Journal of Dairy Science* 75, 3122–3130.

Zinn, S.A., Kazmer, G.W. and Paquin-Platts, D.D. (1993) Milk yield and composition response to a sustained-release formulation of bovine somatotrophin in lactating dairy cows. *Journal of Dairy Science* 76(Suppl. 1), 241 (Abstract).

Zoa-Mboe, A., Head, H.H., Bachman, K.C., Baccari, F., Jr and Wilcox, C.J. (1989) Effects of bovine somatotrophin on milk yield and composition, dry matter intake and some physiological functions of Holstein cows during heat stress. *Journal of Dairy Science* 72, 907–916.

Zullo, P.A. and Martin, J.H. (1990) Effect of bovine somatotrophin on milk composition and product yields: A review. *Cultured Dairy Products Journal* 25, 16–20.

Breeding and Reproduction

The Bovine Gene Map 5

J.E. Womack

Department of Veterinary Pathobiology, College of Veterinary Medicine, Texas A&M University, College Station, Texas 77843–4467, USA

Introduction

Gene mapping in animals is not a new concept. Following the products of meiotic recombination to order genes on the *Drosophila* X-chromosome, Sturtevant (1913) introduced animal biologists to the underlying principles of genomic research. While a few genes were mapped in several laboratory animal species over the ensuing 50–60 years, a comprehensive genetic map was produced for only a single mammal, the laboratory mouse. Both biological and technical restraints were responsible for the lack of proliferation of maps to other mammalian genomes. Meiotic mapping requires (i) segregation of allelic variation and (ii) the analysis of co-segregation of multiple loci in large numbers of meiotic products. Vast collections of mutant phenotypes and early application of immunological and biochemical techniques to gene products in laboratory mice revealed the necessary allelic variation. Crossing, followed by backcrossing and intercrossing of inbred strains provided the material for segregation analysis of meiotic products. The genomics revolution, most apparent in recent developments in human medical genetics but also suddenly emerging as a force in livestock breeding, was made possible by advances on several fronts. First, somatic cell genetics circumvented both the requirements for meiotic segregation and intraspecies allelic variation. Second, molecular techniques revealed DNA level polymorphism that far exceeds the resolution possible in the electrophoretic or immunological analysis of gene products. Following the lead of the human genetics community, animal scientists have come to appreciate the potential value of genome maps in livestock improvement (Soller and Beckmann, 1982; Smith and Simpson, 1986; Beckmann and Soller, 1988). It is the merger of new molecular technologies with the new found appreciation for practical application of genomic information that has led to the development of a substantial bovine map over the last five years.

Goals of Gene Mapping

Chromosomal evolution

Most mammalian gene maps have been constructed with the initial goal of advancing the fundamental knowledge of the species. As biologists, we are curious about the organization and regulation of expression of the complement of genes we refer to as a genome. It is apparent that the well-funded and well-organized human genome initiative will ultimately provide the foundation of our knowledge with respect to mammalian genomes. Generous contributions can be expected from the study of the laboratory mouse. Other mammals, including cattle, have an important role to play, however, especially in unravelling the pathways of mammalian genome evolution. The development of a bovine map for this purpose has been a goal of my laboratory from the beginning and remains a priority among much of the active mapping community.

Marker-assisted selection

The current focus of the bovine genome effort in many laboratories is toward the more practical goal of marker-assisted selection (MAS). Simple DNA tests to identify positive alleles for economic trait loci (ETL), many of which may be quantitative trait loci (QTL), have the potential to rapidly accelerate the efficiency of selective breeding (Soller and Beckmann, 1982; Smith and Simpson, 1986; Weller *et al.*, 1990). Two general approaches to identifying selectable markers of ETL are applicable (Fig. 5.1). The ideal marker is the gene responsible for the trait. Thus, the 'candidate gene' approach seeks allelic variation that is directly responsible for phenotypic variation. This generally requires detailed knowledge of the physiological basis of a trait followed by extensive screening for variation in the candidate genes related to the known physiological and biochemical mechanisms. Once the gene is found, sequence variation related to a positive or negative phenotype must be demonstrated. Ideally, this same variation should become part of the marker assay. As an excellent example, Shuster *et al.* (1992) identified the genetic defect responsible for leucocyte adhesion deficiency (LAD) in Holstein cattle as a missense mutation coding amino acid 128 in CD18. A polymerase chain reaction (PCR) assay was developed to distinguish the mutant and normal alleles, providing the ultimate genetic marker of this economic trait locus. The recessive mutant allele can be eliminated in a single generation by avoiding the use of carriers as breeding stock.

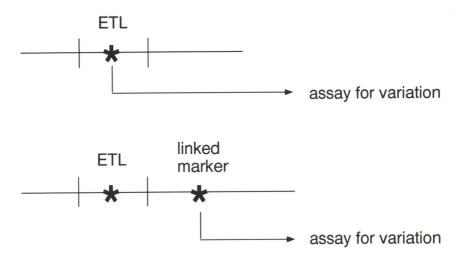

Fig. 5.1. Economic trait loci (ETL) may be followed directly if the gene responsible for the trait is known and the variation can be exploited as a laboratory assay. Linked markers may have no physiological relationship to the trait but simply reside nearby on the same chromosome.

Unfortunately, the genes for most economic traits are unknown. Variation contributing to such traits, even quantitative trait variation, can be genetically mapped, however (Lander and Botstein, 1989; Paterson *et al.*, 1989; Georges *et al.*, 1993a,b). A marker mapped in close proximity to an ETL can be used for selection provided (i) the recombination frequency (map distance) is sufficiently small and (ii) the 'phase' of marker and ETL alleles is known (Fig. 5.2). The frequency of meiotic recombination between linked loci defines the units of a genetic map: 1% recombination represented as one centimorgan (cM). Thus, a marker 10 cM from an ETL can be used to select the ETL with 90% probability of success since 10% of the gametes produced by a parental animal will be recombinants of the marker and ETL alleles. Efficiency is dramatically increased with markers flanking the ETL since recombination in the region spanned by two markers can be detected. One goal of bovine genome mapping is therefore a map of highly polymorphic markers spaced at intervals of 20 cM or less across every cattle chromosome. The total length of the bovine genome is expected to be 2500–3000 cM, thus 125 markers represent the minimum number for a 20 cM map. Since marker mapping is largely a random process, the actual number needed is much larger.

It is obvious from Fig. 5.2 that linked markers must be used with knowledge of phase of the marker allele and the ETL allele. This is not a problem in pedigrees where the phase is initially known and the probability of recombination is small. Linked markers cannot be used at the population

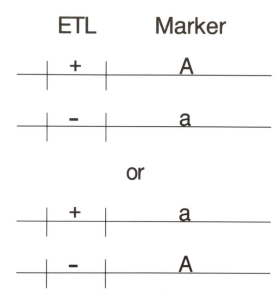

Fig. 5.2. Alleles for ETL and linked markers may be in different phase in different families. Thus, *A* may mark the desirable trait in one pedigree, and *a* in another. Linked markers must therefore be used along with pedigree information.

level (i.e. no pedigree information), however, unless the marker and trait are in linkage disequilibrium. For this reason, a marker in the gene itself is the ideal marker for MAS.

Positional cloning

A third goal of the gene map is to identify and clone genes responsible for ETL via positional cloning. Linked markers have been used successfully in human genetics to clone the genes responsible for disease traits even when the function of the genes and the nature of the gene products is unknown. An excellent example of this 'reverse genetics' approach, now known as positional cloning, comes from the successful search for the cystic fibrosis gene (Rommens *et al.*, 1989). Also exemplified in cystic fibrosis is the need for tightly linked (1 cM or less) markers. Since the marker is merely a starting point for molecular walking and jumping techniques, it must be extremely close in terms of recombination to make cloning feasible. A centimorgan in the 3200 cM human genome is on average about a million base pairs. The cattle genome appears to be slightly smaller in terms of cM while about the same size in base pairs which suggests that a cM may comprise even more base pairs in cattle. A one cM linkage map in cattle is not a foreseeable reality and may never exist except

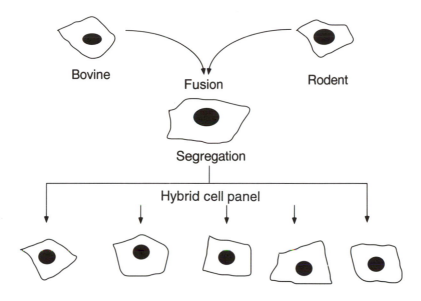

Fig. 5.3. Fusion of cells from different species to produce hybrid somatic cells is facilitated by polyethylene glycol. Complementation of selectable markers permits the recovery of hybrid clones which have independently segregated subsets of the bovine genome. Concordancy of retention and loss of markers establishes synteny.

in isolated chromosomal regions. More realistic possibilities for the use of the map to identify and clone genes will be discussed later.

Methods of Gene Mapping

Synteny mapping

Synteny simply means 'on the same strand' or on the same chromosome. A synteny map is nothing more than a list of genes that are on the same chromosome, even if the chromosome on which they reside is not identified. Since extensive synteny mapping in cattle preceded assignment of genes to specific chromosomes, designations U1, U2, etc. were given to 'unassigned' syntenic groups. These temporary place holders have recently given way to specific chromosomal assignments.

The best approach to synteny mapping is through somatic cell genetics (Fig. 5.3). Cattle-rodent hybrid somatic cells can be constructed in such a configuration that the cattle chromosomes are preferentially lost. Each clone of hybrid cells will have a partial cattle genome along with the complete genome of the rodent cell line. Since chromosome loss is largely random, each clone will have a different subset of cattle chromosomes retained. Analysis of pairs of genes across a panel of such hybrid cell lines will reveal concordance or discordance of retention suggesting their location on the same or different chromosomes, respectively. Gene products or DNA sequences may be mapped by synteny analysis, the only criterion being the ability to detect the presence or absence of the bovine gene or gene product against the rodent background. Enzyme electrophoresis, Southern blotting with cattle or heterologous probes, and PCR have all been used successfully to map cattle genes in panels of hybrid cells.

Somatic cell genetics does not typically assign markers to specific chromosomal regions. Consequently, genes are not ordered in a synteny map. Somatic cell methods employing rearranged chromosomes are an exception to this generalization. Unfortunately, panels of hybrid cell lines with rearranged bovine chromosomes have not yet been systematically used in cattle gene mapping.

In situ *hybridization*

The physical localization of genes or other cloned DNA segments to specific sites on chromosomes is best achieved by *in situ* hybridization. This technique employs the labelling of a probe with a microscopically detectable marker, followed by hybridization of the probe to denatured DNA of the otherwise intact chromosome on a glass slide. The specificity of the reaction is dictated by the uniqueness of the probe utilized (Fig. 5.4). Whole genomes, highly repetitive elements and unique sequences have all been successfully localized by *in situ* hybridization. While radioactive probes predominated in the early application of this technology, fluorescent probes are now widely used. As reviewed by Trask (1991), fluorescence *in situ* hybridization (FISH) has a number of advantages over isotopic labelling. It provides superior spatial resolution and generally requires the visualization of fewer labelled chromosomes. It is faster and the probe is more stable. The sensitivities are similar, each requiring a few kilobase (kb) pairs of uninterrupted sequence as the hybridizing probe. Schemes have been developed which allow multiple probes with different colour signals to be used on the same chromosomes.

Cosmids, cloning vectors that accommodate large DNA sequences (up to 45 kb), can be mapped by FISH. Since these large inserts often contain repetitive DNA, the target DNA must first be pre-hybridized with unlabelled total genomic DNA. This method has been effectively used to anchor the growing bovine linkage map to chromosomes with cosmids that also contain markers useful for linkage mapping (Solinas-Toldo *et al.*, 1993).

The above technologies result in physical maps which describe the localization of genes in physical relationships to other genes and to the chromosomes on which they reside. Higher resolution physical maps which place markers

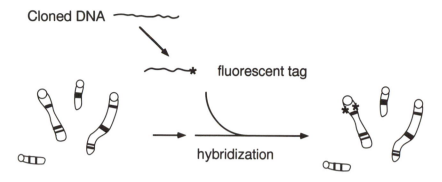

Chromosome spread

Fig. 5.4. Probes for fluorescence *in situ* hybridization (FISH) may be conjugated directly with fluorescent molecules or more often with reporter molecules that bind fluorescent reagents after hybridization. The result of either is a fluorescent signal at the site of probe-chromatin hybridization.

in contiguous clones (contig maps) are forthcoming in livestock but will most likely span small regions of special interest rather than whole chromosomes as is targeted in the human genome initiative and is a prerequisite to total genome sequencing.

Linkage mapping

Linkage mapping defines the relationship of markers to each other in terms of meiotic recombination. Linkage is measurable only if gametic products can be identified. This requires polymorphisms that distinguish maternal and paternal chromosomes in a heterozygous individual. A linkage map is made by determining the percentage of recombinants in the parental arrangements of alleles on a chromosome (Fig. 5.5.). This requires animals that are heterozygous for multiple loci and sufficient numbers of offspring for a statistically meaningful analysis. The segregation of three or more loci permits ordering of genes on the map on the assumption that double recombinants are rare relative to single region recombinants. While the physical distance between two markers is a major factor in the frequency of recombination between them, it is not the only factor. Recombination is not a totally random process but the biological basis for 'hot spots' or 'cold spots' is not known.

Markers for mapping purposes have been classified by O'Brien (1992) as

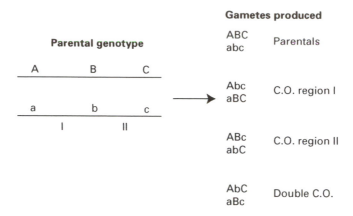

Fig. 5.5. Linkage of markers is determined by an excess of parental allelic combinations in the gametes produced. Map distances are the frequency of recombination as a result of crossing-over (C.O.) between markers. Genes may be ordered in a multi-point analysis on the assumption that double cross-overs will be the combination of alleles observed most infrequently.

class I or II. Class I markers are expressed sequences (genes). They are highly conserved from one species to another and are thus the substances of comparative gene mapping. Unfortunately they are not generally highly polymorphic and therefore difficult to place on linkage maps. Class II markers are the substance of linkage mapping. These are markers whose value lies in their polymorphism. As such, they are not necessarily expressed sequences but often anonymous stretches of DNA.

Two types of polymorphic DNA markers currently predominate linkage mapping. Restriction fragment length polymorphisms (RFLPs) identify loci that differ in the presence or absence of specific restriction sites. These four or six base sites are cut by bacterial enzymes called restriction endonucleases. Electrophoresis of digested DNA is followed by blotting onto a solid membrane which serves as a matrix for hybridization to a labelled probe (Southern, 1975). The size of the fragments that hybridize may reveal a RFLP (Fig. 5.6). The major advantage of RFLPs as linkage markers is that they often define polymorphisms in genes (Type I markers). Thus, they are useful for comparative mapping and are also a potential source of variation in candidate genes for economic traits. Blotting and probe development are laborious and time-consuming, however.

The discovery of microsatellites (Weber and May, 1989; Fries *et al.*, 1990) revolutionized linkage mapping in most animal species, including cattle. These islands of di-, tri-, or tetranucleotide repeats are ubiquitous throughout animal genomes. The number of repeats is often polymorphic, with from two to a dozen or more alleles present in a breeding population. They are defined by

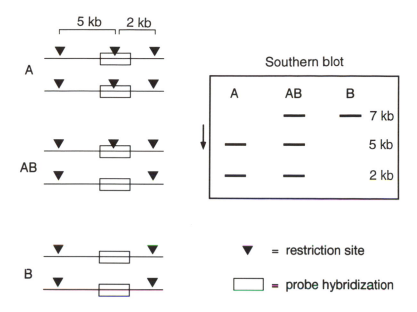

Fig. 5.6. Restriction fragment length polymorphisms (RFLPs) are polymorphisms for the presence or absence of restriction sites which may be detected by DNA fragment length differences after enzyme digestion, Southern blotting, and probe hybridization.

unique flanking sequences which provide primers for amplification by the polymerase chain reaction (PCR). Microsatellites (Fig. 5.7) have become the predominant Type II marker in the growing bovine linkage map.

Other types of markers appear on linkage maps. These may be enzyme polymorphisms, blood group antigens, or phenotypic traits such as a coat colour or horn development. These all add interest and biological significance to the linkage map, but are not sufficiently polymorphic to provide the skeletal framework.

In addition to markers, linkage mapping requires observable meiotic products, usually in the form of offspring from individuals segregating the markers discussed above. Large full sib-ships or half sib-ships are ideal for linkage analysis. Since a comprehensive map requires large numbers of markers scored on a common set of meiotic products, sharing of family material among laboratories is an effective arrangement for generating a linkage map. Such a set of reference families is available in cattle and has been used to generate a foundation linkage map (Barendse *et al.*, 1994).

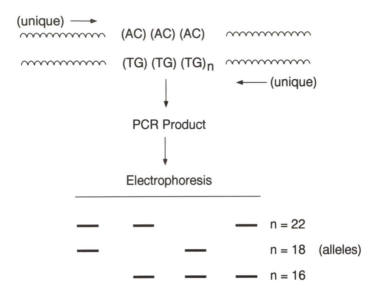

Fig. 5.7. Microsatellites are small islands of di-, tri-, or tetranucleotide repeats embedded in unique sequences of DNA. Amplification of these islands by the PCR often reveals polymorphism in the number of repeat motifs.

Status of the Bovine Map

Physical mapping

More than 400 Type I loci have been mapped in cattle (Fries *et al.*, 1993; O'Brien *et al.*, 1993) primarily through somatic cell genetics. While these markers indicate the boundaries of chromosomal conservation relative to the map-rich genomes of mice and humans, they provide an incomplete comparative map. Conservation of gene order is not addressed by the synteny map.

In situ hybridization, especially FISH, has been used effectively to address the order of Type I loci, to assign syntenic groups to specific chromosomes, and to anchor the rapidly growing linkage map to chromosomes (Fries *et al.*, 1993; Solinas-Toldo *et al.*, 1993; Gallagher *et al.*, 1993). There are presently more than 50 *in situ* localizations of unique sequences on cattle chromosomes. All syntenic groups are now assigned to specific chromosomes and the bovine linkage map is anchored at 35 sites on 26 chromosomes.

Linkage mapping

The published bovine linkage map of 200 markers (Barendse *et al.*, 1994) has grown to almost 500 markers in less than a year (Barendse, personal communication). This was made possible in large measure by international cooperation and the use of a common set of reference families. Combined with the independent development of other maps (Bishop *et al.*, 1994) there are probably more than 800 markers presently assigned to cattle linkage groups. The goal of 20 cM resolution has clearly been achieved over at least 95% of the total genome. The international map includes 65 Type I markers.

Comparative mapping

Approximately 400 loci have been mapped in both cattle and humans. Most of these have also been mapped in mice. Humans and mice are destined to be the 'map rich' species among mammals and much is to be learned about cattle by identifying conserved map regions relative to these species. While extensive conservation of synteny has been observed between cattle and humans (Womack and Moll, 1986; Threadgill *et al.*, 1991), conservation of linkage (conservation of gene order) may not be so prevalent. Barendse *et al.* (1994) incorporated a sufficient number of Type I loci into the linkage map to demonstrate the presence of several rearrangements of gene order within conserved syntenies. O'Brien *et al.* (1993) addressed the value and also some of the problems of comparative gene mapping and proposed a list of comparative anchor loci. Approximately one-half of the 321 anchor loci have been mapped in cattle, primarily by somatic cell genetics. An example of comparative anchor loci is illustrated for human chromosome 12 (Fig. 5.8) where most of the anchor loci that have been mapped in cattle are mapped to chromosome 5. While this map suggests that most genes that are on human 12 will probably be mapped to cattle 5, it does not address evolutionary rearrangements in gene order.

Economic traits

A number of traits of economic significance are beginning to appear on the bovine map. In addition to LAD (Threadgill *et al.*, 1991; Shuster *et al.*, 1992), uridine monophosphate synthase deficiency (UMPS) is also mapped to a specific site on chromosome 1 (Barendse *et al.*, 1993; Schwenger *et al.*, 1993; Ryan *et al.*, 1994). BolLA is associated with susceptibility to leukemia virus infection (Lewin *et al.*, 1988). Georges *et al.* (1993a) have linked the polled locus to microsatellites on chromosome 1. The weaver disease maps to markers on chromosome 4 (Georges *et al.*, 1993b) and also appears to be associated with a quantitative trait for improved milk production. Variation around the prolactin

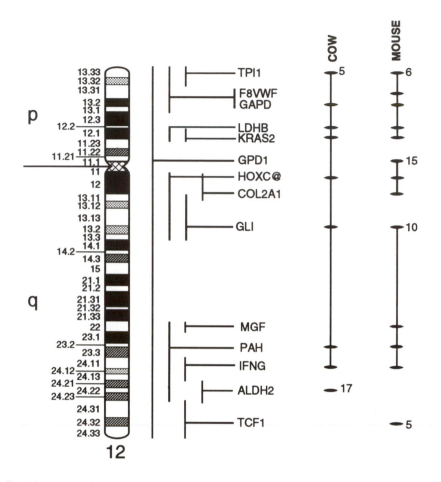

Fig. 5.8. A map of the 14 comparative anchor loci for human chromosome 12. Extensive conservation of synteny with chromosome 5 of cattle is evident from the eight loci that map to that chromosome. Human chromosome 12 homologues are spread over four mouse chromosomes.

gene on chromosome 23 (Cowan *et al.*, 1990) is related to milk production in certain Holstein sire families and Georges *et al.* (1995) have used mapped microsatellites to locate an additional five QTL for milk production. Not intended to be an exhaustive list, these examples illustrate both the candidate gene and the genomic approach to placing ETL on the bovine map.

Future Directions

Sufficient numbers of mapped markers now exist in cattle for extensive genome coverage of families segregating ETL. Unfortunately, different ETL require different segregating families. These resource families are expensive to develop and maintain. None the less, the families are an integral and necessary step in the ultimate application of the gene map to economic improvement. The mapping of ETL to 10–20 cM regions of a chromosome may be followed by high resolution mapping to ultimately identify and clone the responsible genes. Chromosome specific libraries will aid this process.

The increased emphasis on identifying and mapping all the expressed genes of mice and humans places even greater need for comparative mapping in cattle. The mapping of a bovine ETL to a 10 cM region will eventually result in a list of 100–200 candidate genes directly from the human and mouse maps. The efficient use of comparative maps will require an increased emphasis on placing Type I markers on the bovine linkage map.

References

Barendse, W., Armitage, S.M., Ryan, A.M., Moore, S.S., Clayton, D., Georges, M. and Womack, J.E. (1993) A genetic map of DNA loci in bovine chromosome 1. *Genomics* 18, 602–608.

Barendse, W., Armitage, S.M., Kossarek, L.M., Shalom, A., Kirkpatrick, B.W., Ryan, A.M., Clayton, D., Li, L., Neibergs, H.L., Zhang, N., Grosse, W.M., Weiss, J., Creighton, P., McCarthy, F., Ron, M., Teale, A.J., Fries, R., McGraw, R.A., Moore, S.S., Georges, M., Soller, M., Womack, J.E. and Hetzel, D.J.S. (1994) A genetic linkage map of the bovine genome. *Nature Genetics* 6, 227–235.

Beckmann, J.S. and Soller, M. (1988) Detection of linkage between marker loci and loci affecting quantitative traits in crosses between segregating populations. *Theoretical and Applied Genetics* 76, 228–236.

Bishop, M.D., Kappes, S.M., Keele, J.W., Stone, R.T., Sunden, S.L.F., Hawkins, G.A., Solinas-Toldo, S., Fries, R., Grosz, M.D., Yoo, J. and Beattie, C.W. (1994) A genetic linkage map of cattle. *Genetics* 136, 619–639.

Cowan, C.M., Dentine, M.R., Ax, R.L. and Schuler, L.A. (1990) Structural variation around prolactin gene linked to quantitative traits in an elite Holstein sire family. *Theoretical and Applied Genetics* 79, 577–582.

Fries, R., Eggen, A. and Stranzinger, G. (1990) The bovine genome contains polymorphic microsatellites. *Genomics* 8, 403–406.

Fries, R., Eggen, A. and Womack, J.E. (1993) The bovine genome map. *Mammalian Genome* 4, 405–428.

Gallagher, D.S., Threadgill, D.W., Ryan, A.M., Womack, J.E. and Irwin, D.M. (1993) Physical mapping of the lysozyme gene family in cattle. *Mammalian Genome* 4, 368–373.

Georges, M., Drinkwater, R., King, T., Mishra, A., Moore, S.S., Nielsen, D., Sargeant, L.S., Sorensen, A., Steele, M.R., Zhao, X., Womack, J.E. and Hetzel, J. (1993a) Microsatellite mapping of a gene affecting horn development in *Bos taurus*. *Nature Genetics* 4, 206–210.

Georges, M., Dietz, A.B., Mishra, A., Nielsen, D., Sargeant, L., Sorensen, A., Steele, M.R., Zhao, X., Leipold, H. and Womack, J.E. (1993b) Microsatellite mapping of the gene causing weaver disease in cattle will allow the study of an associated quantitative trait locus. *Proceedings of the National Academy of Sciences USA* 90, 1058–1062.

Georges, M., Nielsen, D., Mackinnon, M., Mishra, A., Okimoto, R., Pasquino, A.T., Sargeant, L.S., Sorensen, A., Steele, M.R., Zhao, X., Womack, J.E. and Hoeschele, I. (1995) Mapping genes controlling milk production: towards marker assisted selection in livestock. *Genetics* 139, 907–920.

Lander, E.S. and Botstein, D. (1989) Mapping mendelian factors underlying quantitative traits using RFLP linkage maps. *Genetics* 121, 185–199.

Lewin, H.A., Wu, M.-C., Stewart, J.A. and Nolan, T.J. (1988) Association between Bol*LA* and subclinical bovine leukemia virus infection in a herd of Holstein-Friesian cows. *Immunogenetics* 27, 338–344.

O'Brien, S.J. (1992) Mammalian genome mapping; lessons and prospects. *Current Opinion in Genetics and Development* 1, 105–111.

O'Brien, S.J., Womack, J.E., Lyons, L.A., Moore, K.J., Jenkins, N.A. and Copeland, N.G. (1993) Anchored reference loci for comparative genome mapping in mammals. *Nature Genetics* 3, 103–112.

Paterson, A.H., Lander, E.S., Hewitt, J.D., Peterson, S., Lincoln, S.E. and Tanksley, S.D. (1989) Resolution of quantitative traits into Mendelian factors by using complete linkage map of restriction fragment length polymorphisms. *Nature* 335, 721–726.

Rommens, J.M., Iannuzzi, M.C., Kerem., B.-S., Drumm, M.L., Melmer, G., Dean, M., Rozmahel, R., Cole, J.L., Kennedy, D., Hidaka, N., Zsiga, M., Buchwald, M., Riordan, J.R., Tsui, L.-C. and Collins, F.S. (1989) Identification of the cystic fibrosis gene: chromosome walking and jumping. *Science* 245, 1059–1065.

Ryan, A.M., Gallagher, D.S., Jr., Schöber, S., Schwenger, B. and Womack, J.E. (1994) Somatic cell mapping and *in situ* localization of the bovine uridine monophosphate synthase gene (UMPS). *Mammalian Genome* 5, 46–47.

Schwenger, B., Schöber, S. and Detlef, S. (1993) DUMPS cattle carry a point mutation in the uridine monophosphate synthase gene. *Genomics* 16, 241–244.

Shuster, D.E., Kehrli, Jr., M.E., Ackermann, M.R. and Gilbert, R.O. (1992) Identification and prevalence of a genetic defect that causes leukocyte adhesion deficiency in Holstein cattle. *Proceedings of the National Academy of Sciences USA* 89, 9225–9229.

Smith, C. and Simpson, S.P. (1986) The use of genetic polymorphism in livestock improvement. *Journal of Animal Breeding and Genetics* 103, 205–217.

Solinas-Toldo, S., Fries, R., Steffen, P., Neibergs, H.L., Barendse, W., Womack, J.E., Hetzel, D.J.S. and Stranzinger, G. (1993) Physically mapped, cosmid-derived microsatellite markers as anchor loci on bovine chromosomes. *Mammalian Genome* 4, 720–727.

Soller, M. and Beckmann, J.S. (1982) Restriction fragment length polymorphisms and genetic improvement. *Proceedings of the 2nd World Congress on Genetics Applied to Livestock Production* 6, 396–404.

Southern, E. (1975) Detection of specific sequences among DNA fragments separated by gel electrophoresis. *Journal of Molecular Biology* 98, 503–517.

Sturtevant, A.H. (1913) The linear arrangement of six sex-linked factors in *Drosophila*, as shown by their mode of association. *Journal of Experimental Zoology* 14, 43–59.

Threadgill, D.S., Krause, J.P., Krawetz, S.A. and Womack, J.E. (1991) Evidence for the evolutionary origin of human chromosome 21 from comparative gene mapping in the cow and mouse. *Proceedings of the National Academy of Sciences USA* 88, 154–158.

Trask, B.J. (1991) Fluorescence *in situ* hybridization. *Trends in Genetics* 7, 150–154.

Weber, J.L. and May, P.E. (1989) Abundant class of human DNA polymorphisms which can be typed using the polymerase chain reaction. *American Journal of Human Genetics* 44, 388–396.

Weller, J.L., Kashi, Y. and Soller, M. (1990) Power of daughter and grandaughter designs for determining linkage between marker loci and quantitative trait loci in dairy cattle. *Journal of Dairy Science* 73, 2525–2537.

Womack, J.E. and Moll, Y.D. (1986) A gene map of the cow: Conservation of linkage with mouse and man. *Journal of Heredity* 77, 2–7.

The Application of Genetic Markers in Dairy Cow Selection Programmes

6

J.A.M. Van Arendonk and H. Bovenhuis

Department of Animal Breeding, Wageningen Institute of Animal Sciences, PO Box 338, 6700 AH Wageningen, The Netherlands

Introduction

Animal geneticists have relied largely on quantitative variation in the genetic improvement of livestock, and have developed sophisticated statistical methods to exploit all the information in selection (Henderson, 1984; Kennedy *et al.*, 1988). The breeding values of animals are predicted based on phenotypes of the animal itself and/or those of its relatives. When only observations on the trait of interest are considered, the contribution of observations on relatives to an animal's breeding value depends on the additive genetic relationship, i.e. the proportion of genes shared in common by descent, and the heritability of the trait. In calculating the additive genetic relationship between individuals no knowledge on the actual contribution of a parent to its offspring is used. Instead use is made of Wright's (1922) inbreeding coefficients and the coefficients of relationship between animals. The genetic model underlying current methodology of predicting breeding values assumes a large number of genes affecting the trait each having a small effect (Kennedy, 1988). Identification of genes with a large effect might open new ways for genetic improvement, such as introgression of desirable characteristics from exotic breeds into a commercial population and marker assisted selection within a population. With large numbers of polymorphic marker loci becoming available, marker assisted selection and introgression is expected to improve the exploitation of genes with important effects.

The detection of microsatellites (Jeffreys *et al.*, 1985) and the use of the polymerase chain reaction make it possible to identify differences between individuals in genotype at many genomic sites. These sites are called marker loci, and their alleles are genetic markers. A marker locus is not likely to be a quantitative trait locus (QTL) itself, but it may be linked to a QTL (Soller,

1978). The inheritance of marker alleles can be determined by typing parents and offspring for marker loci. Information on genetic markers and phenotypic observations can be used to determine the location of single genes and of genes affecting quantitative traits (QTL).

Recently, a number of bovine genetic maps have been published (Fries *et al.*, 1993; Bishop *et al.*, 1994; Barendse *et al.*, 1994). Microsatellites seem to be the genetic markers of choice for population studies and marker assisted selection. They appear frequently and apparently randomly throughout the genome, are highly polymorphic and typing can be automated (e.g. Georges, 1991). The published maps cover approximately 90% of the bovine genome and the bovine map is now of sufficient resolution to start looking for genes affecting quantitative traits.

Two steps are required for the implementation of marker assisted selection: (i) location of loci affecting the traits of interest; and (ii) incorporation of that knowledge in the selection strategies. First we will review the experimental designs that can be used to locate QTL. We will then describe possible scenarios to incorporate information on markers in within-breed selection. In this chapter we will talk about a QTL that might be a locus affecting a trait controlled by a large number of genes or a single gene trait.

Detecting Genes of Interest

Loci affecting traits of economic interest (QTL) can be detected through their linkage with marker loci. For detection of a QTL linked to a marker, linkage disequilibrium between the genetic marker and the QTL is required. Linkage disequilibria between marker and QTL across the population are expected in populations that were recently derived from a cross between two divergent populations (F_2 or backcross populations), particularly two inbred lines. The original linkage disequilibrium (D) at formation of the population depends upon genetic differences between the founding populations. This disequilibrium is gradually reduced to $D(1 - r)^n$ by recombination (r) over n generations. It is clear from the above formula that the largest disequilibrium is expected in the first generations after the formation of a population. For detection of linkage in laboratory animals and plants, F_2 or backcrosses between inbred lines are used. This experimental design for detecting QTL is very efficient in terms of number of individuals that need to be typed. We will elaborate on it because it also clearly illustrates the principles underlying detection of a QTL linked to a marker. It is assumed that the two lines are homozygous and carry different alleles for the QTL and marker. In the F_1 population all individuals are heterozygous for the marker as well as for the QTL. An F_1 individual can produce four different alleles (Fig. 6.1). In the F_2 populations individuals will have one of the four different marker genotypes. By comparing the mean performance of the two groups of homozygous individuals, information can be

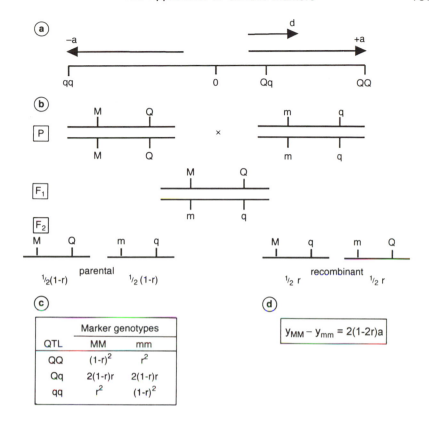

Fig. 6.1. Linkage between marker locus and QTL in the F_2 generation of a cross between inbred lines. M and m, alleles at the marker locus; Q and q, alleles at QTL; a and d, main effect and dominance effect at QTL; r, proportion of recombination between marker locus and QTL; y_{MM} and y_{mm}, mean value of marker genotypes in the F_2 generation. (a) Mathematical scale showing the gene effects at the QTL. (b) Diagrammatic chromosomes of parents, F_1 progeny, and gametes formed by F_1, showing coupling between marker alleles and QTL alleles and frequency of parental and recombinant F_1 gametes. (c) Relative genotype frequencies at the QTL, within F_2 progeny having alternative homozygous marker genotypes. (d) Difference between mean values of alternative homozygous marker genotypes in the F_2 generation (from Soller, 1991).

obtained to infer linkage to a QTL. The expected contrast is a function of the size of the genetic effects at the QTL and the recombination rate between the marker and QTL (Fig. 6.1).

As a result of using inbred lines, all F_1 individuals have the same genotype at all loci and they are heterozygous for all those loci for which the inbred lines were different. The advantage of this design is that information from F_2 individuals can be pooled across families and, secondly, the probability of animals being heterozygous can be maximized by using extreme lines. Producing completely inbred lines, however, is not a viable option for cattle because of the long generation interval and the costs involved. However, populations, in which different alleles are segregating, are available that differ radically for a particular trait. These populations can be expected to differ markedly in allelic frequencies at QTL affecting the trait, approaching fixation for alternative alleles at these loci. With respect to marker loci, however, the populations may share the same marker alleles. Anderson *et al.* (1994) used a cross between European and wild pigs for the detection of QTL in pigs. Pooling of information across families is no longer possible in this case but marker information can be used to classify F_2 offspring as to whether the marked chromosomal region derives from the 'high value' population or the 'low value' population (Beckmann and Soller, 1988). Results of this type of experiment will lead to a better understanding of genetic differences. For practical animal breeding, the results might be of limited value in the short term if favourable alleles are fixed in the commercial population.

For linkage analysis in cattle, existing populations are used in which alleles are segregating. Two types of analysis can be distinguished: an across-population analysis and a within-family analysis. Performing an analysis across the population will only show differences between marker genotypes if overall linkage disequilibrium exists within a population for both marker loci and QTL (Table 6.1). This type of analysis would be particularly appropriate for populations that have been maintained with small effective numbers for many generations, e.g. some of the smaller cattle breeds (Soller, 1991). The majority of microsatellite polymorphisms can be expected to be selectively neutral. Consequently, population-wide linkage disequilibrium between marker loci and QTL in a closed population that has not undergone a crossing episode in its recent history can only be generated by genetic drift, which is primarily a function of effective population size and recombination rate (Hill and Robertson, 1968). With selectively neutral markers, results from an across-population analysis are expected to vary between populations. Also with candidate genes, i.e. genes that are expected to have a direct effect on the trait, results might differ between populations when the candidate gene is (also) linked to a QTL. This has been discussed by Bovenhuis *et al.* (1992) for associations between milk protein polymorphisms and milk production traits.

For most traits of interest, populations are generally polymorphic at QTL as well as at the marker loci. In this case, the degree of linkage disequilibrium that can be generated by crossing populations is limited. In such populations, mapping can be based on the disequilibrium necessarily found within individual families within a single population. In dairy cattle populations large sire half-sib families are available and much information on quantitative traits is often collected routinely. For mapping of a QTL with respect to a marker locus the sire needs to be heterozygous at both loci. Markers can be preselected on the degree of polymorphism to minimize the proportion of sires homozygous for

Table 6.1. Overview of expected marker contrast and individuals informative with respect to linkage in different experimental designs for detecting a QTL with two alleles (Q^1 and Q^2) using a marker with two alleles (M^1 and M^2).

Design	Expected marker contrast[1]	Informative individuals
Inbred lines		
F-2	$2(1 - 2r)\, a$	All
Backcross	$(1 - 2r)\,(a + d)$	All
Segregating populations: across population[2]	$\left[\dfrac{2D}{(s + t)\,(u + v)}\right] a$	All
Segregating populations: within families		
Daughter design	$(1 - 2r)\,[a + (q - p)d]$	Only double heterozygous sires
Grand-daughter design	$\frac{1}{2}(1 - 2r)\,[a + (q - p)d]$	Only double heterozygous grandsires

[1] The expected marker contrast $M^1M^1 - M^2M^2$. The following parameters are used: θ is the recombination fraction between marker and QTL, a is half the difference between Q^1Q^1 and Q^2Q^2 genotypes, d is the dominance deviation, and p is frequency of Q^1 allele ($q = 1 - p$).
[2] Linkage disequilibrium at the population level is $D = sv - tu$ where s, t, u and v is the haplotype frequency of M^1Q^1, M^1Q^2, M^2Q^1 and M^2Q^2, respectively.

the marker but the level of heterozygosity at the QTL can not be changed. Not all sires in the experiment will be informative with respect to linkage. For the analysis it is important to realize that linkage relationship between the marker and QTL alleles may be different for different sires.

Consider a sire with the following genotype M^1Q^1/M^2Q^2 where M and Q refer to alleles for the marker and the QTL, respectively, and M^1 is linked to Q^1 and M^2 is linked to Q^2. Half of the sire's daughters will inherit the M^1 allele, and the other half will inherit the M^2 allele. When there is no recombination, the daughters inheriting the M^1 allele will also inherit the Q^1 allele. Similarly, the daughters inheriting the M^2 allele will also inherit the Q^2 allele. Therefore, comparing differences between the mean performance of daughters inheriting the M^1 and the M^2 allele provides information about a linked QTL (Table 6.1). To detect a QTL with a substitution effect of 0.1–0.3 phenotypic standard deviation, it is necessary to determine genotypes of thousands of daughters (Weller *et al.*, 1990). The number of daughters available on a single sire might result in a too-low power of detection. In addition, in a number of cases a single sire would not be heterozygous at both alleles and consequently be uninformative with respect to linkage. Thus, progeny of several sires needs to be analysed jointly, and it is necessary to analyse marker effects within sires.

To reduce the number of offspring scored for markers at the expense of increased numbers scored for the quantitative trait is to progeny test the offspring in order to decrease the error variance of the quantitative trait evaluation. This can be readily applied to a dairy population's progeny testing scheme. In this design, which is referred to as a 'grand-daughter design', sons, rather than daughters, of a heterozygous elite sire are scored for markers. Compared to the daughter design, the contrast between sons inheriting different marker alleles is reduced by a factor of one-half in the grand-daughter design (Table 6.1). However, due to the reduction in environmental variance the scheme is more efficient for detecting QTL for a given number of individuals scored for markers.

Table 6.1 gives an overview of different designs that can be used for detecting linkage between a marker and a QTL. The table shows the expected difference between the individuals carrying different marker alleles, which is a good indication of the power of the design to detect QTL, and the individuals that provide information about the QTL linked to the marker. In the case of no dominance the difference between the M^1M^1 and M^2M^2 in the F_2 design is twice the difference that can be observed in the daughter design. Further, all individuals are double heterozygous whereas in the daughter design it is not known whether the sire is heterozygous or homozygous for the QTL. These two factors make the F_2 cross between inbred lines more efficient for detecting QTL. As compared to the daughter design, in the grand-daughter design the difference between the individuals inheriting the M^1 and the M^2 allele is further reduced by a factor of one half. However, because in this design grand-daughter means are used, the error variance is reduced, making this design more efficient. This is especially true for low heritability traits.

Identified QTL Effects

The detection of QTL in bovines goes back a long time. Most of the early studies in this area were at the population level using polymorphic blood groups, milk proteins or enzymes as genetic markers. When these loci do not have a direct effect on the trait of interest, the estimated effects originate from population-wide linkage disequilibrium between marker loci and QTL. The forces that induce linkage disequilibrium might differ from one population to another and consequently results of these types of studies might lack consistency.

Results on within-family studies have been reported by Neimann-Sorensen and Robertson in 1961 and, more recently, by Geldermann *et al.* (1985), Cowan *et al.* (1990), Hoeschele and Meinert (1990), and Bovenhuis and Weller (1994). These studies searched for QTL using only a limited number of genetic markers.

Hoeschele and Meinert (1990) reported a significant linkage of the weaver locus, a defective gene in Brown Swiss, with a QTL affecting milk and fat yield. Carrier cows produced approximately 700 kg of milk and 26 kg of fat more than non-carrier cows.

Bovenhuis and Weller (1994) studied the effects of milk protein loci and QTL linked to milk protein loci on milk production traits using a maximum likelihood method. Besides significant direct effects of several milk protein genes on milk production traits, a significant effect of a QTL linked to the casein genes on fat percentage was found. The QTL explained about 28% of the additive genetic variance for fat percentage.

The first genome wide scan in dairy cattle was performed by Georges *et al.* (1995). They used 159 microsatellite markers and a grand-daughter design to detect genes affecting milk production traits. The data consisted of 14 grandsires with in total 1518 progeny tested sons. Milk production traits studied were milk yield, fat yield, protein yield, fat percentage and protein percentage. The analysis was performed within the 14 families. In total, significant QTL effects were detected on five chromosomes: a QTL on chromosome 9 increased the amount of milk, fat and protein produced, the QTL on chromosomes 6 (casein) and 20 on the other hand tended to increase milk yield but decrease fat and protein percentage, QTLs on chromosome 1 and 10 seemed to have differential effects on milk composition. The identified QTLs would explain between 38 and 179% of the expected mendelian sampling variation of the sire and the authors conclude that the estimated effects of the QTLs are probably overestimates.

Results from these studies demonstrate that QTL with reasonable effects are still segregating in populations selected for several generations. More detailed experiments are needed to obtain reliable estimates of position and effects of QTL. Recently one experiment has started involving 35 US grandsires (Da *et al.*, 1994) and one involving sires progeny tested in The Netherlands and New Zealand (Georges, personal communication). Identification of QTL explaining a significant fraction of the genetic variance is a first step towards the application of marker assisted selection for quantitative trait loci.

Besides genes affecting quantitative traits, a number of genes have been mapped that are known to be controlled by single genes: the double muscling locus (Georges *et al.*, 1990), the weaver locus (Georges *et al.*, 1993a) and the polled locus (Georges *et al.*, 1993b).

Utilizing Linkage Associations in Breeding Programmes

Marker assisted selection is expected to increase genetic response by affecting time, intensity and accuracy of selection (Soller and Beckman, 1983; Smith and Simpson, 1986; Weller and Fernando, 1991). Marker genotypes can be scored at a very young age or even before an embryo is implanted in the recipient cow when multiple ovulation and embryo transfer (MOET) is used. This opens ways to reduce generation interval. In addition, marker genotypes can be obtained on all selection candidates, which is an advantage for sex-limited traits and carcass traits. In the next section a method is described to incorporate marker information in the prediction of genetic merit of selection candidates. In addition, the effects of using marker information on the accuracy of predicting the genetic merit will be described. In the last section, studies on the consequences of marker assisted selection in dairy cattle will be summarized.

Incorporating marker information

Best linear unbiased prediction (BLUP) methods are currently used to predict breeding values of animals in a large number of countries and species. Selection in most cases is across generations and age classes. Consequently information on markers needs to be combined with BLUP breeding values into a single selection criterion (Hoeschele and Romano, 1993). The same data are used to predict polygenic and marker linked effects, which makes it necessary to predict both genetic effects simultaneously.

The prediction of an animal's breeding value using BLUP is based on phenotypes of the animal itself or those of its relatives. When only observations on the trait of interest are considered, the contribution of relatives to an animal's breeding value depends on the additive genetic relationship, i.e. the proportion of genes shared in common by descent, and the heritability of the trait. In building the numerator relationship matrix or its inverse no knowledge on the actual contribution of a parent to its offspring is used. Fernando and Grossman (1989) and Van Arendonk *et al.* (1994a) showed that information on a single marker could be used in an animal model by fitting additive effects for alleles at a QTL linked to the marker and additive polygenic effects for alleles at the remaining quantitative trait loci. The total additive genetic variance (σ_a^2) is equal to $\sigma_a^2 = \sigma_u^2 + 2\sigma_v^2$ where is σ_u^2 is the polygenic variance and σ_v^2 is the variance of gametic effects at the marked QTL. Let $\alpha_u = \sigma_e^2/\sigma_u^2$ and $\alpha_v = \sigma_e^2/\sigma_v^2$, then the mixed model equations to predict genetic effects are:

$$\begin{bmatrix} \mathbf{X/X} & \mathbf{X/Z} & \mathbf{X/W} \\ \mathbf{Z/X} & \mathbf{Z/Z} + \mathbf{A}_u^{-1}\alpha_u & \mathbf{Z/W} \\ \mathbf{W/X} & \mathbf{W/Z} & \mathbf{W/W} + \mathbf{G}_{v|r}^{-1}\alpha_v \end{bmatrix} \begin{bmatrix} \hat{\beta} \\ \hat{\mathbf{u}} \\ \hat{\mathbf{v}} \end{bmatrix} = \begin{bmatrix} \mathbf{X/y} \\ \mathbf{Z/y} \\ \mathbf{W/y} \end{bmatrix} \qquad (1)$$

where:

y is the vector of observations on the trait of interest,
β is the vector with fixed effects,
u is the vector with random additive polygenic effects,
v is the vector with random gametic effects at the marked QTL,
e is the vector of random residual effects.

The matrices **X**, **Z** and **W** are incidence matrices relating observations to fixed effects, polygenic effects and gametic QTL effects, respectively. $\mathbf{A}_u$ is the numerator relationship matrix for polygenic effects and $\mathbf{G}_{v|r}$ is the gametic relationship matrix for the QTL linked to the marker. To build $\mathbf{G}_{v|r}$ information is used on marker allele transmission and the recombination rate (r) between marker and QTL. Efficient algorithms have been developed to obtain the inverse of $\mathbf{G}_{v|r}$. These algorithms differ in the way they distinguish parental origin of gametic effects (Bink and Van Arendonk, 1994). In the mixed model, information on several generations of relatives is used to predict effects at the QTL.

In most studies it has been assumed that all animals have genotypic information. In the most likely scenario for marker assisted selection, marker genotypes will be available only on a limited number of individuals. Many of the animals in a population may have unknown genotypes because several generations of ancestors are included in the animal model genetic evaluation and most of these ancestors will have unknown genotypes. Van Arendonk *et al.* (1994a) and Wang *et al.* (1995) have presented procedures to incorporate information from animals whose genotypes are unknown. Information on several unlinked marker loci, each linked to a different QTL, can be used by including an effect for each marked QTL. The number of equations per animal in this case is $2m + 1$ where m is the number of marked QTL. The number of equations per animal can be reduced to 1 by combining information on all marked QTL and polygenes and the use of a combined numerator relationship matrix (Van Arendonk *et al.*, 1994a). The total number of equations can also be reduced by using a reduced animal model (Cantet and Smith, 1991; Goddard, 1992) or absorbing QTL equations for individuals whose marker genotypes are unknown (Hoeschele, 1993). Depending on the costs and benefits of typing for markers, only elite animals of the current generation and their offspring will have known marker genotypes. In this case, marker information will only affect the structure of the inverse of the combined numerator relationship matrix for some animals, whereas for others (e.g. ancestors and offspring with unknown genotypes) the structure will be the same as without markers. This feature makes application of the combined model attractive in populations where marker genotypes are available on a limited number of animals.

Accuracy
The predicted breeding value of an animal can be written as:

$$a_i = v_i^p + v_i^m + u_i$$

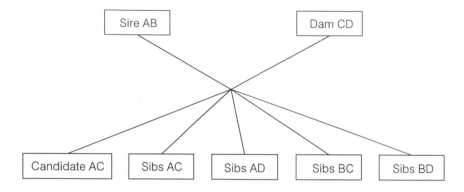

Fig. 6.2. Full sib pedigree structure: accuracy of predicting effects at marked QTL is calculated for selection candidate. A, B, C and D refer to alleles at marker locus. Phenotypes are only on full sibs.

where v_i^p and v_i^m are the predictions of the gametic effects at the marked QTL and u_i is the predicted additive polygenic effect. Information on transmission of alleles at the marker locus, and the recombination rate between the marker and the QTL, is used in building the gametic relationship matrix $\mathbf{G}_{v|r}$. Prediction of effects at the marked QTL is based on the observations on relatives. The accuracy of prediction of gametic effects is studied for a two-generation full sib structure (Fig. 6.2) and a three-generation half sib structure (Fig. 6.3).

Two-generation full sib structure. Phenotypic observations are available on full sibs of the selection candidate. For the marked QTL the relation between the selection candidate and a full sib depends on the marker genotype. The full sib that inherited the same marker alleles (A and B) has a relationship with the selection candidate of $(1 - r)^2 + r^2$ where r is the recombination rate. The relationship with sibs AD and BC, which have only one marker allele in common with the selection candidate, is ½. For the sib that inherited a different marker allele from both parents the relationship is $2(1 - r)r$. With r at 0.1 the relationship of the full sibs ranges from 0.18 to 0.82. The contribution of an observation to the predicted genetic effects at the marked QTL clearly depends on r and the marker genotype of the selection candidate and the individual with the observation. The accuracy of prediction of the genetic effect at the marked QTL ($v_i^p + v_i^m$) is calculated dependent on the number of observations on fulls sibs, r and the fraction of variance explained by the marked QTL (Table 6.2).

Based on phenotypic observations on full sibs (no markers) it is not possible to predict the extent to which the genetic value of the selection candidate deviates from the parental average, i.e. the within-family deviation. Genetic variation within a full sib family amounts to 50% of the additive genetic variance. As a consequence the maximum accuracy of prediction of the genetic

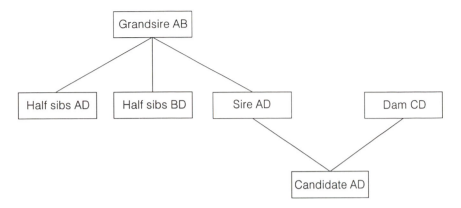

Fig. 6.3. Three-generation half sib structure. Observations are on sibs of grandsire only. A, B and D denote alleles at the marker (D refers to allele transmitted by dam).

value of the selection candidates based on phenotypic information on parents and sibs is $\sqrt{0.5} = 0.701$ in a situation without marker information. With the help of markers it is possible to explain part of the within-family deviations which enables a within-family selection. With markers, the accuracy of predicting the genetic value of selection candidates can be increased substantially. With an observation on one full sib in each marker genotype ($N_{fs} = 1$) the accuracy increased from 0.426 for r at 0.5 (which corresponds to a situation without markers) to 0.553 for r at 0.02 when the fraction of phenotypic variance explained by QTL (f_{QTL}) is 0.25. The results for N_{fs} at 100 illustrates the extent to which markers can be used to explain within-family deviations. When

Table 6.2. Accuracy of prediction of genetic value at the marked QTL of selection candidate based on observations on full sibs dependent on the fraction of phenotypic variance explained by the marked QTL (f_{QTL}), recombination rate (r) and number of full sibs within marker genotype [a].

f_{QTL}^{b}	r	Number of observations				
		1	2	5	10	100
0.25	0.50	0.426	0.516	0.609	0.652	0.701
	0.10	0.497	0.609	0.735	0.803	0.892
	0.02	0.553	0.677	0.813	0.883	0.969
0.10	0.50	0.295	0.385	0.506	0.582	0.691
	0.10	0.332	0.437	0.586	0.689	0.871
	0.02	0.365	0.481	0.647	0.760	0.948

[a] This is the number of individuals with the same marker genotype, i.e. total number of full sibs of selection candidate is four times this number.
[b] Fraction of phenotypic variance explained by the marked QTL.

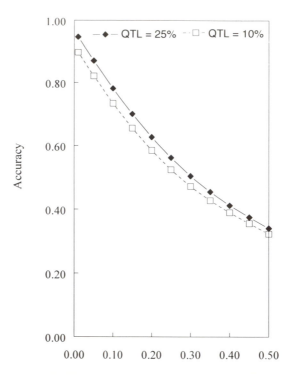

Fig. 6.4. Influence of recombination rate on accuracy of predicting contribution of sire to candidate in three-generation full sib structure for single QTL which explains 10 and 25% of variance.

the marked QTL explains a smaller fraction of the genetic variance, the accuracies of prediction are lower but the benefit of using markers remains.

Three-generation full sib structure. The number of full sibs with phenotypic information which is available in practical dairy breeding schemes is restricted. This implies that we need to look at other sources for the prediction of effects. One option would be to use half sibs within the same generation. For example, paternal half sisters can be used for the prediction of the contribution of the sire (i.e. 50% of the additive genetic variance). Another option is to look at individuals from the previous generation. In Fig. 6.3 a three generation half sib structure is given for which accuracies have been calculated. Observations are available on 200 half sib offspring (100 for each marker genotype) of the paternal grandsire and the selection candidate inherited one marker allele from that grandsire. In Fig. 6.4 the accuracy of predicting the contribution of the sire at the marked QTL is given for a range of recombination rates. The accuracy increased from 0.34 for r at 0.5 to 0.95 for r at 0.01 when the f_{QTL} is 0.25. The accuracy reduced slightly when the f_{QTL} is 0.10 but the effect of

f_{QTL} will depend on the number of half sib offspring for which observations are available.

In the above example, the selection candidate inherited the paternal marker allele from the sire and consequently its gametic effect was highly related to one of the gametic effects in the grandsire. Gametic effects in the grandsire can be predicted accurately due to the large number of offspring. In 50% of the cases, the selection candidate will inherit a marker allele coming from the paternal granddam. Accuracy of prediction will be much lower in that case. It is important to realize that for a single marker, selection candidates produced by a given sire can only be divided in two groups, i.e. those which received the marker allele coming from the grandsire and those which inherited the marker allele coming from the granddam. This restricts the selection differential that can be realized within a sire family in a situation where marker linked effects are predicted based on offspring from the grandsire. The situation changes when there are multiple marked QTL, in which case the number of favourable alleles transmitted by the sire will vary between offspring.

Until now we have only considered the accuracy of predicting the paternal gametic effect in the three-generation structure. Accuracy of the gametic effect transmitted by the dam is expected to be the same when the dam transmits the marker allele coming from its sire. In that case the marker linked effects can also be linked to offspring of a grandsire.

The situation of a single marked QTL with a large effect has been described so far, but a situation with multiple marked QTL is very likely. In this case, equations (1) need to be extended to accommodate genetic effects at each marked QTL and selection of animals would be on the sum of the predicted genetic effects (Van Arendonk *et al.*, 1994a). It is expected that the situation with a single marked QTL with a large effect will be similar to the situation with multiple marked QTL each with a smaller effect.

It is important to remember that mixed equations (1) will combine all available information to predict the genetic effects. We have looked at two very distinct situations in order to provide insight into the accuracies that can be achieved by using markers in predicting breeding values. To obtain optimum gain from using markers it is important to look at different population structures in order to capitalize on this new technology.

Simulation studies

Zhang and Smith (1992) determined the value of marker assisted selection using linkage disequilibrium between markers and QTL across the whole population. To simulate linkage disequilibrium, four base populations were created by 2, 5, 10 and 20 generations of random mating from a cross between two inbred lines. Linkage disequilibrium allows the use of marked QTL effects across a population. Initial linkage disequilibrium had a large influence on the genetic gain from using markers. In outbreeding populations of livestock that have been largely closed for many generations, linkage disequilibrium will be limited with loosely linked markers. For populations which are close to linkage equilibrium, marker assisted selection must be within family, requiring relatively large amounts of data to determine marker linked effects. A different situation would arise if it were possible to score QTL genotypes directly since associations can be utilized across the population in that case.

Bovenhuis and De Boer (1994) have studied the value of using milk protein genetic variants as additional selection criteria in a closed adult breeding scheme in which multiple ovulation and embryo transfer (MOET) was used on adult cows. In that study, animals were scored for the genotypes that had a direct effect on the traits of interest. A base population of 64 donor cows and 16 sires was simulated. Each sire was mated at random to four donor cows resulting in eight progeny. Selection of males and females was after the 256 females had one phenotypic observation and all selection candidates were scored for the QTL genotypes. To restrict inbreeding, the number of males selected per full sib group was restricted to one. As is illustrated in the previous section, without genotypic information all full sib males have the same breeding value and random selection is practised. Typing of animals would enable the selection of sires with the most favourable genotype from within the full sib group, i.e. increasing the selection intensity. Genotypic information hardly affected the selection criteria for females because selection was after they had completed the first lactation. Annual genetic response during the first seven generations increased by 2.4–4.8% compared with a situation without markers. Additional gain was due to an increased selection intensity on the male side. The effect on accuracy of selection was negligible. For a progeny testing scheme, it would be expected that increased genetic gain could be obtained if including milk protein genotypes results in an increased selection intensity. When considering a situation where MOET is used to obtain young bulls, and only a limited number of males of each full sib family is progeny tested, milk protein genotypes can be used to increase selection intensity. However, if only one full sib male is available or if all full sibs are allowed to be progeny tested, no additional gain from selection for milk protein genotypes is expected.

Gomez-Raya and Gibson (1993) compared selection strategies to increase the frequency of favourable alleles in dairy cattle using information on single locus genotypes. The fastest increase was found when all cows and sires are genotyped, because this strategy may not be appropriate for a large dairy cattle population because of the high costs associated with it. It could be applied, however, in a nucleus selection scheme. In a progeny testing scheme, one could genotype all bull sires, bull dams and young bulls, and select young bulls based on their genotype. Gomez-Raya and Gibson (1993) showed that increase in the frequency of the favourable allele was slower under random mating than under a negative assortative mating.

Gibson (1994) determined the short-term and long-term response of selection with or without additional information on genotype at a single locus. Without genotypic information, the favourable allele moves to fixation due to its selective advantage mediated through its effect on phenotype. The use of genotypic information increased the rate of response in the early generations, but this advantage was eventually lost, resulting in a lower long term response. At the time of selection, both phenotypic and genotypic information on all selection candidates was available. Bovenhuis and De Boer (1994) used genotypic information for within-full-sib selection in sires before progeny test results were available. They did not find a reduction in long-term response. This adds to the arguments for using genotypic information at a separate stage of selection (Gibson, 1994).

The above strategies of including genotypic information in a dairy cattle breeding scheme can also be applied when information is available on marker rather than QTL genotypes. In the above situation estimates for QTL genotypes could be determined and utilized across the population. For markers linked effects within families will have to estimated. This makes it very important to look at the amount of data that is available to estimate the marker linked effects. In the three-generation half sib design we used information on grandsires while in the full sib design phenotypic information on full sibs in the same generation was used. The latter information can be generated when MOET is used but it is necessary to wait until lactation data is available before selecting sires. An alternative in a progeny testing scheme would be to type the first crop of daughters of selected sires (M. MacKinnon personal communication). This information can be obtained before the first sons are born.

Kashi *et al.* (1990) determined the benefit of using genetic markers within families to select young bulls prior to progeny testing and calculated the effect of preselection on the frequency of favourable QTL alleles. With a proportion preselected of 0.25, the average genetic merit of bulls increased by $0.30\sigma_a$ to $0.53\sigma_a$. Meuwissen and Van Arendonk (1992) predicted the fraction of within-family variance that can be explained by tracing markers transmitted from grandsire to grandsons. A 20% increase in annual genetic change was found in an open nucleus scheme in which sires were selected at 2 years of age, and markers explained 20% of the within-family genetic variance. The fraction of within-family variance that can be explained by markers is of critical importance. To predict 10% of the within-family variance in grand-offspring, informative genotypes on 500 daughters of both grandsires are needed for a large number of markers. In the latter study variance reduction at the marked QTL was accounted for, which was not the case in Kashi *et al.* (1990). The large differences in results between studies illustrates the need for more research in this area. In particular, ways to optimize the design of the scheme to make optimal use of this new technology need to be studied.

Concluding Remarks

It is clear that molecular genetics cannot replace traditional selection methods but should be integrated with them to obtain maximum rates of improvement. Most characters of interest to cattle breeders are quantitative in nature, and it is most likely that there are a great number of loci affecting the trait, each with a small effect (Shrimpton and Robertson, 1988). Genes with a small effect are very difficult to map and therefore it seems likely that for part of the quantitative genetic variation traditional selection methods have to be applied whereas for other parts marker assisted selection might be used. Due to recombination, associations between a marker and a QTL have to be used within families and such associations will erode over time. Closer linked markers will help to predict allelic effects at QTL more accurately. Smith and Smith (1993) concluded that the main advantage of closer markers is an increase in linkage disequilibrium between markers and QTL, which would allow marker assisted selection across the population. Linkage disequilibrium across the population rather than within specific families would make evaluation and selection simple and general. The question, however, is whether a small recombination rate implies linkage disequilibrium across the population (Van Arendonk *et al.*, 1994b). In a population recently derived from a cross of distinct lines, which is the situation studied by Lande and Thompson (1990), this might be realistic. But within a segregating population such a relationship can only result from random drift. Close linkage is expected to be advantageous but is not expected to lead to across population linkage disequilibrium within most livestock populations (Van Arendonk *et al.*, 1994b).

A different situation arises, however, when we know the genotype for the QTL itself. The detection of markers linked to the Booroola gene for fecundity is a recent example of using comparative maps and modern mapping tools in livestock (Montgomery *et al.*, 1993). Isolation of the gene using positional cloning strategies will be retarded by the lack of candidate genes in the homologous human region and by the lack of high density linkage maps for sheep (Hetzel, 1993). Most biologically and economically important traits are polygenic. Detection of genes responsible for differences in these traits is important both for a better understanding of the underlying processes as well as for practical selection purposes. But we are more likely to find markers than the genes and it is therefore worthwhile to invest also in more sophisticated methods to determine genotypes of animals and to use this information in selection programmes.

References

Anderson, L., Haley, C.S., Ellegren, H., Knott, S.A., Johansson, M., Andersson, L., Andersson-Eklund, K., Edfors-Lilja, I., Fredholm, M., Hansson, I., Håkansson, J. and Lundström, K. (1994) Genetic mapping of quantitative trait loci for growth and fatness in pigs. *Science* 263, 1771–1774.

Barendse, W., Armitage, S.M., Kossarek, L.M., Shalom, A., Kirkpatrick, B.W., Ryan, A.M., Clayton, D., Li, L., Neibergs, H.L., Zhang, N., Grosse, W.M., Weiss, J., Creighton, P., McCarthy, F., Ron, M., Teale, A.J., Fries, R., McGraw, R.A. and Moore, S.S. (1994) A genetic linkage map of the bovine genome. *Nature Genetics* 6, 227–235.

Beckmann, J.S. and Soller, M. (1988) Detection of linkage between marker loci and loci affecting quantitative traits in crosses between segregating populations. *Theoretical and Applied Genetics* 76, 228–236.

Bink, M.C.A.M. and Van Arendonk, J.A.M. (1994) Marker assisted prediction of breeding values in dairy cattle populations. In: *Proceedings of the 5th World Congress on Genetics Applied to Livestock Production*, Guelph 19, 311–318.

Bishop, M.D., Kappes, S.M., Keele, J.W. and Stone, R.T. (1994) A genetic linkage map for cattle. *Genetics* 136, 619–639.

Bovenhuis, H. and de Boer, I.J.M. (1994) The potential contribution of milk protein loci to improvement of dairy cattle. In: *Proceedings of the 5th World Congress on Genetics Applied to Livestock Production*, Guelph 19, 311–318.

Bovenhuis, H. and Weller, J.I. (1994) Mapping and analysis of dairy cattle quantitative trait loci by maximum likelihood methodology using milk protein genes as genetic markers. *Genetics* 137, 267–280.

Bovenhuis, H., Van Arendonk, J.A.M. and Korver, S. (1992) Associations between milk production polymorphisms and milk production traits. *Journal of Dairy Science* 75, 2549–2559.

Cantet, R.J.C. and Smith, C. (1991) Reduced animal model for marker assisted selection using best linear unbiased prediction. *Genetics Selection Evolution* 23, 221–233.

Cowan, C.M., Dentine, M.R., Ax, R.L. and Schuler, L.A. (1990) Structural variation around prolactin gene linked to quantitative traits in an elite Holstein sire family. *Theoretical and Applied Genetics* 79, 577–582.

Da, Y., Ron, M., Yanai, A., Band, M., Everts, R.E., Heyen, D.W., Weller, J.I., Wiggans, G.R. and Lewin, H.A. (1994) The dairy bull DNA respository: a resource for mapping quantitative trait loci. In: *Proceedings of the 5th World Congress on Genetics Applied to Livestock Production*, Guelph 21, 229–232.

Fernando, R.L. and Grossman, M. (1989) Marker assisted selection using best linear unbiased prediction. *Genetics Selection and Evolution* 21, 467–477.

Fries, R., Eggen, A. and Womack, J.E. (1993) The bovine genome map. *Mammalian Genome* 4, 405–428.

Geldermann, H., Pieper, U. and Roth, B. (1985) Effects of marked chromosome sections on milk performance in cattle. *Theoretical and Applied Genetics* 70, 138–146.

Georges, M. (1991) Hypervariable minisatellites and their use in animal breeding. In: Shook, L.B., Lewin, H.A. and McLaren, D.G. (eds) *Gene Mapping Techniques and Applications*. Marcel Dekker, Inc., New York, pp. 89–112.

Georges, M., Lathrop, M., Hilbert, P., Marcotte, A., Schwers, A., Swillens, S., Vassart, G. and Hanset, R. (1990) On the use of DNA fingerprints for linkage studies in cattle. *Genomics* 6, 461–474.

Georges, M., Dietz, A.B., Mishra, A., Nielsen, D., Sargeant, L.S., Sorensen, A., Steele, M.R., Zhao, X., Leipold, H., Womack, J.E. and Lathrop, M. (1993a) Microsatellite mapping of the gene causing Weaver disease in cattle will allow the study of an associated quantitative trait locus. *Proceedings of the National Academy of Science USA* 90, 1058–1062.

Georges, M., Drinkwater, R., King, T., Mishra, A., Moore, S.S., Nielsen, D., Sargeant, L.S., Sorensen, A., Steele, M.R., Zhao, X., Womack, J.E. and Hetzel, J. (1993b) Microsatellite mapping of a gene affecting horn development in *Bos taurus*. *Nature Genetics* 4, 206–210.

Georges, M., Nielsen, D., Mackinnon, M., Mishra, A., Okimoto, R., Pasquino, A.T., Sargeant, L.S., Sorensen, A., Steele, M.R., Zhao, X., Womack, J.E. and Hoeschele,

I. (1995) Using a complete microsatellite map and the grandaughter design to locate polygenes controlling milk production. *Genetics* 139, 907–920.

Gibson, J.P. (1994) Short-term gain at the expense of long-term response with selection of identified loci. In: *Proceedings of the 5th World Congress on Genetics Applied to Livestock Production*, Guelph 21, 201–204.

Goddard, M.E. (1992) A mixed model for analyses of data on multiple genetic markers. *Theoretical and Applied Genetics* 83, 878–886.

Gomez-Raya, L. and Gibson, J.P. (1993) Within-family selection at an otherwise unselected locus in dairy cattle. *Genome* 36, 433–439.

Henderson, C.R. (1984) *Applications of Linear Models in Animal Breeding*. University of Guelph.

Hetzel, J. (1993) Livestock genome research. *Nature Genetics* 4, 327–328.

Hill, W.G. and Robertson, A. (1968) Linkage disequilibrium in finite populations. *Theoretical and Applied Genetics* 38, 226–331.

Hoeschele, I. (1993) Elimination of quantitative trait loci equations in an animal model incorporating marker data. *Journal of Dairy Science* 76, 1693–1713.

Hoeschele, I. and Meinert, T.R. (1990) Association of genetic defects with yield and type traits: The Weaver locus effect on yield. *Journal of Dairy Science* 73, 2503–2515.

Hoeschele, I. and Romano, E.O. (1993) On the use of marker information from grandaughter design. *Journal of Animal Breeding and Genetics* 110, 429–449.

Jeffreys, A., Wilson, V. and Thein, S.L. (1985) Hypervariable minisatellite regions in human DNA. *Nature* 314, 67–73.

Kashi, Y., Hallerman, E. and Soller, M. (1990) Marker-assisted selection of candidate bulls for progeny testing programmes. *Animal Production* 51, 63–74.

Kennedy, B.W. (1988) Genetic properties of animal models. *Journal of Dairy Science* 71(suppl 2), 17–26.

Kennedy, B.W., Verrinder Gibbins, A.M., Gibson, J.P. and Smith, C. (1988) Coalescence of molecular and quantitative genetics for livestock improvement. *Journal of Dairy Science* 73, 2619–2627.

Lande, R. and Thompson, R. (1990) Efficiency of marker-assisted selection in the improvement of quantitative traits. *Genetics* 124, 743–756.

Meuwissen, T.H.E. and Van Arendonk, J.A.M. (1992) Potential improvements in rate of genetic gain from marker-assisted selection in dairy cattle breeding schemes. *Journal of Dairy Science* 75, 1651–1659.

Montgomery, G.W., Crawford, A.W., Penty, J.M., Dodds, K.G., Ede, A.J., Henry, H.M., Pierson, C.A., Lord, E.A., Galloway, S.M., Schmack, A.E., Sise, J.A., Swarbrick, P.A., Hanrahan, V., Buchanan, F.C. and Hill, D.F. (1993) *Nature Genetics* 4, 410–414.

Neimann-Sorensen, A. and Robertson, A. (1961) The associations between blood groups and several production characteristics in three Danish cattle breeds. *Acta Agricultura Scandinavica* 11, 163.

Shrimpton, A.E. and Robertson, A. (1988) The isolation of polygenic factors controlling bristle score in *Drosophila melanogaster*. 2. Distribution of third chromosome bristle effects within chromosome sections. *Genetics* 118, 445–459.

Smith, C. and Simpson, S.P. (1986) The use of genetic polymorphisms in livestock improvement. *Animal Production* 20, 1–10.

Smith, C. and Smith, D.B. (1993) The need for close linkages in marker-assisted selection for economic merit in livestock. *Animal Breeding Abstracts* 61, 197–204.

Soller, M. (1978) The use of loci associated with quantitative effects in dairy cattle improvement. *Animal Production* 27, 133–179.

Soller, M. (1991) Mapping quantitative trait loci affecting traits of economic importance in animal populations using molecular markers. In: Schook, L.B., Lewin, H.A. and McLaren, D.G. (eds) *Gene Mapping Techniques and Applications*. Marcel Dekker, Inc., New York, pp. 21–49.

Soller, M., and Beckmann, J.S. (1983) Genetic polymorphism in varietal identification and genetic improvement. *Theoretical and Applied Genetics* 67, 25–33.

Van Arendonk, J.A.M., Tier, B. and Kinghorn, B.P. (1994a) Use of multiple genetic markers in prediction of breeding values. *Genetics* 137, 319–329.

Van Arendonk, J.A.M., Bovenhuis, H., Van der Beek, S. and Groen, A.F. (1994b) Detection and exploitation of markers linked to quantitative traits in farm animals. In: *Proceedings of the 5th World Congress on Genetics Applied to Livestock Production*, Guelph 21, 193–200.

Wang, T., Fernando, R.L., van der Beek, S., Grossman, M. and van Arendonk, J.A.M. (1995) Covariance between relatives for a marked quantitative trait locus. *Genetics Selection and Evolution* 27, 251–274.

Weller, J.I. and Fernando, R.L. (1991) Strategies for the improvement of animal production using marker-assisted selection. In: Schook, L.B., Lewin, H.A. and McLaren, D.G. (eds) *Gene Mapping Techniques and Applications*. Marcel Dekker, Inc., New York, pp. 305–328.

Weller, J.I., Kashi, Y. and Soller, M. (1990) Power of daughter and grandaughter designs for determining linkage between marker loci and quantitative trait loci in dairy cattle. *Journal of Dairy Science* 73, 2525–2537.

Wright, S. (1922) Coefficients of inbreeding and relationship. *American Naturalist* 56, 330–338.

Zhang, W. and Smith, C. (1992) Computer simulation of marker-assisted selection utilizing linkage disequilibrium. *Theoretical and Applied Genetics* 83, 813–820.

Breeding for Longevity in Dairy Cows

7

E. Strandberg

Department of Animal Breeding and Genetics,
Swedish University of Agricultural Sciences, PO Box 7023,
S-75007 Uppsala, Sweden

Introduction

The length of life of a dairy cow has substantial impact on the economic performance. The largest effect is probably that a longer life decreases the cost of replacement per year. Also, a longer average life will lead to a higher proportion of cows in later high-producing lactations. An increased length of productive life from about three to four lactations increased milk yield per lactation or profit per year by 11–13% (Renkema and Stelwagen, 1979; Essl, 1984).

There may also be additional benefits if the lengthened life is due to less culling for disease due to lower disease incidence. This would lead to lower treatment costs for diseases. However, given an improvement of disease resistance, it might be more beneficial to increase voluntary culling instead of increasing average length of life. Van Arendonk (1985, 1986) and Rogers *et al.* (1988) showed that when involuntary culling decreased it was best to increase voluntary culling so that the total effect on productive life was less than expected.

These factors point to some of the complexities surrounding breeding for longevity. The aim of this chapter is to summarize the knowledge on how to define a breeding goal which takes account of longevity, how to calculate economic weights for longevity, and how to evaluate animals in order to facilitate improvement in longevity.

The Breeding Goal

One of the first steps taken when starting a breeding programme should be to establish the breeding goal. However, even before that, the overall objective needs to be established. For the dairy farmer it is probably to achieve the greatest profit per year possible and for the society it could be to produce food as efficiently as possible, e.g. measured as cost per unit of energy, protein or per kg of milk. These are not breeding goals but objectives on a higher (and phenotypic) level. These overall objectives are sometimes called 'profit functions', keeping in mind though that they need not specify 'economic profit'.

The breeding goal in itself is not longevity nor does it even have to include longevity as a trait *per se*, as will be illustrated using a simple example from Goddard (1989). Assuming that cows live a maximum of two years, the profit per cow and year (the profit function) is:

$$P = \frac{1}{1+S}P_1 + \frac{S}{1+S}P_2^\star - \frac{1}{1+S}c_r$$

where: P_1 is mean profit from cows during the first year (lactation),
       $P_2^\star$ is mean profit from cows retained for the second year,
       S is mean stayability, proportion kept for a second year,
       c_r is net replacement cost (heifer cost minus cull cow value).

Culling is on predicted profitability during the second year, C. Therefore $P_2^\star = P_2 + i\sigma_c$, where P_2 is mean profit in second year if no culling occurred, i is the standardized selection differential and σ_c is the standard deviation of C. As culling increases, i increases and thus $P_2^\star$; however, S will decrease and the replacement cost per year will increase.

If profit is affected by milk yield (M) only and if the average is the same in both years (if no culling), the culling index $C = v_M M$ and $\sigma_c = v_m \sigma_M$, where v_M and σ_M are the value and the standard deviation of milk, respectively. The equation above now becomes:

$$P = \frac{1}{1+S}v_M M + \frac{S}{1+S}v_M(M + i\sigma_M) - \frac{1}{1+S}c_r$$

Now assume that the herd is replaced with cows that in the current situation have breeding values ΔM higher for milk yield and ΔS higher for stayability. The farmer can use this improvement in one of two ways: either he culls the same proportion as before (i.e. S and i will be constant), or he can keep the same cut-off level below which cows are culled. In the first case the economic weight of M will be v_M and that of S will be zero. In the second case, S and i will change and i is now dependent on the new milk yield, and both economic weights will be non-zero. The weight for stayability contains one term relating to increased proportion of second-year cows and one term relating to the reduction in replacement costs. Because M and S are 100% correlated, both ways of calculating economic weights will rank animals in the same order (Goddard, 1989).

The conclusion from these examples is that if the current culling rate is optimal, the correct economic weights are calculated by assuming either a constant culling rate or a constant cutoff point below which cows are culled (Goddard, 1989).

The simple model above was extended to include several traits and both voluntary culling (on *C*) and involuntary culling, which Goddard (1989) defined as 'culling for traits whose only effect on profit is through stayability'. Here it is necessary to assume constant cutoff levels for all traits, and the involuntary culling traits need not be included in the breeding goal. Their contribution to the profit function is accounted for by their relationship with stayability.

Although this method seems theoretically tempting it may be a bit difficult to find traits that only affect profit through stayability. Goddard (1989) suggested 'broken down udder' as such a trait. Other traits might be various injuries (udder, feet and legs). However, most other diseases or disorders usually have costs associated with them, even if they, in severe cases, lead to culling. Therefore, it is difficult to assign them to only one of these categories (Beard and James, 1993).

Another solution would be to express the profit function in terms of voluntary and involuntary culling and assume that voluntary culling is constant (Goddard, 1989; Rogers and McDaniel, 1989; Beard and James, 1993). The traits in the breeding goal would then be traits used for voluntary culling (e.g. milk yield) and the involuntary culling part of longevity, where the involuntary culling now is defined as culling not explained by the other traits in the breeding goal (Goddard, 1989). In this situation, economic weights for ordinary traits and the involuntary part of longevity are calculated assuming a constant voluntary culling (Rogers and McDaniel, 1989; Beard and James, 1993).

As already mentioned when discussing the calculation of economic weights, it can be useful to divide the breeding goal into 'ordinary traits', such as milk yield, and involuntary culling, which is defined as culling not explained by the other traits in the breeding goal. This is also important when estimating and interpreting genetic parameters.

Estimates of the genetic correlation between first lactation milk yield and observed productive life have ranged from 0.3 to 0.9 (Strandberg, 1985; Harris *et al.*, 1992; Short and Lawlor, 1992; Visscher and Goddard, 1994). However, several simulation studies have shown that these correlation estimates are severely biased due to the voluntary culling on milk yield and do not correctly describe the expected genetic response to selection (Essl, 1989; Strandberg, 1991; Dekkers, 1993; Strandberg and Håkansson, 1994).

To avoid this problem, length of productive life adjusted for milk yield has been suggested as the trait of choice, often termed *functional* productive life. When productive life is adjusted for level of milk yield, estimates of genetic correlation with milk yield have been much lower, usually around zero (Ducrocq *et al.*, 1988b; Sölkner, 1989; Harris *et al.*, 1992; Short and Lawlor, 1992). Therefore, whenever examining an estimate of genetic correlation between milk yield and longevity, one should note whether the longevity measured is adjusted for milk yield or not.

The economic weight for productive life adjusted for milk yield usually ranges from 30% to 70% of that for milk yield, expressed per genetic standard deviation (Rogers et al., 1988; Allaire and Gibson, 1992; Harris and Freeman, 1993), with the exception of Van Arendonk (1991) where the weight of productive life was higher than that of milk yield, mainly due to the very high phenotypic standard deviation of productive life (33 months). The relative economic values are affected by the net replacement cost, the average length of life, the level of production (Allaire and Gibson, 1992), and the feed costs (Harris and Freeman, 1993).

Traits in the Goal and in the Selection Criteria

Having decided how to handle voluntary and involuntary culling and how to calculate economic weights, the next question is what traits to use, in the breeding goal and in the selection criteria. Here one can distinguish between traits that actually determine longevity, i.e. traits that are part of the culling criteria, and traits that are only, or mainly, indicators of longevity. For another review, and an excellent one, see Dekkers and Jairath (1994).

Determinants of longevity

Health traits
Health traits (diseases and disorders) are naturally one of the main groups of traits affecting longevity of the dairy cow. Due to the lack of health trait registration in most countries there is a shortage of studies showing exactly how important diseases are in impairing longevity. One way to get an approximation is to study the culling reasons reported by the farmers. Shook (1989) concluded in a review that mastitis is the third most common reason for culling, after milk yield and reproduction. Dentine et al. (1987a) reported that of all cows culled, about one-third were culled for diseases, one-third of which were culled for mastitis. In Sweden 15–25% of those culled are culled for mastitis and another 15% are culled for other diseases (Svensk Husdjursskötsel, 1992).

The economic values of health traits have been reviewed by Rogers (1994) and are summarized in Table 7.1. Most health traits have economic values of about 10% of that for milk yield, if expressed per genetic standard deviation. Mastitis is the exception with an economic value of about 25% of that for milk yield. This value includes cost for drugs, labour, veterinary treatment, discarded milk and premature culling. Heuven (1987) found an even higher value for mastitis in later lactations (70% of that of milk yield).

Even if mastitis is the trait included in the breeding goal, selection could be on somatic cell counts (SCCs) instead, if measurements of clinical mastitis are not available. In addition, heritabilities of SCCs have usually been

Table 7.1. Genetic standard deviations (σ_A) on a per lactation basis and approximate net economic values (US$) under US conditions for some breeding goal traits (after Rogers 1994).

Trait	σ_A	Value per unit	Value per σ_A	Relation to milk
Milk yield (ME), kg	650	0.14	91.00	1.00
Days open	6	1.5	9.00	0.10
Clinical mastitis	0.15[1]	150[2]	22.50	0.25
Ketosis[1]	0.06	100[2]	6.00	0.07
Milk fever[1]	0.08	100[2]	8.00	0.09
Displaced abomasum[1]	0.04	100	4.00	0.04
Laminitis[1]	0.10	100[2]	10.00	0.11

[1] Measured as a discrete trait (0 or 1).
[2] At least.

somewhat higher than those for clinical mastitis, even if the latter is correctly treated as a binomial trait (Emanuelson, 1988; Lin *et al.*, 1989; Shook, 1989; Lyons *et al.*, 1991). Including mastitis or SCCs in the selection index improved the total response in the breeding goal by only 0.4–0.8% (Strandberg and Shook, 1989; Duval and Colleau, 1993). However, the breeding goal consisted only of milk yield and mastitis traits, not longevity.

Dystocia (calving difficulty) is most common in first-calf heifers (Philipsson *et al.*, 1979; Erb *et al.*, 1985). Heritability estimates have usually been higher for first calvings (0.03–0.20) compared with later calvings (0.00–0.08) (Philipsson *et al.*, 1979). Due to the commonly found negative genetic correlation between direct and maternal effects, it is recommendable to evaluate bulls both as sires and as maternal grandsires (e.g. Philipsson, 1976; Thompson *et al.*, 1981).

Milk fever is very rare in first-lactation cows, but when it occurs it may increase the risk of culling by more than five times (Gröhn *et al.*, 1986a). Heritability estimates for milk fever have been close to zero (Emanuelson, 1988) and for the first parities very difficult to estimate (Gröhn *et al.*, 1986b). Ketosis is also more frequent in later lactations and usually heritability estimates are low. For both traits, the low heritability and the low frequency in first lactation limits the use of these traits in selection.

Several of these diseases are unfavourably genetically correlated with milk yield and would therefore need to be included in the breeding goal and the selection criteria, even if selection does not aim at improving longevity itself. Mastitis and SCCs are extensively studied and almost invariably a genetic antagonism with milk yield has been found (Emanuelson, 1988; Banos and Shook, 1990; Simianer *et al.*, 1991; Boettcher *et al.*, 1992; Weller *et al.*, 1992; Welper and Freeman, 1992). A similar situation seems to exist for ketosis (Emanuelson, 1988).

Reproduction

Reproduction is the largest or second largest culling reason in dairy cattle, 15–40% of those culled are culled for inadequate reproductive performance (Shook, 1989; Svensk Husdjursskötsel, 1992). The economic value for days open (calving to conception interval) is approximately 10% of that of milk yield (Table 7.1).

Estimates of heritability for reproductive measures have been low, below 10% (Schaeffer and Henderson, 1972; Berger *et al.*, 1981; Philipsson, 1981; Hansen *et al.*, 1983a,b; Freeman, 1984, 1986; Strandberg and Danell, 1989; Lyons *et al.*, 1991). Nevertheless, there is a substantial genetic coefficient of variation, 3–18% (Philipsson, 1981; Strandberg and Danell, 1989), which indicates that it is possible to distinguish between sire progeny groups. Furthermore, there is evidence of genetic antagonism with milk yield (Berger *et al.*, 1981; Philipsson, 1981; Hansen *et al.*, 1983a; Strandberg and Danell, 1989; Oltenacu *et al.*, 1991) suggesting that reproduction will deteriorate if selection is on milk yield only.

Indicators of longevity for genetic evaluation

Stayability

Stayability is probably the most commonly used indicator of longevity. This trait is measured as the survival (0 or 1) to start a certain lactation (Robertson and Barker, 1966; Schaeffer and Burnside, 1974), up to a certain age (Everett *et al.*, 1976a,b; Hudson and Van Vleck, 1981; DeLorenzo and Everett, 1982; Van Doormaal *et al.*, 1985) or up to a certain time after first calving (Van Doormaal *et al.*, 1985). The ages used have been 36, 48, 60, 72 and 84 months, probably chosen assuming 24 months at first calving and a 12-month calving interval. Van Doormaal *et al.* (1985) used a different approach to determine the time when to measure stayability. They studied the hazard (the relative risk of culling) during the cow's life and chose the periods of low risk of culling as the time to measure stayability. This resulted in stayability up to 42, 54, 66, and 78 months of age, and up to 17, 30, 43, and 55 months from first calving. DeLorenzo and Everett (1986) chose 41 and 54 months, being the average ages in the middle of the dry period in the first two calving intervals, respectively.

Although stayability is a binary trait, most studies have used linear models. One exception is DeLorenzo and Everett (1986) who used a logistic linear model. They found quite substantial differences in ranking of sires between the logistic model and the linear model; rank correlations were 0.6–0.7. Nevertheless, no-one seems to have used this model further, perhaps due to computational limitations.

Heritability estimates for stayability have been low, usually less than 0.05 (Strandberg, 1985; Dentine *et al.*, 1987b; Brotherstone and Hill, 1991a; Short and Lawlor, 1992; Visscher and Goddard, 1995). Unfortunately for selection purposes, there was also a tendency to lower heritabilities for earlier measures, the ones mainly used in selection. The only exception was DeLorenzo and Everett (1986) who estimated heritabilities of around 0.27 for survival to 41 or 54 months of life using a logistic linear model.

Survival within each lactation
Instead of using stayability Madgwick and Goddard (1989) used a series of survival scores, S_i, where $S_i = 1$ if the cow survived from year i to year $i + 1$ after first calving and $S_i = 0$ otherwise.

Heritabilities for survival scores were low, the highest were between 0.028 and 0.053 (for S_0). They found very high correlations between sire solutions based on linear analysis of survival scores and a nonlinear analysis. Genetic correlations among survival scores were around 0.8 for S_0 to S_3. The genetic correlations between production in first lactation and survival scores were highest for S_0 to S_3, between 0.26 and 0.63, depending on the production trait. One of the 'problems' with their analysis was that it was performed on Australian cattle, which have very long productive lives: the average in their study was 5.8–6.6 years. The very low culling from one lactation to the next might have affected the performance of the method. Visscher and Goddard (1994) also found low heritabilities for survival scores in Australian Friesians, but somewhat higher in Jerseys (0.07–0.08).

Length of productive life
Length of productive life in itself is not a useful measure for selection because of the long time before it is realized. For many animals one would only know that length of life is at least as long as a certain time. Such records are called censored. Different methods have been used to accommodate censoring.

Failure time analysis. Failure time analysis has been used before in the fields of medicine and engineering, e.g. to study the effect of various medical treatments on relapse, recovery or survival (e.g. Kalbfleisch and Prentice, 1980). This approach models the actual survival times or usually (because it is more convenient) the hazard, the risk of failure at a certain time given that the individual survived up to that time. The observations used are the failure times (e.g. death) or the censoring times (e.g. still alive at the end of data collection), combined with an indicator of whether the measure is censored or not.

Probably the most commonly used model is the proportional hazards model (e.g. Kalbfleisch and Prentice, 1980). The hazard $\lambda(t; z)$ for time t and a set of known covariates z:

$$\lambda(t;\ z) = \lambda_0(t)e^{z'\beta}$$

where the hazard is the product of a time-dependent term $\lambda_0(t)$, called the baseline hazard, and a time-independent term $e^{z'\beta}$. The baseline hazard is related to the general ageing process and the term $e^{z'\beta}$ depends on the covariates in z, e.g. effect of treatment, animal, etc. Two of the most common assumptions about the baseline hazard are (i) that it is constant ($\lambda_0(t) = \lambda$), in which case the survival times follow an exponential distribution, and (ii) that it is $\lambda_0(t) = \lambda p(\lambda t)^{p-1}$, for some values of the parameters λ and p, in which case the survival times follow a Weibull distribution. Examples of the two distributions are given in Fig. 7.1.

Cox (1972) simplified the equation above by showing that $\lambda_0(t)$ need not be specified. Smith (1983) and Smith and Quaas (1984) studied age at disposal using an extension of the Cox's regression model:

$$\lambda_{jklm}(t) = \lambda_{0j}(t) \exp \{ h_{jk} + g_l + s_{lm} \}$$

where: $\lambda_{0j}(t)$ is a piecewise constant baseline hazard, stratified by year of birth j; h_{jk} is the kth herd effect nested within the jth year of birth; g_l is the lth genetic group; and s_{lm} is the mth sire within the lth group. In contrast to traditional failure time analysis where all effects are considered fixed, here both herd and sire were assumed to be random.

Another new feature of this model was the stratification, i.e. they divided the material into strata, here based on year of birth. Each stratum was allowed to have a different baseline hazard. This method of stratification can be used if a certain factor does not seem to act multiplicatively on the hazard as it would if added to the model component in the exponent (e.g. Kalbfleisch and Prentice, 1980).

The next large study using failure time analysis was the thesis of Ducrocq (1987; summarized in Ducrocq et al., 1988a,b). Smith (1983) and Smith and Quaas (1984) used age as the time scale. Because there is virtually no culling for almost two years, it is almost impossible for any parametric distribution to fit that survival or hazard function well. To avoid this problem, Ducrocq (1987) and Ducrocq et al. (1988a,b) used length of productive life (time from first calving to disposal). Also, instead of deciding a priori to use an exponential distribution, they estimated the parameters p and λ of the Weibull distribution (of which the exponential is a special case) to get the best fit possible.

Another major extension was the inclusion of time-dependent effects. In the above equation the stratification is according to year of birth and the herd effect (h_{jk}) is constant throughout the life of the cow. However, Ducrocq (1987) argued that it is more likely that each production year has a specific effect on the hazard of all cows alive at that time. For instance, perhaps the herd size is increasing during a few years. The hazard of cows alive during that period, regardless of when they were born, is lower than for cows during other periods. To accommodate this, Ducrocq (1987) included a herd effect that was piecewise constant, changing every new year.

To account for a changing herd size, Ducrocq (1994) also included a time-dependent effect which was a combination of season and class of change in herd size the previous year. The effects were assumed to change at 1 March and 1 December each year. The date 1 March was chosen because the quota period

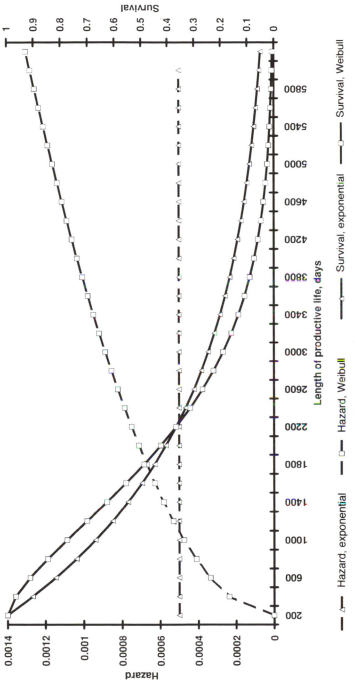

Fig. 7.1. Example of survival and hazard as a function of length of productive life (time scale in days) for two distributions: (i) the exponential with baseline hazard 0.0005, and (ii) the Weibull distribution with parameters $p = 1.5$ and $\lambda = 0.0005$.

ended 31 March. The risk of culling was higher in the months just before the end of the quota period compared with during the rest of the year. Also, in herds of decreasing size, the hazard was much higher than in herds of increasing size.

The effect of within-herd class of milk yield was included in an attempt to estimate length of productive life adjusted for voluntary culling, termed by Ducrocq (1987) 'functional productive life'. The correlation between sire estimates for length of productive life (actually relative risk of culling) and estimated breeding values for milk yield changed from favourable (-0.28) to slightly unfavourable (0.13) with the adjustment for within-herd class of milk yield. This result was not repeated in the study by Ducrocq (1994): both correlations were -0.40. One possible explanation put forward was that the breed used in Ducrocq (1994), Normande, is a dual-purpose breed with voluntary culling for other traits as well.

Estimating heritabilities using failure time analysis is not as straightforward as for ordinary linear models. Due to the methods and the censoring it is not possible to estimate the phenotypic variance in an unambiguous way. However, making some simplifying assumptions, Ducrocq *et al.* (1988b) estimated a 'pseudo-heritability' of 0.085.

The future developments in this area should focus on incorporating the relationship between sires into the model in a similar manner as for the ordinary mixed models. Theoretically, it should be possible to set up an animal model also for this type of model; however, the predictive power and use of such individual animal breeding values might be limited. Ducrocq and Sölkner (1994) are currently working to create a general computer program to be used for failure time analysis of animal breeding data. So far only the Cox's regression is available, but development of a Weibull model is under way.

Extension of censored records. VanRaden and Klaaskate (1993) tried to accommodate censored records while avoiding the complexity of the nonlinear methods used in failure time analysis. Extension of incomplete (censored) records is a common method when studying 305-day milk yield, and the idea here was to use a similar procedure for length of productive life. The predicted records were expanded to give the same variance as for completed records and the predicted records were given a lower weight in the mixed model analysis later. The R^2-values for predicting their longevity measure from earlier ages were quite low except at 72 months of age. However, the genetic correlations between longevity from completed records only and longevity from combined projected and completed records were all quite high, above 0.92.

Properties of the different methods of genetic evaluation
One important property of methods used for genetic evaluation of longevity is the ability to account for censoring. For failure time analysis, censoring is accounted for directly within the method but not in the other methods.

In the first two methods the problem with censoring is at least partially avoided by the definition of the trait. For genetic evaluation using stayability a short period is chosen, e.g. 17 months of productive life, and all animals are given the opportunity to live that long and no animals are censored. Although this procedure avoids censoring it means that not all information in the data is utilized. Cows culled 1 day or 1 year before the limit are treated as identical. Danner *et al.* (1993) compared sire breeding values estimated using a Cox's regression model and sire breeding values for stayability to different ages using an ordinary linear model. In both models relationships were included, and adjustment for within-herd milk yield deviation was carried out. The full data set was divided according to year of first calving and then truncated to simulate different amounts of censoring. The rank correlation between breeding values estimated on the full data set or on the censored data set using the Cox's regression model were higher than when applying the corresponding stayability model. The difference was largest when censoring was high, i.e. for stayability to 16 months of productive life and its corresponding censored data set including two years of data.

The ability to account for systematic environmental effects is also different for the described methods. In the analysis of stayability and extended censored records of longevity, the environmental effects included usually pertain to first calving and are assumed constant regardless of the actual longevity. Both with failure time analysis and the traits survival within each lactation (Madgwick and Goddard, 1989), it is possible to have time-dependent environmental effects, which should improve the possibility to account for a changing environment.

The genetic model is somewhat different for the various methods. In failure time analysis the implicit assumption is that survival is the same throughout life. In the other methods, it is possible to study each measure (stayability, survival through different lactations) as separate genetic traits, which may be an advantage.

The adjustment for milk yield to achieve a measure of functional productive life should be possible in all methods, either by preadjustment or by inclusion of milk yield in the model. Because culling is within herd and year, the milk yield should be deviated from the herd-year mean.

The traditional linear models use the relationships among animals. Due to computational limitations, this is not yet possible for the failure time analysis methods, except for the sire model using Cox's regression.

Type traits – conformation – workability
Several authors have studied type traits, and also the relationship with longevity. However, here the focus will be on longevity adjusted for milk yield. Quite naturally, it is difficult to compare type traits measured in different ways in different countries and breeds. However, even within country and breed there can be difficulties in interpreting correlations between type traits and longevity because length of productive life seems to be a different trait in registered

(pedigree) and grade (non-pedigree) cows. Dentine *et al.* (1987b) found genetic correlations between the same longevity measures in registered and grade cows to be considerably lower than unity. The explanation is that grade cows to a certain extent are culled for different reasons than are registered cows (Dentine *et al.*, 1987a).

DeLorenzo and Everett (1982) found that the predicted difference (PD, half the breeding value) for type traits did not explain any variation in PD for stayability, even if milk yield was excluded and type was the only variable in the equation, except for the top 10% of milk producers. Other authors have found that PD for type had moderate to high genetic correlation with longevity for registered herds but low for grade herds (Table 7.2).

Quite a few studies have found several udder traits (especially fore udder attachment and udder support) to be consistently and usually moderately genetically related to stayability adjusted for milk yield (Table 7.2). These traits also seem to behave fairly consistently between registered and grade cows.

Body traits (e.g. body depth, stature) have positive correlations with longevity in registered cows (about 0.15) but negative (about -0.15) in grade cows (Table 7.2). A similar situation exists for foot angle, which is positively correlated (about 0.25) with productive life in registered cows but basically uncorrelated in grade cows.

Heritability estimates of type traits vary with population (country, registered, grade), type of measurement and category of trait. It seems that body traits have higher heritabilities (0.25–0.45) than udder traits (0.2–0.3) and feet and legs (0.15–0.25).

Expected response to selection

Indirect prediction of functional productive life using type traits in grade cows resulted in maximum correlations between index and breeding goal, r_{IH}, of 0.5–0.75 (Boldman *et al.*, 1992; Short and Lawlor, 1992; Weigel and Cassell, 1994; Dekkers *et al.*, 1994). However, indirect prediction gave a higher r_{IH} than direct prediction when the effective number of progeny was less than 75 ($r_{IH} \approx 0.67$) (Boldman *et al.*, 1992). Again, it is important to distinguish between registered and grade cows, as the ability to predict productive life from type traits is larger in registered cows (Weigel and Cassell, 1994).

Some authors have studied the ability to predict a breeding goal consisting of yield and involuntary culling. Rogers and McDaniel (1989) found that r_{IH} was increased from 0.567 to 0.608 (by about 7%), when they included udder depth, teat placement, and foot angle, compared with only including milk yield. Veerkamp *et al.* (1994) similarly showed an increase of about 5% in r_{IH} when including udder depth, teat length, foot angle, and angularity over and above yield traits. When studying these r_{IH} one should remember that such correlations calculated when many traits are included in the index are somewhat overestimated (Sales and Hill, 1976; Visscher, 1994).

Table 7.2. Estimates of genetic correlations between functional productive life and type traits.

Reference	Reg. or grade	PDT[5]	Stature	Body depth	Foot angle	Fore udder	Udder depth	Udder support	Fore/rear teat
Rogers et al., 1989[1]	g	0.03	-0.18	-0.23	0.07	0.10	0.20	0.13	0.20
Rogers et al., 1991[2]	r	0.73[3]	0.14	–	–	0.36	0.31	–	0.06
Brotherstone and Hill, 1991	r	0.43-0.65[3]	0.02-0.08	0.15-0.33	0.22-0.35	0.28-0.45	0.08-0.24	0.12-0.27	0.32-0.40
Boldman et al., 1992	g		-0.21	-0.20	-0.12	0.46	0.47	0.22	0.17
Klassen et al., 1992	r		0.23					0.07	-0.02
Short and Lawlor, 1992	r		0.18	-0.03	0.26	0.47	0.50	0.43	0.42
	g		-0.19	-0.32	-0.06	0.24	0.39	0.18	0.08
Harris et al., 1992[4]	r		-0.69	-0.35	-0.14	0.35	0.04	0.41	0.11
Bagnato, 1993		0.01[3]	-0.15	-0.30	-0.05	0.20	0.31	0.22	0.12
Dekkers et al., 1994	r		0.15	0.09	0.21	0.26		0.24	
	g		0.03	-0.08	0.06	0.19		0.06	

[1] Stayability to 54 months of age.
[2] Jerseys, survival to 20 months of productive life.
[3] Final score.
[4] Guernsey.
[5] Predicted difference for type.

Allaire and Gibson (1992) showed that ignoring selection on longevity could decrease overall genetic response by 5–9%. The higher losses occurred when the heritability of functional productive life was higher (0.10 instead of 0.05). Rekik and Allaire (1993) similarly calculated an expected increase in the breeding goal response of 2–3% when including one stayability measure besides milk yield. The inclusion of stayability measures was more beneficial when correlations between milk yield and involuntary culling were unfavourable. In both studies the breeding goal consisted of milk yield and involuntary culling.

Dekkers (1993) showed that when biased estimates between milk yield and longevity were used (i.e. without adjusting longevity for milk yield), the response in the breeding goal decreased by about 4% compared with the optimal situation, which was when the true correlation was known.

Conclusions

Although the title of this chapter is 'breeding for longevity', our aim should not be to improve longevity in itself. Our aim should be to improve the lifetime profit, efficiency or some other measure of utility. In doing so, we will probably also improve the animal's ability to live longer by improving traits that determine longevity. However, as pointed out before, the actually observed longevity may not change at all or may not change as much as expected from the changes in the other traits.

What breeding goal is best if one wants to improve lifetime profit of dairy cows? There is no general answer to that question, but a few main points can be outlined. First of all, milk production traits should be included – this is the main reason for keeping dairy cows. After that, traits that influence the costs of production and determine longevity should be included, i.e. broadly speaking health and reproductive traits. It is sometimes argued that one cannot include these traits because they are not recorded in the population. However, it is quite possible to have, for example, mastitis in the breeding goal without recording it routinely, as long as one has estimates of genetic correlation between somatic cell counts and mastitis.

The health and reproductive traits to be included in the goal must be decided for each specific population. Nevertheless, as a minimum, mastitis and some measure of reproduction, e.g. days open, should be included. All possible health and reproductive traits should not be included, because in practice, the more traits one needs to estimate variance components for, the more likely one is to get parameter estimates that are invalid.

Type traits should, in general, not be included in the breeding goal, because in commercial herds, no income or cost is directly associated with the type trait. The importance of type traits lies in their ability to indicate health problems or decreased longevity.

A tempting compromise is to include production traits, mastitis, days open, and involuntary culling in the breeding goal. Involuntary culling should then, at least, be adjusted for the culling due to production.

The traits in the selection criterion could be the same as those in the goal. However, now the availability of records is important. If somatic cell counts are available one could use that trait: it might be as good for predicting mastitis as mastitis itself. Of the type traits, udder traits seem the most promising. It would be interesting to know more about their relationship to, for example, mastitis, rather than only their relationship to longevity. For prediction of days open, it might be valuable to use both calving to first insemination interval and number of inseminations, because they describe two, sometimes opposing, sides of days open.

One should be somewhat cautious of including all measured traits in the selection criterion. If the true parameters are known, the precision of evaluation always increases with increasing number of traits. However, parameters are estimated with error. As pointed out by Sales and Hill (1976) and Visscher (1995), the precision might even decrease if traits are included that have weak true correlations to the goal but that have estimates (with large error variances) that indicate otherwise.

If no or few health and reproduction traits are recorded in the population, one solution could be to use evaluations of longevity as an overall measure of health, reproduction, and involuntary culling. However, these measures should be adjusted for production. If possible, one should use failure time analysis because of its ability to fully use the information in lifetime data, and to account for censoring and changing systematic environmental effects.

References

Allaire, F.R. and Gibson, J.P. (1992) Genetic value of herd life adjusted for milk production. *Journal of Dairy Science* 75, 1349–1356.

Bagnato, A. (1993) Herdlife in the Italian Holstein Friesian. *Proceedings of the Interbull Annual Meeting*, Aarhus, Denmark, Aug 19–20. Bulletin no. 8.

Banos, G. and Shook, G.E. (1990) Genotype by environment interaction and genetic correlations among parities for somatic cell count and milk yield. *Journal of Dairy Science* 73, 2563–2573.

Beard, K.T. and James, J.W. (1993) Including the involuntary component of longevity in a breeding objective. Paper G2.1 presented at the annual meeting of the European Association of Animal Production, I: 74–75. Aarhus, Denmark.

Berger, P.J., Shanks, R.D., Freeman, A.E. and Laben, R.C. (1981) Genetic aspects of milk yield and reproductive performance. *Journal of Dairy Science* 64, 114–122.

Boettcher, P.J., Hansen, L.B., VanRaden, P.M. and Ernst, C.A. (1992) Genetic evaluation of Holstein bulls for somatic cells in milk of daughters. *Journal of Dairy Science* 75, 1127–1137.

Boldman, K.B., Freeman, A.E., Harris, B.L. and Kuck, A.L. (1992) Prediction of sire transmitting abilities for herd life from transmitting abilities for linear type traits. *Journal of Dairy Science* 75, 552–563.

Brotherstone, S. and Hill, W.G. (1991) Dairy herd life in relation to linear type traits and production. 1. Phenotypic and genetic analyses in pedigree type classified herds. *Animal Production* 53, 279–287.

Cox, D.R. (1972) Regressions models and life tables (with discussion). *Journal of the Royal Statistical Society*, B34, 187–220.

Danner, C., Sölkner, J. and Essl, A. (1993) Prediction of breeding values for longevity: comparison of proportional hazards analysis for culling rate and BLUP for stayability. Proceedings of the annual meeting of the European Association of Animal Production, Aarhus, Denmark, Aug 16–19 (Mimeo).

Dekkers, J.C.M. (1993) Theoretical basis for genetic parameters of herd life and effects on response to selection. *Journal of Dairy Science* 76, 1433–1453.

Dekkers, J.C.M. and Jairath, L.K. (1994) Requirements and uses of genetic evaluations for conformation and herd life. *Proceedings of the 5th World Congress on Genetics Applied to Livestock Production* 17, 61–68. Department of Animal and Poultry Science, University of Guelph, Guelph, Ontario, Canada.

Dekkers, J.C.M., Jairath, L.K. and Lawrence, B.H. (1994) Relationships between sire genetic evaluations for conformation and functional herd life of daughters. *Journal of Dairy Science* 77, 844–854.

DeLorenzo, M.A. and Everett, R.W. (1982) Relationships between milk and fat production, type and stayability in Holstein sire evaluation. *Journal of Dairy Science* 65, 1277–1285.

DeLorenzo, M.A. and Everett, R.W. (1986) Prediction of sire effects for probability of survival to fixed ages with a logistic linear model. *Journal of Dairy Science* 69, 501–509.

Dentine, M.R., McDaniel, B.T. and Norman, H.D. (1987a) Comparison of culling rates, reasons for disposal, and yields for registered and grade Holstein cattle. *Journal of Dairy Science* 70, 2616–2622.

Dentine, M.R., McDaniel, B.T. and Norman, H.D. (1987b) Evaluation of sires for traits associated with herdlife of grade and registered Holstein cattle. *Journal of Dairy Science* 70, 2623–2634.

Ducrocq, V. (1987) An analysis of length of productive life in dairy cattle. PhD thesis, Cornell University, Ithaca, NY.

Ducrocq, V. (1994) Statistical analysis of length of productive life for dairy cows of the Normande breed. *Journal of Dairy Science* 77, 855–866.

Ducrocq, V. and Sölkner, J. (1994) 'The Survival Kit' – a FORTRAN package for the analysis of survival data. *Proceedings of the 5th World Congress on Genetics Applied to Livestock Production* 22, 51–52. Department of Animal and Poultry Science, University of Guelph, Guelph, Ontario, Canada.

Ducrocq, V., Quaas, R.L., Pollak, E.J. and Casella, G. (1988a) Length of productive life of dairy cows: 1. Justification of a Weibull model. *Journal of Dairy Science* 71, 3061–3070.

Ducrocq, V., Quaas, R.L., Pollak, E.J. and Casella, G. (1988b) Length of productive life of dairy cows: 2. Variance component estimation and sire evalutation. *Journal of Dairy Science* 71, 3071–3079.

Duval, E. and Colleau, J.J. (1993) Selecting for mastitis resistance in dairy cattle. Paper GC 1.4 presented at the annual meeting of the European Association of Animal Production, I, 16–17. Aarhus, Denmark.

Emanuelson, U. (1988) Recording of production diseases in cattle and possibilities for genetic improvement: a review. *Livestock Production Science* 20, 89–106.

Erb, H.N., Smith, R.D., Oltenacu, P.A., Guard, C.L., Hillman, R.G., Powers, P.A., Smith, M.C. and White, M.E. (1985) Path model of reproductive disorders and performance, milk fever, mastitis, milk yield and culling in Holstein cows. *Journal of Dairy Science* 68, 3337–3349.

Essl, A. (1984) Zusammenhang zwischen Leistungszucht und Nutzungsdauer bei Kuhen. *Zuchtungskunde* 56, 337–343.

Essl, A. (1989) Estimation of the genetic correlation between first lactation milk yield and length of productive life by means of a half-sib analysis: a note on the estimation bias. *Journal of Animal Breeding and Genetics* 106, 402–408.

Everett, R.W., Keown, J.F. and Clapp, E.E. (1976a) Relationships among type, production and stayability in Holstein cattle. *Journal of Dairy Science* 59, 1505–1510.

Everett, R.W., Keown, J.F. and Clapp, E.E. (1976b) Production and stayability trends in dairy cattle. *Journal of Dairy Science* 59, 1532–1539.

Freeman, A.E. (1984) Secondary traits: sire evaluation and the reproductive complex. *Journal of Dairy Science* 67, 449–458.

Freeman, A.E. (1986) Genetic control of reproduction and lactation in dairy cattle. *Proceedings of the 3rd World Congress on Genetics Applied to Livestock Production.* XI, 3–13.

Goddard, M.E. (1989) The use of stayability in the selection objective for dairy cows. Unpublished mimeograph, University of New England, Armidale, Australia.

Gröhn, Y., Saloniemi, H. and Syväjärvi, J. (1986a) An epidemiological and genetic study on registered diseases in Finnish Ayrshire cattle. I. The data, disease occurrence and culling. *Acta Veterinaria Scandinavica* 27, 182–195.

Gröhn, Y., Saloniemi, H. and Syväjärvi, J. (1986b) An epidemiological and genetic study on registered diseases in Finnish Ayrshire cattle. III. Metabolic diseases. *Acta Veterinaria Scandinavica* 27, 209–222.

Hansen, L.B., Freeman, A.E. and Berger, P.J. (1983a) Yield and fertility relationships in dairy cattle. *Journal of Dairy Science* 66, 293–305.

Hansen, L.B., Freeman, A.E. and Berger, P.J. (1983b) Association of heifer fertility with cow fertility and yield in dairy cattle. *Journal of Dairy Science* 66, 306–314.

Harris, B.L. and Freeman, A.E. (1993) Economic weights for milk yield traits and herd life under various economic conditions and production quotas. *Journal of Dairy Science* 76, 868–879.

Harris, B.L., Freeman, A.E. and Metzger, E. (1992) Analysis of herd life in Guernsey dairy cattle. *Journal of Dairy Science* 75, 2008–2016.

Heuven, H.C.M. (1987) Diagnostic and genetic analysis of mastitis field data. PhD thesis, University of Madison, Wisconsin.

Hudson, G.F.S. and Van Vleck, L.D. (1981) Relationships between production and stayability in Holstein cattle. *Journal of Dairy Science* 64, 2246–2250.

Kalbfleisch, J.D. and Prentice, R.L. (1980) *The Statistical Analysis of Failure Time Data.* Wiley, New York, NY.

Klassen, D.J., Monardes, H.G., Jairath, L.K., Cue, R.I. and Hayes, J.F. (1992) Genetic correlations between lifetime production and linearized type in Canadian Holstein. *Journal of Dairy Science* 75, 2272–2282.

Lin, H.K., Oltenacu, P.A., Van Vleck, L.D., Erb, H.N. and Smith, R.D. (1989) Heritabilities of and genetic correlations among six health problems in Holstein cows. *Journal of Dairy Science* 72, 180–186.

Lyons, D.T., Freeman, A.E. and Kuck, A.L. (1991) Genetics of health traits in Holstein cattle. *Journal of Dairy Science* 74, 1092–1100.

Madgwick, P.A. and Goddard, M.E. (1989) Genetic and phenotypic parameters of longevity in Australian dairy cattle. *Journal of Dairy Science* 72, 2624–2632.

Oltenacu, P.A., Frick, A. and Lindhé, B. (1991) Relationship of fertility to milk yield in Swedish dairy cattle. *Journal of Dairy Science* 74, 264–268.

Philipsson, J. (1976) Studies on calving difficulty, stillbirth and associated factors in Swedish dairy cattle. III. Genetic parameters. *Acta Agriculturae Scandinavica* 26, 211–220.

Philipsson, J. (1981) Genetic aspects of female fertility in dairy cattle. *Livestock Production Science* 8, 307–319.

Philipsson, J., Foulley, J.L., Lederer, J., Liboriussen, T. and Osinga, A. (1979) Sire evaluation standards and breeding strategies for limiting dystocia and stillbirth. *Livestock Production Science* 6, 111–127.

Rekik, B. and Allaire, F.R. (1993) Contribution of stayability records to the accuracy of selection for improved production value and herd life. *Journal of Dairy Science* 76, 2299–2307.

Renkema, J.A. and Stelwagen, J. (1979) Economic evaluation of replacement rates in dairy herds. I. Reduction of replacement rates through improved health. *Livestock Production Science* 6, 15–27.

Robertson, A. and Barker, J.S.F. (1966) The correlations between first lactation milk production and longevity in dairy cattle. *Animal Production* 8, 241–252.

Rogers, G.W. (1994) Requirements and uses of evaluations for health and reproductive traits. *Proceedings of the 5th World Congress on Genetics Applied to Livestock Production* 17, 81–88. Department of Animal and Poultry Science, University of Guelph, Guelph, Ontario, Canada.

Rogers, G.W. and McDaniel, B.T. (1989) The usefulness of selection for yield and functional type traits. *Journal of Dairy Science* 72, 187–193.

Rogers, G.W., Van Arendonk, J.A.M. and McDaniel, B.T. (1988) Influence of involuntary culling on optimum culling rates and annualized net revenue. *Journal of Dairy Science* 71, 3463–3469.

Rogers, G.W., McDaniel, B.T., Dentine, M.R. and Funk, D.A. (1989) Genetic correlations between survival and linear type traits measured in first lactation. *Journal of Dairy Science* 72, 523–527.

Rogers, G.W., Hargrove, G.L., Cooper, J.B. and Barton, E.P. (1991) Relationships among survival and linear type traits in Jerseys. *Journal of Dairy Science* 74, 286–291.

Sales, J. and Hill, W.G. (1976) Effect of sampling errors on efficiency of selection indices. 2. Use of information on associated traits for improvement of a single important trait. *Animal Production* 23, 1–14.

Schaeffer, L.R. and Henderson, C.R. (1972) Effects of days dry and days open on Holstein milk production. *Journal of Dairy Science* 55, 107–112.

Schaeffer, L.R. and Burnside, E.B. (1974) Survival rates of tested daughters of sires in artificial insemination. *Journal of Dairy Science* 57, 1394–1400.

Shook, G.E. (1989) Selection for disease resistance. *Journal of Dairy Science* 72, 1349–1362.

Short, T.H. and Lawlor, T.J. (1992) Genetic parameters of conformation traits, milk yield, and herd life in Holsteins. *Journal of Dairy Science* 75, 1987–1998.

Simianer, H., Solbu, H. and Schaeffer, L.R. (1991) Estimated genetic correlations between disease and yield traits in dairy cattle. *Journal of Dairy Science* 74, 4358–4365.

Smith, S.P. (1983) The extension of failure time analysis to problems of animal breeding. PhD thesis, Cornell University, Ithaca, NY. *Dissertation Abstracts International*, B43(12), 3848.

Smith, S.P. and Quaas, R.L. (1984) Productive lifespan of bull progeny groups: failure time analysis. *Journal of Dairy Science* 67, 2999–3007.

Sölkner, J. (1989) Genetic relationships between level of production in different lactations, rate of maturity and longevity in a dual purpose cattle population. *Livestock Production Science* 23, 33–45.

Strandberg, E. (1985) Estimation procedures and parameters for various traits affecting lifetime milk production: a review. Report 67, Department of Animal Breeding and Genetics, Swedish University of Agricultural Sciences (also in Strandberg, 1991).

Strandberg, E. (1991) Breeding for lifetime performance in dairy cattle. PhD thesis. Report 96, Department of Animal Breeding and Genetics, Swedish University of Agricultural Sciences.

Strandberg, E. and Danell, B. (1989) Genetic and phenotypic parameters for production and days open in the first three lactations of Swedish dairy cattle. *Acta Agriculturae Scandinavica* 39, 203–215.

Strandberg, E. and Shook, G.E. (1989) Genetic and economic responses to breeding programs that consider mastitis. *Journal of Dairy Science* 72, 2136–2142.

Strandberg, E. and Håkansson, L. (1994) Effect of culling on the estimates of genetic correlation between milk yield and length of productive life in dairy cattle. *Proceedings of the 5th World Congress on Genetics Applied to Livestock Production* 17, 77–80. Department of Animal and Poultry Science, University of Guelph, Guelph, Ontario, Canada.

Svensk Husdjursskötsel (1992) Å Arsstatistik, SHS 1991/92. [Statistics from production and disease recording systems]. Swedish Association of Livestock Breeding and Production, Hållsta, S-63184 Eskilstuna. Medd. 172.

Thompson, J.R., Freeman, A.E. and Berger, P.J. (1981) Age of dam and maternal effects for dystocia in Holsteins. *Journal of Dairy Science* 64, 1603–1609.

Van Arendonk, J.A.M. (1985) Studies on the replacement policies in dairy cattle. II. Optimum policy and influence of changes in production and prices. *Livestock Production Science* 13, 101–123.

Van Arendonk, J.A.M. (1986) Economic importance and possibilities for improvement of dairy cow herd life. *Proceedings of the 3rd World Congress on Genetics Applied to Livestock Production* IX, 9–100.30, September 1994.

Van Arendonk, J.A.M. (1991) Use of profit equations to determine relative economic value of dairy cattle herd life and production from field data. *Journal of Dairy Science* 74, 1101–1107.

Van Doormaal, B.J., Schaeffer, L.R. and Kennedy, B.W. (1985) Estimation of genetic parameters for stayability in Canadian Holsteins. *Journal of Dairy Science* 68, 1763–1769.

VanRaden, P.M. and Klaaskate, E.J.H. (1993) Genetic evaluation of length of productive life including predicted longevity of live cows. *Journal of Dairy Science* 76, 2758–2764.

Veerkamp, R.F., Brotherstone, S., Stott, A.W., Hill, W.G. and Simm, G. (1994)

Combining transmitting abilities for yield and linear type in an index for selection on production and longevity. *Proceedings of the 5th World Congress on Genetics Applied to Livestock Production* 17, 69–72. Department of Animal and Poultry Science, University of Guelph, Guelph, Ontario, Canada.

Visscher, P.M. and Goddard, M.E. (1995) Genetic parameters for milk yield, survival, workability, and type traits for Australian dairy cattle. *Journal of Dairy Science* 78, 205–220.

Visscher, P.M. (1995) Bias in multiple genetic correlation from half-sib designs. *Genetics Selection Evolution* 27, in press.

Weigel, D.J. and Cassell, B.G. (1994) Differences in multiple trait prediction of transmitting abilities for herdlife due to data source. *Proceedings of the 5th World Congress on Genetics Applied to Livestock Production* 17, 73–76. Department of Animal and Poultry Science, University of Guelph, Guelph, Ontario, Canada.

Weller, J.I., Saran, A. and Zeliger, Y. (1992) Genetic and environmental relationships among somatic cell count, bacterial infection, and clinical mastitis. *Journal of Dairy Science* 75, 2532–2540.

Welper, R.D. and Freeman, A.E. (1992) Genetic parameters for yield traits of Holsteins, including lactose and somatic cell score. *Journal of Dairy Science* 75, 1342–1348.

Effect of Draught Work on the Metabolism and Reproduction of Dairy Cows

8

E. Zerbini[1], Takele Gemeda[2], Alemu Gebre Wold[2]
and Azage Tegegne[1]

[1]*International Livestock Research Institute (ILRI), PO Box 5689, Addis Ababa, Ethiopia;* [2]*Institute of Agricultural Research (IAR), PO Box 2003, Addis Ababa, Ethiopia*

Introduction

Due to increasing population and livestock pressure on the land, farmers in developing countries may not be able to continue maintaining draught oxen specifically for work purposes. The use of dairy cows for traction could benefit total farm output and incomes through increased milk production, while alleviating the need to feed draught oxen year-round and to maintain a follower herd to supply replacement oxen (Gryseels and Goe, 1984; Gryseels and Anderson, 1985; Matthewman, 1987; Barton, 1991). Besides contributing to a better utilization of scarce feed resources, the use of dairy cows for draught would allow males to be fattened and sold younger, and could also lead to greater security of replacements. More productive animals on farm could result in a reduction of stocking rates and overgrazing, thus contributing to the establishment of a more productive and sustainable farming system.

Even though cows are used for draught in many parts of the developing world (Matthewman, 1987), limited research has been conducted on the specific effects of work on reproduction. Also, the available information is not consistent due to differences in animal genotype, environment, amount and duration of work and feeding regimes. Reproductive performance of cows is an important factor that determines whether cows are adopted for draught power. Working cows could perform at higher levels of efficiency than oxen, but only if nutrient inputs are adequate to meet their greater requirements and milk production and reproduction are kept at levels comparable to non-working cows (Mathers *et al.*, 1985; Matthewman *et al.*, 1994). Energy deficits during the working season could result in body weight and body condition losses, thus affecting the production and reproduction efficiency of cows (Teleni and Hogan, 1989; Teleni *et al.*, 1989; Zerbini *et al.*, 1993a). As well, the specific effect of work on energy metabolites is an important factor in evaluating the overall productive efficiency of cows used for draught work.

ILRI and the Ethiopian Institute of Agricultural Research (IAR) have researched different aspects of the use of dairy cows for draught work. This chapter deals with the current knowledge on the metabolism and reproduction of dairy cows used for draught purposes. Information generated from ILRI and IAR research are used to elucidate the interplay of factors affecting work output, metabolism, physiological responses and the reproductive physiology and performance of dairy cows used for draught. Data from research on other species have also been included where information is lacking for cattle.

Work Output and Efficiency of Draught Dairy Cows

Determination of the optimum workload that dairy cows should efficiently undertake is an essential component for successful adoption of cow traction technologies into smallholder mixed farming systems.

Results of our investigation with crossbred cows in the Ethiopian Highlands (Zerbini *et al.*, 1992) show that dairy cows were able to work at a rate of about 500 W. This work rate, at a speed of $0.75 \, m \, s^{-1}$, implies that the sustainable horizontal draught force was roughly 670 N. This represented about 14% of mean body weight, in line with what would be expected (e.g. Barwell and Ayre, 1982 and CEEMAT/FAO, 1972). The work efficiency of cows increased from about 7% to 26% as the workload increased to its maximum (Fig. 8.1). Cardiorespiratory measurements indicated that during work each additional heart beat transported approximately 72 ml of oxygen which is in turn equivalent to $1 \, kJ \, s^{-1}$ of additional mechanical work output (Zerbini *et al.*, 1992).

Results from our investigation (Gemeda *et al.*, 1995) show that over a period of three years, work output of dairy cows averaged more than 200 MJ per cow of net energy which was equivalent or above that required by farmers for land cultivation.

Energy Metabolism in Working Dairy Cows

In working cows, glucose is not only an important energy source for muscles but also a major substrate for lactose synthesis and an essential energy source for reproductive activity (Young, 1977; Kamiya and Daigo, 1988; Butler and Smith, 1989; Gaines, 1989). Fatty acids are precursors of milk fat, but also influence neuroendocrine activities regulating the synthesis and release of reproductive hormones (Shillo, 1992).

Plasma concentrations of non-esterified fatty acids (NEFAs) and β-hydroxybutyrate have been found to increase, and concentrations of glucose, magnesium and inorganic phosphorus decrease in exercised cows (Matthewman,

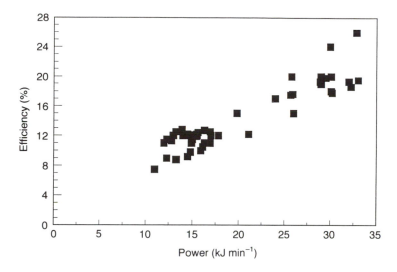

Fig. 8.1. Work efficiency of F₁ crossbred cows used for draught. Adapted from Zerbini *et al.* (1992).

1990). The response of blood metabolites to exercise is influenced by diet and some adaptation to exercise has been observed over a three week working period. These changes are indicative of energy deficits during exercise and are comparable to changes in blood metabolite concentrations observed in fasting animals. Similarly, the effect of work on energy metabolites has been examined using 40 crossbred dairy cows (Zerbini *et al.*, 1995). In working cows, plasma glucose was on average 16% lower during working hours than in non-working cows, apparently indicating a substantial drain of glucose from blood to muscle (Fig. 8.2). The partial increase in blood glucose concentrations during the resting and feeding hours could be due to a lower demand for glucose in the muscle and/or a greater supply of glucose to the bloodstream originating from gluconeogenesis in the liver (Young, 1977).

During working hours a 150% increase was observed in plasma NEFA concentration of working cows and plasma NEFA and glucose were negatively related (Fig. 8.2). A rapid decrease of plasma NEFA observed in working cows during the resting and feeding period corresponded to an increase in plasma glucose concentrations (Zerbini *et al.*, 1995). Similarly, Nachtomi *et al.* (1991) showed that increased energy intake lowered plasma NEFA and β-hydroxybutyrate concentrations and raised glucose level in dairy cows.

Gemeda *et al.* (1995) reported a greater loss of body weight in working non-supplemented compared to working supplemented cows. This suggests that in non-supplemented cows during working hours the decrease in blood glucose concentrations stimulated NEFA release from adipose tissue and their

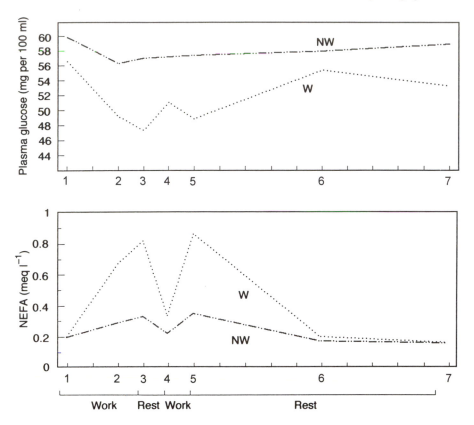

Fig. 8.2. Plasma glucose and NEFA concentrations of F$_1$ crossbred cows. Working cows on working days (W) vs non-working cows (NW). Adapted from Zerbini *et al.* (1995).

utilization by muscle. Ketone bodies could also have served as oxidizable substrates in skeletal muscle and the lactating mammary gland (Heitmann *et al.*, 1987).

Greater utilization of glucose during working hours has been related to a greater concentration of lactate in blood (Zerbini *et al.*, 1995). In a study with working oxen, Updhyay and Madan (1985) reported that muscle lactate concentrations increased during and after work, and the increase in lactate during working hours was likely to be due to increasing muscle tissue hypoxia. This could have happened in spite of adjustments in cardiac output and in metabolite concentrations occurring during work (Weber *et al.*, 1987; Zerbini *et al.*, 1992).

When animals are working, there is a compulsory diversion of energy yielding substrates to contracting muscles. During these periods free fatty acids are mobilized from fat depots even when the animals are reasonably well fed (Teleni and Hogan, 1989). Zerbini *et al.* (1995) and Matthewman (1990) found that plasma glucose concentration was affected by both energy intake and work output, while responses of NEFA, β-hydroxybutyrate and lactate were more related to energy expenditure during working hours. The NEFA response to work suggests that NEFAs are the principal fuel in the muscle tissue of working dairy cows, since muscle uptake of free fatty acids is directly proportional to the plasma concentration of this substrate (Norris *et al.*, 1978). However, depletion of muscle glycogen stores has been correlated with muscle fatigue and a rapid and sustained increase in the uptake of glucose by the exercising muscle, while the uptake of free fatty acids tends to increase throughout the exercise period (Pethick *et al.*, 1991). Utilization of glucose by skeletal muscle has been shown to be insensitive to changes in physiological state (pregnancy and lactation) in sheep, suggesting that there is an obligatory requirement by the tissue for glucose (Pethick *et al.*, 1991). However, the release of catecholamines and the reduction of insulin in circulating blood during work periods stimulates hydrolysis of triglycerides to free fatty acids from fat depots and their utilization in muscles. These events indicate the dominant role of free fatty acids in energy supply to contracting muscles. Teleni and Hogan (1989) reported that when cattle and buffaloes are worked, there is an increase in glucose entry rate of approximately 84% and an increase in free fatty acids entry rate of approximately 150%. In pregnant and/or lactating working cows or buffaloes, the increased entry rate of glucose could be accounted for by the gravid uterus or the mammary gland uptake. The role of working muscles in the utilization of glucose in late pregnant or early lactating cows is not yet clearly understood.

Reports by Zerbini *et al.* (1995) indicated that the average glucose concentrations of non-supplemented, supplemented, non-working and working cows were 50.0, 55.5, 54.9 and 46.0 mg dl^{-1}, respectively. The lower plasma glucose concentrations in non-supplemented cows whether working or not, and of working cows during working hours (in relation to a greater depletion of body reserves), could have been critical in determining the minimum blood glucose concentration required to trigger ovarian activity and related reproductive functions. These results confirm those of an earlier study (Zerbini *et al.*, 1993a) indicating a decrease of the probability of conception by a factor of 4 in non-supplemented compared with supplemented cows and by a factor of 2 in working compared with non-working cows. Lower plasma glucose concentrations in working cows only during working days compared with a continuously lower plasma concentration in non-supplemented cows suggest a more transient effect of work than feeding.

Energy Metabolism, Reproductive Physiology and Work Stress

Endocrine responses to stressful stimuli represent one of the most important homeostatic adaptations to environmental changes (Petraglia *et al.*, 1990). The correlation between endocrine response to stress and reproductive function is of great interest because exposure to chronic stress may affect reproductive functions. Several reports (Jagger *et al.*, 1987; Kirkwood *et al.*, 1987; Kawate *et al.*, 1991) have shown an influence of stress-induced responses on physiological changes in the reproductive system in different species. The length and intensity of exposure to the stressors may influence these changes.

Periods of energy restriction affect reproductive performance at the hypothalamic or pituitary level, either by inhibiting gonadotrophin-releasing hormone (GnRH) secretion and/or reducing pituitary sensitivity to GnRH. Energy restriction had no effect on the magnitude of luteinizing hormone (LH) response to GnRH but did delay response time. Glucose infusion did not alter concentrations of plasma LH, the response to exogenous GnRH, or the number of cows exhibiting ovarian activity by day 53 post-partum (McCaughey *et al.*, 1988). Glycogen increases progressively in uterine arteries during pregnancy and then decreases gradually after parturition; little glycogen has been found in post-partum cycling cows. These results suggest that storage of glycogen in the uterine artery may serve as an important source of energy for the recovery of the uterus and its arteries to their non-pregnant states (Kamiya and Daigo, 1988). Kirkwood *et al.* (1987) indicated that glucose status affects the rate of reproductive development and in extreme situations of hypoglycaemia, ovarian activity will be inhibited in ruminants apparently by inhibition of GnRH secretion. However, other pathways also must be involved because bypassing the 'GnRH pulse generator' was not an effective treatment for all cows. Conceivably the primary cause of anoestrus is different for different stages of anoestrus and the mediating mechanisms for anoestrus are at least partially involved with blood glucose and the endogenous opioid peptide system (Petraglia *et al.*, 1986).

The reduced LH pulse frequency in post-partum cows in low body condition is a function of reduced frequency of GnRH pulses from the hypothalamus rather than impaired pituitary function. Body condition appears to have a direct effect on the hypothalamus, although the possibility of body condition also altering the sensitivity of the hypothalamus to oestradiol cannot be excluded. Peters and Perera (1989) have shown that ovarian follicles are responsive to LH pulses by releasing oestradiol pulses early in the post-partum period. The sensitivity of the pituitary to the negative feedback effect of oestradiol could decline with time post-partum (Wright *et al.*, 1990).

Cows that exhibited preovulatory LH surges in response to GnRH treatment had significantly higher plasma oestradiol-17β and lower FSH (follicle stimulating hormone) concentrations before treatment (Jagger *et al.*, 1987) suggesting that LH response to GnRH treatment is dependent on follicular status in the immediate pretreatment period. Kawate *et al.* (1991) showed that the number of GnRH receptors in the anterior pituitary decreased from the late-mid to the early luteal phase in cows. Oestrogen exerts a highly selective effect on the gonadotrophin secretory process, and successive GnRH stimuli will result in an increase in the maximal rate and mass of secretion of biologically active LH (Urban *et al.*, 1991).

The mechanisms whereby poor nutrition decreases plasma gonadotrophin is not known. However, it has been shown that the oestrogen negative feed back is more sensitive in underfed animals. This increased sensitivity will in turn limit gonadotrophic stimulation of the ovary and hence delay or prevent ovulation. For example, a correlation has been observed between plasma gonadotrophin levels and percentage ideal body weight in women (Kirkwood *et al.*, 1987; Loucks *et al.*, 1992).

A further mechanism indicated by Kirkwood *et al.* (1987), whereby nutrition may modulate the effects of oestrogen, is through changes in steroid metabolic clearance rate. The effectiveness of the oestrogen feedback signal normally resulting in ovulation will be limited if poor nutrition results in the maintenance of a high oestradiol metabolic clearance rate. Exercise seems to have the opposite effect as indicated by Keizer *et al.* (1981) who reported a sharp decrease of the metabolic clearance rate of oestradiol during bicycle ergometer work.

Reproductive dysfunction is common in low body weight women runners and others who exercise strenuously. Amenorrhoeic runners had a lower percentage body fat than normally menstruating runners who were, in turn, leaner than non-exercising women (Kirkwood *et al.*, 1987). There are marked differences in menstrual patterns, total body weight, weight loss over time, percentage body fat, exercise patterns, and some serum hormone levels among female long distance runners, joggers and controls, all of which may influence hypothalamic function and be expressed as alterations in menstrual patterns (Dale *et al.*, 1979).

It has been suggested that adipose tissue may be a significant extragonadal source of plasma oestrogens. However, whether fatness influences plasma oestrogen levels, or whether plasma oestrogen levels influence fatness, remains to be determined. Plasma steroids can influence voluntary feed intake which may eventually result in changes in body composition.

Exercise-induced hyperprolactinaemia may also interfere with normal hypothalamic-pituitary-ovarian function and result in decreased ovarian oestrogen production (Boyden *et al.*, 1983; Nelson *et al.*, 1986; Lloyd *et al.*, 1987; Yahiro *et al.*, 1987). Both exercise and the accompanying changes in body composition may influence normal reproductive function in women athletes as well as in other female species. Results from Nnakwe (1990) support the role of body composition-related factors in the aetiology of exercise-induced amenorrhoea. Petraglia *et al.* (1986) proposed that an 'energy drain' by increased energy output during exercise and a decrease in body fatness, augments prolactin release and increases central nervous system β-endorphin activity thereby affecting GnRH.

Reproduction Performance of Dairy Cows Used for Draught

Resumption of post-partum oestrus and conception

When breeding females are used for draught purposes, the impact of work stress on fertility should be ascertained and incidence and type of anoestrus should be determined (Jainudeen, 1985). Bamualim *et al.*, (1987) indicated that in buffalo cows, work might reduce reproductive performance. On the other hand, the study by Winugroho and Situmorang (1989) suggested that work, *per se*, was not a major factor influencing ovarian activity if energy reserves were adequate. Feed supplementation of thin working buffaloes induced a return to normal ovarian activity. Reh and Host (1985) reported that fertility was 6–7% lower in working than in non-working cows. However, research conducted in India, with working and non-working Red-Sindhi cows, over two lactations, showed no significant differences in milk production and length of lactation (Reh and Host, 1985). Agyemang *et al.* (1991) reported that the reproductive and productive performance of draught and non-draught cows were similar. However, the work done was lower than the amount required by a small-holder farmer.

Zerbini *et al.* (1993a) reported that diet supplementation significantly decreased days to first oestrus and days to conception in non-working and working cows. When work treatment was superimposed on non-supplemented treatment, the effect on reproduction was deleterious. Differences in the first 200 days post-partum, in onset of oestrus and conception between treatment groups, seem to be related to work in the first post-partum period but related to a greater extent to diet supplementation if a 365-day period was considered, suggesting a longer term effect of the supplementation than work (Fig. 8.3). In supplemented cows, work significantly delayed days to conception. However, by 365 days post-partum, conception rate was similar for supplemented non-working and supplemented working cows. For occurrence of first oestrus, the diet effect was considerably larger than the work effect (a probability factor of 5 versus a probability factor of 2). This was less pronounced for conception. This is contrary to results from other studies (Wells *et al.*, 1981; Spicer *et al.*, 1990) which indicated that supplementary feeding did not influence interval from calving to first ovulation, conception rate or interval from calving to conception. Body condition at calving significantly affected post-partum reproductive ability of non-working and working cows (Fig. 8.4). This indicates that cows with greater body reserves at calving, and the ability to use these reserves during the post-partum period, can partly overcome the negative effect of dietary energy restrictions on oestrus onset and conception (Zerbini *et al.*, 1993a).

The results of this study indicate a decrease in percentage of cows

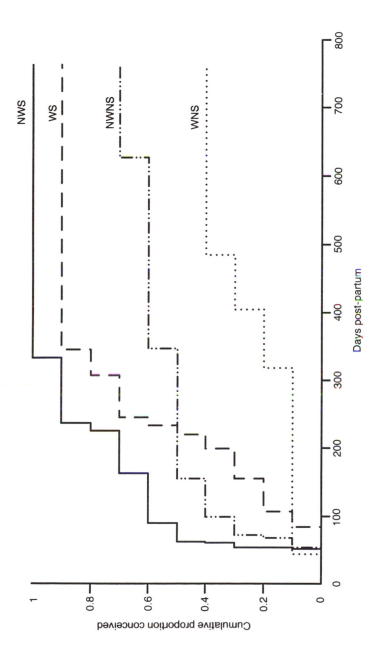

Fig. 8.3. Distribution function of interval to conception after calving of F_1 crossbred cows. NWNS = Non-working non-supplemented; NWS = Non-working supplemented; WNS = Working non-supplemented; WS = Working supplemented. Adapted from Zerbini *et al.* (1993a).

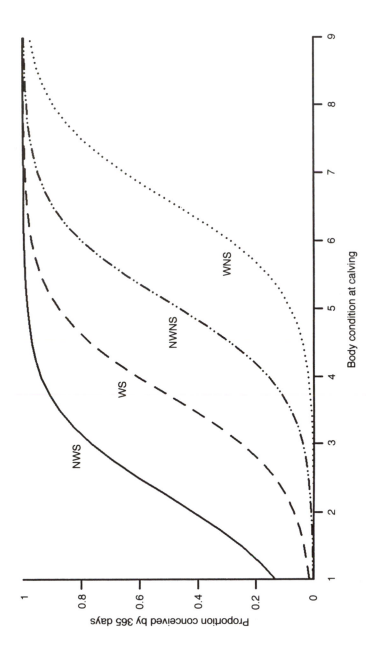

Fig. 8.4. Predicted conception rate at 365 days vs initial body condition of F_1 crossbred cows. NWNS = Non-working non-supplemented; NWS = Non-working supplemented; WNS = Working non-supplemented; WS = Working supplemented. Adapted from Zerbini *et al.* (1993a).

showing oestrus and in conception rate when work was applied to non-supplemented cows. Similar data on the onset of post-partum oestrus and conception rates of non-supplemented non-working and supplemented working cows suggests that diet supplementation seems to overcome the negative effect of work on post-partum reproductive performance.

Post-partum anoestrous interval was extended in a larger proportion of working than in non-working cows. Work did not influence conception rate in supplemented cows, but had a substantial influence in non-supplemented cows. The significant delay of conception for supplemented working cows compared with supplemented non-working cows indicated that work output of cows might be associated with longer calving intervals and the economic trade-offs between the two factors should be examined in detail. Once pregnancy was established there was no effect of work on maintenance of pregnancy. A greater proportion of supplemented working cows cycled between 120 days and 1-year post-partum indicating that work applied soon after calving delayed, but did not suppress oestrus and conception in subsequent resting or working periods (Zerbini *et al.*, 1993a,b).

Ovarian function

The effects of work and diet supplementation on progesterone secretion and the incidence of short luteal phases and ovulations without oestrus has been investigated in F_1 crossbred dairy cows (Zerbini *et al.*, 1993b). Based on plasma progesterone concentrations, ovulation started 62 days earlier than onset of behavioural oestrus. Of the total ovulations, 36% were not associated with behavioural signs of oestrus and occurred in 33% of cows. The incidence of ovulations without oestrus was higher in working (42%) than in non-working (30%) cows and in non-supplemented (42%) than in supplemented (33%) cows (Figs 8.5 and 8.6). Short luteal phases occurred in 33% of the cows before the establishment of normal oestrous cycles. More supplemented-working cows had ovulations without oestrus before normal oestrous cycles were established. The incidence of short luteal phases or ovulations without oestrus did not influence pregnancy in subsequent normal oestrus periods.

All the ovulations without oestrus were accompanied by formation of functional corpora lutea. The lower total progesterone output and peak progesterone concentrations in ovulations without oestrus than in ovulations accompanied by manifestation of behavioural oestrus shown by Zerbini *et al.* (1993b) agrees with earlier reports (Schams *et al.*, 1978; Eger *et al.*, 1988; Tegegne *et al.*, 1993). This may show inadequate corpora lutea formation or inefficient function to induce the manifestation of behavioural oestrus. Such a phenomenon is usually associated with stressful conditions such as nutritional inadequacy, workload and high intrauterine temperatures which could result in hormonal imbalances and affect reproductive functions. Such corpora lutea senesce prematurely and

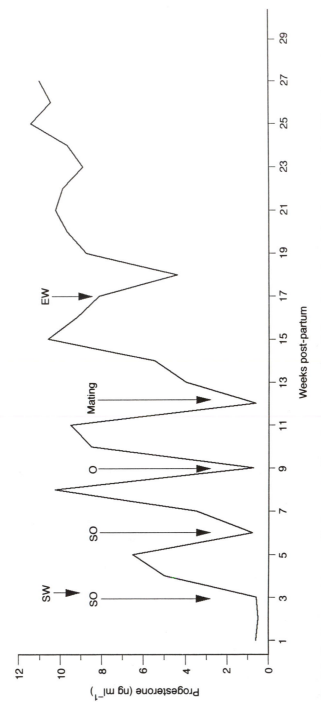

Fig. 8.5. Plasma progesterone profile in a cow with early ovarian cyclic activity and conceived during the work period. SW = start work; EW = end work; O = oestrus; SO = ovulation without oestrus. Adapted from Zerbini *et al.* (1993b).

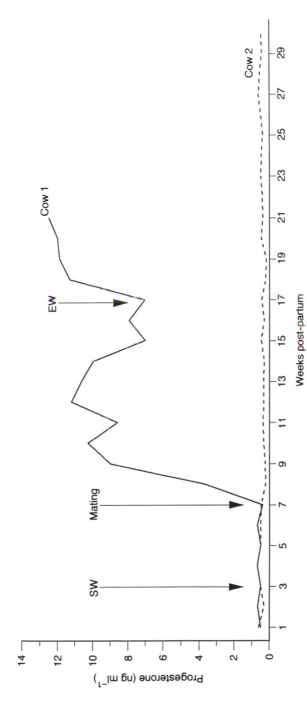

Fig. 8.6. Plasma progesterone profiles in cows with early ovarian cyclic activity followed by conception (cow 1) and by an extended post-partum anoestrus (cow 2). SW = start work; EW = end work. Adapted from Zerbini *et al.* (1993b).

the mechanisms involved could be due to deficiencies in the maturational process within the preovulatory follicle and/or inadequacies of the ovulatory stimulus, shortcomings in the support of the corpora lutea once they have formed, or a premature activation of the luteolytic process such as an early release of prostaglandin $F2\alpha$ from the uterus (Cooper *et al.*, 1991; Lishman and Inskeep, 1991). In the present study, however, fertility was not impaired in cows that experienced short luteal phases and ovulations without oestrus prior to the establishment of normal oestrous cycles. This is in agreement with other studies (Mukasa-Mugerwa *et al.*, 1991; Tegegne *et al.*, 1993) and indicates that these phenomena may not be indicative of reproductive abnormality.

The failure of cows to express the overt signs of oestrus and subsequently accept the bull while having ovulations without oestrus could be a mechanism of adaptation to postpone possible pregnancies until cows are physically and physiologically ready to conceive and maintain the pregnancy to full term. Therefore, specific intervention options on the amount of work and/or feeding regimes sufficient to induce behavioural oestrus during the early post-partum period need to be established.

In Zerbini *et al.* (1993a), five supplemented working cows were mated more than once and their oestrous cycles were normal, probably suggesting early embryonic mortality. Although a number of factors have been implicated, the causes for embryonic losses in working cows during the first three weeks of pregnancy is not clear (Jainudeen, 1985). The extent to which heat generated during work affects the chain of reproductive events and the physiological mechanisms involved are also not yet properly understood. Work load increases body temperature and the heart and respiration rates (Zerbini *et al.*, 1992) and could contribute to embryonic losses. Elevated intrauterine temperature could affect the availability of water, electrolytes, nutrients and hormones, which could then influence ovarian function and fertility (Thatcher *et al.*, 1983). An increase of 0.5°C above the mean uterine temperature of 38.6°C on the day of insemination and 38.4°C on the day after insemination has been found to decrease conception rates by 12.8 and 6.9%, respectively (Gwazdauskas *et al.*, 1973). One possible cause for the reduced conception rate could be heat damage to the spermatozoa that fertilize the ovum. This could lead to the production of abnormal embryos which subsequently die (Burfening and Ulburg, 1968).

Extended post-partum anoestrus, high incidence of ovulations without oestrus and short luteal phases in the working cows could be partly due to work stress. Work *per se* could affect the functions of the hypothalamic–pituitary–ovarian axis. The regulation of the functions of this axis consists of the interplay of several neurotransmitter, hypothalamic, pituitary and gonadal hormones acting through stimulatory and inhibitory pathways and positive and negative feedbacks. Work stress can modify the metabolism of these transmitters and hormones (Collu *et al.*, 1984), reduce the duration and intensity of oestrus and under severe conditions will induce anoestrus in the cow (Bond and McDowell, 1972). In athletic women, for instance, a causal relationship between athletic activity and an increased incidence of menstrual dysfunction, as well as a shortening of the luteal phase of the menstrual cycle, have been observed (Collu *et al.*, 1984). Other factors, such as glucocorticoids have been shown to greatly increase during work stress and are capable of exerting a direct inhibitory effect on gonadal hormone production (Bambino and Hsueh, 1981). However, data on the relationship of the neurotransmitter to work-induced modification of the hypothalamic–pituitary–gonadal function and other physiological mechanisms involved in regulation of reproductive functions of post-partum cows used for draught work are not available.

In another experiment conducted on crossbred dairy cows in Ethiopia (unpublished), work was applied when cows were already cycling regularly, to estimate possible effects of work *per se* when nutrition was not a limiting factor. In this study cows below medium body condition score (<5; range 1–9) were not able to show oestrus, but after attaining medium body condition (=5) they started cycling and continued to cycle, indicating that a certain body energy reserve is critical to trigger ovarian cyclic activity. Data on milk progesterone concentrations indicated that once ovarian activity was initiated, cows continued to cycle regularly even during a working period, indicating that work *per se* had no effect on post-partum ovarian activity when energy intake was not a limiting factor. However, oestrus cycle length was significantly longer during working compared with pre-work periods, indicating the occurrence of silent ovulations even when nutrient supply in the diet was above requirements (Fig. 8.7). Similarly, observations on the effect of work on ovarian activity of swamp buffalo cows indicated that although there was evidence of ovarian activity in some of the working buffalo cows, there was a tendency for ovarian activity to decrease in this group compared with the non-working animals (Bamualim *et al.*, 1987).

Nutritional constraints combined with draught work activities could cause losses in body condition and lead to post-partum anoestrus (Jainudeen, 1985). Teleni and Hogan (1989) indicated that cows may stop cycling if they lose approximately 17% of their calving body weight. In mature swamp buffaloes, work reduced ovarian activity from 75 to 31% (Bamualim *et al.*, 1987). It is possible that the depletion of body reserve nutrients to certain critical levels signals metabolic controls to switch off processes such as ovarian function that are not necessary for the immediate survival of the animal. Winugroho and Situmorang (1989) reported that work *per se* had little effect on the ovarian activity of swamp buffaloes. Nineteen percent of total cycles in working group animals showed abnormal patterns compared with 12% in the control group animals. It is likely that this was due to poor body condition or low feed energy intake or both. Body weight loss due to work was reported to be greater in heavier working buffalo in better body condition. This implies that thinner animals require less energy than fatter animals to complete the same amount of work because less energy is spent moving their body mass. Therefore, a clear definition of the body weight/condition at the start of the work season and the rate of weight loss compatible with normal ovarian activity is desirable.

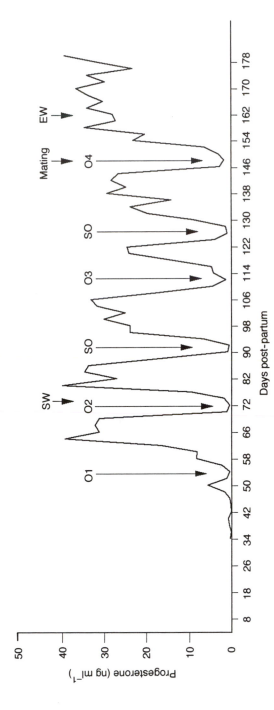

Fig. 8.7. Pattern of milk progesterone profile in a cow with silent oestrus during the working period. O1, O2, O3, O4 = oestrus; SO = ovulation without oestrus; SW = start work; EW = end work. Adapted from Abate (1994).

Table 8.1. Cumulative work output, body weight change (kg), milk yield, completed lactations and conceptions of F_1 crossbred cows over a period of two years.

Treatment	n	Work output (kg)	Body weight change (kg)	Milk yield (kg)	Completed lactations	Conceptions
NWNS	10	-	−76	1226	1.3	0.8
NWS	10	-	−9	3186	1.6	1.7
WNS	10	499.4	−86	927	1.1	0.4
WS	10	508.2	−23	3044	1.1	1.2
s.e.	10	6.3	9.7	219	-	-
F test						
Work		-	NS	NS		
Supplement		NS	***	***		

NWNS = Non-working non-supplemented.
NWS = Non-working supplemented.
WNS = Working non-supplemented.
WS = Working supplemented.
Source: Gemeda *et al.* (1995)

Conceptions over multiple lactations

Gemeda *et al.* (1995) found that the number of working supplemented cows that conceived in year one was similar to that of non-working supplemented cows. However, over a period of two and three years, the number of working supplemented cows that conceived were 29 and 20% lower than those of non-working supplemented cows. Over a period of three years, number of calves born followed the same pattern of number of conceptions. The results indicate that if work and inadequate nutrition exist for as long as one year, milk production and number of conceptions could be reduced by 54 and 78%, respectively, compared with those of adequately fed cows (Table 8.1).

Body weight losses have been reported to impair ovarian activity in female buffaloes and cows (Teleni *et al.*, 1989; Agyemang *et al.*, 1991). As well, over a period of two years, supplementary feeding reduced body weight loss of cows from 0.11 to 0.02 kg day^{-1} and was associated with a 59 and 63% increase in the number of conceptions and parturitions, respectively, compared with a non-supplemented diet. In particular, supplementation of working cows reduced body weight loss from 0.12 to 0.03 kg day^{-1} and doubled the number of conceptions and parturitions compared with working non-supplemented cows (Gemeda *et al.*, 1995). Body condition score followed a similar pattern to that of body weight change over the three-year period. These results indicate that the probability of conception is not greater than 20% in cows with body condition score lower than 3 (range 1–9) and with body weight losses of about 15% from calving body weight (average of 412 kg).

In year one the number of calves born from working supplemented cows was 80% lower (1 out of 10 compared with 5 out of 10) than that of non-working supplemented cows, despite the fact that the number of working supplemented cows which conceived was similar to that of non-working supplemented cows. This is consistent with the delay in conception after parturition reported for working-supplemented compared with non-working supplemented cows by Zerbini *et al.* (1993a). Relatively fewer lactations and parturitions, and greater days in milk of working supplemented cows, over a period of three years, reflects the delayed conception in working supplemented compared with non-working supplemented cows. This is due to both a direct effect of work *per se* and to a deficit of energy yielding substrates, particularly during the working/ lactating periods. Over a period of three years, diet was the main factor that affected reproduction of dairy cows used for draught work (Zerbini *et al.*, 1995).

Reports of Momongan (1985) and Jabbar (1983) indicated that age at first calving, service period and calving interval are much greater in draught cows than in non-draught cows. The delay in service period and calving interval was attributed to longer post-partum anoestrus in draught breeds, to deliberately delayed mating in draught breed cattle and higher incidence of embryonic mortality in draught cows than in dairy cows.

Recovery after work: long-term effects

After a two-year period, non-supplemented working and non-working cows had lost on average 20% of their initial body weight, were dry and not cycling. These cows were supplemented to estimate the time required to re-establish regular reproductive activity. The major problem was resumption of ovarian activity in low body weight, low body condition cows. Preliminary results indicate that cows could resume reproductive activity and conceive again in less than three months of supplementation. The average body weight gain necessary to resume oestrus in these cows was 44 kg. These results have important implications for on-farm feed budgeting and use for optimal production output of multipurpose cows in situations of fluctuating feed quality and availability.

Economic Implications

The potential of the use of crossbred cows for milk production and traction was substantiated by simulating the production parameters and investment returns over a three-year period using the ILRI bio-economic herd model (Shapiro *et al.*, 1994). The Incremental Internal Rate of Return (IIRR) of supplemented working cows over supplemented non-working cows under on-station conditions is greater than 70%. The IIRR is very high because the incremental investment cost is very low while the benefits of work are large.

The effect over time of introducing crossbred dairy cows into a typical farm herd of local cattle for work and milk production were also simulated and compared with using the local cows for milk production and local oxen for traction. Then, the financial implications were investigated using incremental benefit/cost analysis. The incremental benefit/cost ratio of having supplemented working cows over the traditional system of local cows and oxen is about 3.5 and the IIRR is 78%. The incremental benefit/cost ratio is high because of the very high productivity of the crossbred cows (5–6 times milk yield) relative to local cows.

The value of work more than compensated for the small reduction in milk production and longer calving interval found in working cows when supplementation took place to ensure adequate nutrition. The greater returns to investment in supplemented working crossbred cows was thus mainly a result of the higher value of the work output, in spite of the higher feed costs and lower off-take (milk, calves).

Conclusions

With appropriate feeding regimes dairy cows could be used for draught purposes without any detrimental effects on fertility, but calving intervals would be extended. Work *per se* does not influence post-partum ovarian activity when the energy reserve is adequate, but work does delay the interval from calving to conception in dairy cows.

Work increased the incidence of ovulations without oestrus and short luteal phases. However, these events did not influence pregnancy in subsequent normal oestrus periods.

Economic analysis of ILRI data shows that the greater returns to investment in supplemented working dairy cows compared with non-working cows or with a traditional system, was mainly a result of the higher value of the work output, in spite of the higher feed costs and relatively lower off-take of milk and calves.

There is a need to quantify the energy partition to different functions by working, lactating and breeding cows. The nutrient demand of the multipurpose cow is complex and the success of a nutritional management strategy will depend largely on the level of feed intake and, over short periods, on the level of body reserves (Preston and Leng, 1987; Egan and Dixon, 1993). The mechanism by which body reserves contribute to the energy expenditure of working cows is not clear. Future research priorities should include defining minimal nutrient requirements for pregnant and/or lactating working animals to allow for an optimal reproductive performance.

The physiological mechanisms involved in anoestrus and ovulations without oestrus, and the significance of such phenomena in affecting post-partum reproductive performance and fertility in working cows require further detailed investigations. Detailed physiological studies on post-partum reproductive functions in relation to hormonal changes, nutrition, work and their interaction, are warranted. In addition, the influence of work *per se* on the neuroendocrine regulation of anoestrus, ovulations without oestrus, fertilization, implantation, embryonic survival and development need further studies. As well, length of the oestrus cycle, duration and intensity of oestrus and time of ovulation need to be established and the incidence and type of anoestrus in the draught animal should be determined.

To optimize the post-partum anoestrus period, draught dairy cows must regain weight during lactation and farmers must have management skills to integrate strategically physiological events such as pregnancy and lactation with draught work requirements.

References

Abate, E. (1994) Effect of work and energy supplementation on the oestrus cycle length and ovarian activity in crossbred cows used for draught. MSc thesis. Alemaya University of Agriculture, PO Box 138, Dire Dawa, Ethiopia.

Agyemang, K., Abiye Astatke, Anderson, F.M. and Worldeab W. Mariam (1991) A study on the effects of work on reproductive and productive performance of crossbred dairy cows in the Ethiopian highlands. *Tropical Animal Health and Production* 23, 241–248.

Bambino, T.H. and Hsueh, A.J.W. (1981) Direct inhibitory effect of glucocorticoids upon testicular luteinizing hormone receptor and steroidogenesis '*in vivo*' and '*in vitro*'. *Endocrinology* 108, 2142–2150.

Bamualim, A., Ffoulkes, D. and Fletcher, I.C. (1987) Preliminary observations on the effect of work on intake, digestibility, growth and ovarian activity of swamp buffalo cows. Draught Animal Power Project, DAP Project Bulletin, No. 3, pp. 6–10. James Cook University, Townsville, Australia.

Barton, D. (1991) The use of cows for draught in Bangladesh. Australian Centre for International Agricultural Research. Draught Animal Bulletin, No. 1, pp. 14–26. James Cook University of North Queensland, Australia.

Barwell, I. and Ayre, M. (1982) *The Harnessing of Draught Animals*. Intermediate Technology Publications, Ardington, Oxen, UK.

Bond, J. and McDowell, R.E. (1972) Reproductive performance and physiological responses of beef females as affected by prolonged high environmental temperature. *Journal of Animal Science* 19, 374–382.

Boyden, T.W., Pamenter, R.W., Stanforth, P., Rotkis, T. and Wilmore, J.H. (1983) Sex steroids and endurance running in women. *Fertility and Sterility* 39, 629–632.

Burfening, P.J. and Ulberg, L.C. (1968) Embryonic survival subsequent to culture of rabbit spermatozoa at 38°C and 40°C. *Journal of Reproduction and Fertility* 15, 87–95.

Butler, W.R. and Smith, R.D. (1989) Interrelationships between energy balance and postpartum reproductive function in dairy cattle. *Journal of Dairy Science* 72, 767–783.

Collu, R., Gibb, W. and Ducharme, J.R. (1984) Effects of stress on the gonadal function. *Journal of Endocrinology Investigations* 7, 529–537.

Cooper, D.A., Carver, D.A., Villeneuve, P., Silvia, W.J. and Inskeep, E.K. (1991) Effects of progestagen treatment on concentrations of prostaglandins and oxytocin in plasma from the posterior vena cava of post-partum beef cows. *Journal of Reproduction and Fertility* 91, 411–421.

Dale, E., Gerlach, D.H. and Wilhite, A.L. (1979) Menstrual dysfunction in distance runners. *Obstetrics and Gynecology* 54, 47–53.

Egan, A.R. and Dixon, R.M. (1993) Feed resources and nutrition needs for draught power. In: Pryor, W.J. (ed.) *Draught Animal Power in the Asian-Australasian Region.* Australian Centre for International Agricultural Research. Proceedings No. 46, pp. 93. James Cook University, Townsville, Australia.

Eger, S., Shemesh, M., Schindler, H., Amir, S. and Foote, R.H. (1988) Characterization of short luteal cycles in the early postpartum period and their relation to reproductive performance of dairy cows. *Animal Reproduction Science* 16, 215–224.

Gaines, J. (1989) The relationship between nutrition and fertility in dairy herds. *Veterinary Medicine* October 1989, pp. 997–1002.

Gemeda, T., Zerbini, E., Wold, A.G. and Demissie, D. (1995) Effect of draught work on performance and metabolism of crossbred cows. 1. Effects of work and diet on body weight change, body condition, lactation and productivity. *Animal Science* 60, 361–367.

Gryseels, G. and Anderson, F. (1985) Use of crossbred dairy cows as draught animals: experiences from the Ethiopian highlands. In: *Research Methodology for Livestock on Farm Trials.* Proceedings of a workshop held March 25–28 at Aleppo, Syria, pp. 237–258. International Development Research Centre, Ottawa.

Gryseels, G. and Goe, M.R. (1984) Energy flows on smallholder farms in the Ethiopian highlands. International Livestock Centre for Africa, Addis Ababa, Ethiopia. *ILCA Bulletin* 17, 2–9.

Gwazdauskas, F.C., Thatcher, W.W. and Wilcox, C.J. (1973) Physiological, environmental and hormonal factors at insemination which may affect conception. *Journal of Dairy Science* 56, 873–881.

Heitmann, R.N., Dawes, D.J. and Sensing, S.C. (1987) Hepatic ketogenesis and peripheral ketone body utilization in the ruminant. *Journal of Nutrition* 117, 1174–1180.

Jabbar, M.A. (1983) Effect of draught use of cows on fertility, milk production and consumption. In: Davis, C.H., Preston, T.R., Dolberg, F., Haque, M. and Saadullah, M. (eds) *Maximum Livestock Production from Minimum Land.* Proceedings of the fourth seminar held in Bangladesh 2–4 May 1983, pp. 71–85. Bangladesh Agricultural University, Mymensingh. Department of Animal Science.

Jainudeen, M.R. (1985) Reproduction in draught animals: Does work affect female fertility? In: Copland, J.W. (ed.) *Draught Animal Power for Production.* Australian Centre for International Argricultural Research. Proceedings No. 10, pp. 130–133. James Cook University, Townsville, Queensland, Australia.

Jagger, J.P., Peters, A.R. and Lamming, G.E. (1987) Hormone response to low-dose GnRH treatment in postpartum beef cows. *Journal of Reproduction and Fertility* 80, 263–269.

Kamiya, S. and Daigo, M. (1988) Prepartum and postpartum glycogen accumulation in bovine uterine arteries. *Animal Reproduction Science* 16, 191–198.

Kawate, N., Inaba, T. and Mori, J. (1991) Gonadotropin-releasing hormone receptors in anterior pituitaries of cattle during the oestrous cycle. *Animal Reproduction Science* 24, 185–191.

Keizer, H.A., van Shaik, F.W., de Beer, E.L., Schiereck, P., van Heeswijk, G. and Poortman, J. (1981) Exercise-induced changes in estradiol metabolism and their possible physiological meaning. *Journal of Applied Physiology* 48, 765–769.

Kirkwood, R.N., Cumming, D.C. and Aherne, F.X. (1987) Nutrition and puberty in the female. *Proceedings of the Nutrition Society* 46, 177–192.

Lishman, A.W. and Inskeep, E.K. (1991) Deficiencies in luteal function during re-initiation of cyclic breeding activity in beef cows and in ewes. *South African Journal of Animal Science* 21, 59–76.

Lloyd, T., Buchanan, J.R., Bitzer, S. Waldman, J.C., Myers, C. and Ford, B.G. (1987) Interrelationship of diet, athletic activity, menstrual status, and bone density in collegiate women. *American Journal of Clinical Nutrition* 46, 681–684.

Loucks, A.B., Laughlin, G.A., Mortola, J.F., Girton, L., Nelson, J.C. and Yen, S.S.C. (1992) Hypothalamic-pituitary-thyroidal function in eumenorreheic and amenorrheic athletes. *Journal of Clinical Endocrinology and Metabolism* 75, 513–518.

McCaughey, W.P., Rutter, L.M. and Manns, J.G. (1988) Effect of glucose infusion on metabolic and reproductive function in postpartum beef cows. *Canadian Journal of Animal Science* 68, 1079–1087.

Mathers, J.C., Pearson, R.A., Sneddon, C.J., Matthewman, R.W. and Smith, A.J. (1985) The use of draught cows in agricultural systems with particular reference to their nutritional needs. In: Smith, A.J. (ed.) *Milk Production in Developing Countries*, pp. 476–496. University of Edinburgh Press, Edinburgh.

Matthewman, R.W. (1987) Role and potential of draught cows in tropical farming systems: a review. *Tropical Animal Health and Production* 19, 215–222.

Matthewman, R.W. (1990) Effect of sustained exercise on milk yield, milk composition and blood metabolite concentrations in Hereford × Friesian cattle. PhD thesis. Centre for Tropical Veterinary Medicine, University of Edinburgh, Roslin, Midlothian EH25 9RG.

Matthewman, R.W., Dijkman, J.T. and Zerbini, E. (1994) The management and husbandry of male and female draught animals: Research achievements and needs. In: Lawrence, P.R., Lawrence, K., Dijkman, J.T. and Starkey, P.H. (eds) *Research for Development of Animal Traction*. Proceedings of the West African Animal Traction Network, held in Kano, 9–13 July 1990, pp. 125–136. International Livestock Centre for Africa, Addis Ababa, Ethiopia.

Momongan, V.G. (1985) Reproduction in draught animals. In: Copland, J.W. (ed.) *Draught Animal Power for Production*. Australian Centre for International Agriculture Proceedings No. 10, pp. 123–129. James Cook University, Townsville, Queensland, Australia.

Mukasa-Mugerwa, E., Tegegne, A. and Ketema, H. (1991) Patterns of postpartum oestrus onset and associated plasma progesterone profiles in *Bos indicus* cows in Ethiopia. *Animal Reproduction Science* 24, 73–84.

Nachtomi, E., Halevi, A., Bruckental, I. and Amir, S. (1991) Energy-protein intake and its effect on blood metabolites of high-producing dairy cows. *Canadian Journal of Animal Science* 71, 401–407.

Nelson, E.M., Fisher, E.C., Catsos, P.D., Meredith, C.N., Tursksoy, R.N. and Evans, W.J. (1986) Diet and bone status in amenorrheic runners. *American Journal of Clinical Nutrition* 43, 910–916.

Nnakwe, N. (1990) Menstrual irregularities, nutritional patterns and mineral intake and excretion of female athletes. *Nutrition Research* 10, 23–30.

Norris, B., Shade, D.S. and Eaton, R.P. (1978) Effects of altered free fatty acids mobilization on the metabolic response to exercise. *Journal of Clinical Endocrinology and Metabolism* 46, 254–259.

Peters, A.R. and Perera, B.M.A.O. (1989) Pulsatile secretion of oestradiol-17B in postpartum dairy cows. *Animal Production* 49, 335–338.

Pethick, D.W., Miller, C.B. and Harman, N.G. (1991) Exercise in merino sheep – the relationships between work intensity, endurance, anaerobic threshold and glucose metabolism. *Australian Journal of Agricultural Research* 42, 599–620.

Petraglia, F., Vale, W. and Rivier, C. (1986) Opioids act centrally to modulate stress-induced decrease in luteinizing hormone in the rat. *Endocrinology* 119, 2445–2450.

Petraglia, F., Monzani, A., Fabbri, G., De Ramundo, Di Domenica, P., Saletti, C., Genazzani, A.D., Volpe, A., Bernasconi, S. and Genazzani, A.R. (1990) Reproductive function and endocrine responses to stress. In: Nappi, G., Genazzani, A.R., Mortignoni, E., and Petraglia, F. (eds) *Stress and The Ageing Brain*. Raven Press, Ltd., New York, pp. 51–61.

Preston, T.R. and Leng, R.A. (1987) *Matching Ruminant Production Systems with Available Resources in the Tropics and Subtropics*. Preamble Books, Armidale, Australia.

Reh, I. and Host, P. (1985) Beef production from draught cows in small scale farming. *Quarterly Journal of International Agriculture* 24, 38–47.

Shapiro, B., Zerbini, E. and Gemeda, T. (1994) The returns to investment in dual use of crossbred cows for milk production and draught work in the Ethiopian Highlands. Paper presented at the First Workshop of the Animal Traction Network for Ethiopia, held in Addis Ababa, 27–28 January, 1994.

Schams, D., Schallenberger, E., Menzer, C.H., Stangl, J., Zottmeier, K., Hottmann, B. and Karg, H. (1978) Profiles of LH, FSH and progesterone in postpartum dairy cows and their relationship to commencement of cyclic functions. *Theriogenology* 10, 453–467.

Spicer, L.J., Tucker, W.B. and Adams, G.D. (1990) Insulin-like growth factor-I in dairy cows: relationship among energy balance, body condition, ovarian activity, and oestrus behaviour. *Journal of Dairy Science* 73, 929–937.

Shillo, K.K. (1992) Effects of dietary energy on control of luteinizing hormone secretion in cattle and sheep. *Journal of Animal Science* 70, 1271–1282.

Teleni, E., Boniface, A.N., Sutherland, S. and Entwistle, K.W. (1989) The effect of depletion of body reserve nutrients on reproduction in *Bos indicus* cattle. Draught Animal Power Project Bulletin No. 8, pp. 1–10. James Cook University, Townsville, Australia.

Teleni, E. and Hogan, J.P. (1989) Nutrition of draught animals. In: Hoffmann, D., Nari, J. and Petheran, R.J. (eds) *Draught Animals in Rural Development*. Australian Centre for International Agricultural Research Proceedings No. 27. James Cook University, Townsville, Australia, pp. 118–133.

Tegegne, A., Geleto, A. and Kassa, T. (1993) Short luteal phases and ovulations without oestrus in primiparous Borana (*Bos indicus*) cows in the central highlands of Ethiopia. *Animal Reproduction Science* 31, 21–31.

Thatcher, W.W., Badinga, L., Collier, R.J., Head, H.H. and Wilcox, C.J. (1983) Thermal stress effects on the bovine conceptus: early and late pregnancy. In: *Reproduction des Ruminants en Zone Tropicale*, Pointe-a-Pitre (F.W.I.), 8–10 Juin,

1983. Institute National de Recherche Agronomique Publ., 1984 (Les Colloques de l'INRA, No. 20).

Updhyay, R.C. and Madan, M.L. (1985) Studies on blood acid-base status and muscle metabolism in working bullocks. *Animal Production* 40, 11–16.

Urban, J.R., Veldhuis, J.D. and Dufau, M.L. (1991) Oestrogen regulates the gonadotropin-releasing hormone-stimulated secretion of biologically active luteinizing hormone. *Journal of Clinical Endocrinology and Medicine* 72, 660–668.

Wells, P.L., Hopley, J.D.H. and Holness, D.H. (1981) Fertility in the Afrikaner cow. 1. The influence of concentrate supplementation during the post-partum period on ovarian activity and conception. *Zimbabwe Journal of Agricultural Research* 19, 13–21.

Weber, J.M., Parkhouse, W.S., Dobson, G.P., Harman, J.C., Snow, D.H. and Hochacha, P.W. (1987) Lactate kinetics in exercising thoroughbred horses: regulation of turnover rate in plasma. *American Journal of Physiology* 253, 896–903.

Winugroho, M. and Situmorang, P. (1989) Nutrient intake, workload and other factors affecting reproduction of draught animals. In: *Draught Animals in Rural Development*. Australian Centre for International Agricultural Research, Proceedings No. 27, pp. 186–189. James Cook University, Townsville, Australia.

Yahiro, J., Glass, A.R., Fears, W.B., Ferguson, E.W. and Vigersky, R.A. (1987) Exaggerated gonadotropin response to luteinizing hormone-releasing hormone in amenorrheic runners. *American Journal of Obstetrics and Gynecology* 156, 586–591.

Young, J.W. (1977) Gluconeogenesis in cattle. Significance and methodology. *Journal of Dairy Science* 60, 1–15.

Zerbini, E., Gemeda, T., Franceschini, R., Sherington, J. and Wold, A.G. (1993a) Reproductive performance of F1 crossbred dairy cows. Effect of work and diet supplementation. *Animal Production* 57, 361–369.

Zerbini, E., Gemeda, T., O'Neill, D.H., Howell, P.J. and Schroter, R.C. (1992) Relationships between cardio-respiratory parameters and draught work output in F_1 crossbred dairy cows under field conditions. *Animal Production* 55, 1–10.

Zerbini, E., Gemeda, T., Tegegne, A., Wold, A.G. and Franceschini, R. (1993b) Effects of work and diet on progesterone secretion, short luteal phases and ovulations without oestrus in postpartum F_1 crossbred dairy cows. *Theriogenology* 43, 571–584.

Zerbini, E., Gemeda, T., Wold, A.G., Nokoe, S. and Demissie, D. (1995) Effect of draught work on performance and metabolism of crossbred cows. 2. Effect of work on roughage intake, digestion, digesta kinetics and plasma metabolites. *Animal Science* 60, 369–378.

Health Control

Control of Mastitis 9

J.E. Hillerton
*Institute for Animal Health, Compton,
Berkshire RG20 7NN, UK*

Introduction

Milk from the large majority of mammary glands of cows is sterile and no culturable organisms, parasitic or commensal, are recoverable normally. However, as with most animal tissues, foreign materials are frequently introduced and these induce a defensive inflammatory response, a mastitis. The extent of this is extremely variable according to the nature of the stimulus and the responsiveness of the cow. Mild responses, either due to low stimulation, poor recognition of the irritant by the host defences or rapid effectiveness in removing the stimulus are undetectable without detailed investigation. These are subclinical mastitis. More severe responses, clinical mastitis, result in changes in the rheology of the secretion, usually clotting of milk, and possibly changes in its colour. The udder tissue also appears abnormal, becoming oedematous, reddened and painful to the touch. This clinical mastitis may lead to systemic effects, especially elevated temperature and in severe cases death from any of a number of possible physiological effects.

Short term mastitis may result from physiological disturbance of the gland following minor physical injury or invasion by inert materials. Mastitis is usually caused by bacterial infection although many other microorganisms including fungi, algae, mycoplasmas and viruses can cause disease. At least 137 separate possible causes are known (Watts, 1988). However, the vast majority of cases result from a group of no more than ten 'common' pathogens. Microbiological investigation of clinical mastitis reveals a causative agent in 75–95% cases (Wilson and Kingwill, 1975; Wilesmith *et al.*, 1986, Hillerton et al., 1993b) with good evidence that microorganisms have been involved in almost all cases.

Mastitis can occur in the mammary gland at all stages of the cow's, and bull/steer's, life. Usually clinical disease occurs during lactation but particular forms of mastitis are also common in non-lactating animals either before first calving or in the dry period between lactations for the adult animal. These dry period cases include clinical mastitis fairly infrequently but sub-clinical infections are common. It has been shown that up to 18% of quarters can be infected in dairy heifers at first calving (Pankey *et al.*, 1991).

The occurrence, persistence and consequences of these mastitic infections vary greatly with the cause and hence an understanding of the aetiology is important in the investigation and implementation of strategies to control mastitis.

Mastitis

There are two groups of bacteria that are responsible for the vast majority of infections. Globally, the most important group are the contagious organisms which are readily transmitted between and within animals. This group comprises *Staphylococcus aureus, Streptococcus agalactiae* and *Streptococcus dysgalactiae*, all capable of causing clinical mastitis, and a series of minor pathogens including other staphylococcal species and *Corynebacterium bovis*. The sub-group of minor pathogens is usually only responsible for sub-clinical infections or very mild clinical disease. The other major group, which may be relatively more important in developed dairy industries, comprises opportunistic, environmentally associated bacteria, often faecal organisms, such as *Streptococcus uberis*, and similar streptococci, *Escherichia coli* and various other Gram-negative bacteria (IDF, 1987).

Actinomyces pyogenes is generally associated with teat damage. Its importance is that in both the milking and dry cow it causes very significant tissue damage and loss of secretory tissue. The relative importance of the different pathogens varies geographically, with the local structure of the industry and in response to control measures in application (Fig. 9.1).

It is commonly assumed that all pathogens enter the mammary gland via the teat duct, although the mode and timing of entry may differ markedly. Some bacteria, *S. agalactiae, S. aureus, S. dysgalactiae* and *A. pyogenes* and the minor pathogen *C. bovis*, can quite obviously colonize the teat duct orifice and/or lining (Bramley *et al.*, 1979b) although this might not always be a precursor to invasion of the gland. There is no obvious evidence of colonization by *S. uberis* or *E. coli* and they appear to enter directly and therefore possibly in smaller numbers (Bramley *et al.*, 1981). These mechanisms are not exclusive and the various modes of transmission of bacteria show this. It is part of the complexity of preventing mastitis or intervening in clinical episodes to understand the specific mechanisms involved so that they can be targeted in applying control.

Often very small numbers of bacteria lead to disease (Newbould and Neave, 1965). Bacteria grow well inside the mammary gland. They may, however, have to express certain virulence factors to stay there and to achieve a certain growth

No. of cases

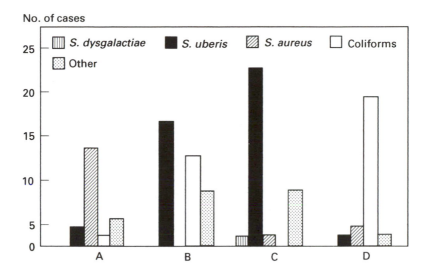

Fig. 9.1. Occurrence of clinical mastitis caused by different pathogens in one year on four dairy farms from a group of twenty. The four farms were selected for their wide disparity in causes of mastitis, demonstrating that they have very different problems possibly requiring quite different solutions. Data from the study reported by Hillerton *et al.* (1995).

rate. Many bacteria are removed by milking and milking frequency may be important in limiting infection rate (Hillerton and Winter, 1992). Frequent emptying of the glands has been a traditional, and not unsuccessful, means of treatment. To avoid removal by milking out at the very next milking bacteria (*S. aureus* and *S. agalactiae*) may adhere to epithelial cells (Frost *et al.*, 1977); grow rapidly (*E. coli*) to prevent removal by cells of the immune system; or be resistant to killing (*S. aureus*) despite phagocytosis (Craven and Anderson, 1979). *S. uberis* does not adhere (Thomas *et al.*, 1995) nor grow quickly and seems to benefit by being poorly recognized by the host and by producing a toxin that inhibits phagocytosis by neutrophils (Leigh and Field, 1994).

The variety of these pathologies shows the complexity of mastitis but also reveals opportunities for intervention by biological strategies once the mechanisms are fully known.

Study of the epidemiology of the disease is important in control of mastitis as it allows identification of risk factors to specific types of infection. This can allow targeting of preventive control techniques and is becoming increasingly important where control measures have been fairly successful but problems remain or there are failures of usually successful schemes.

The incidence and frequency of mastitis has been shown to vary with animal factors, including breed, age of animal, stage of lactation, position of teat and production performance; environmental factors, including season of the year, potential vectors (varying from the milking machine to flies); and the farm management system, especially housing and animal handling. The influence of all of these varies with the different pathogens causing the infection.

Control Strategy

The basis of all disease control is in recognizing the problem, understanding what it is, having the motivation to respond to it and having effective tools to use in response. Much of the incentive for mastitis control has come from the need for increased efficiency in dairy production, to compete for market or to increase product value and profit. These may often be combined. The drive for increased efficiency has resulted both from the recognition of the losses resulting from both clinical and sub-clinical mastitis, perhaps 20% of potential production (Beck *et al.*, 1992), and from market restriction, e.g. as imposed by quotas on output. These have applied in the past and led to today's developed dairy industries. Similar motivations are apparent in the developing industries where there may be targets, e.g. to improve human nutrition to WHO targets (500 g milk per person per day), which necessitates increased output, and quality drives to be able to produce this milk and make it available to urban populations.

Various approaches have been suggested and tried with respect to controlling bovine mastitis. When effective antibiotics became available at the end of the Second World War it was believed that they could be used to achieve eradication of infection within the herd and possibly complete eradication for some pathogens (Plastridge, 1958). It was realized that this might only be achievable for *S. agalactiae*, the relatively obligate parasite (Wilson, 1963). Effectively, eradication of *S. agalactiae* has become possible, not from cost inefficient broad-spread use of lactational antibiotic therapy (Stableforth *et al.*, 1949) but from a total control approach. Significant progress in reducing the level of infection by *S. aureus* and *S. dysgalactiae* by this approach has also been made.

It is clear that complex problems like mastitis cannot be controlled by too simple approaches. Control can be applied at a local level, the cow or herd, or at a national herd level. It is more efficient to have a standardized control scheme that can be defined broadly rather than individual schemes for each herd. Implementation and monitoring of the control schemes can also be made either locally or more centrally.

Whatever the strategy there should be a means of comparison to evaluate success. Morris (1975) considered that to be relevant this should be a direct indicator of mastitis, perhaps level of infection in the herd as percentage of quarters infected or rate of clinical mastitis. He argued against milk cell count, as only 25–38% of the variation in this is due to mastitis (Westgarth, 1971; Morris, 1975). In Scandinavia cell count was used as the primary screen for infection and nationally those countries achieved low cell counts for bulk milk,

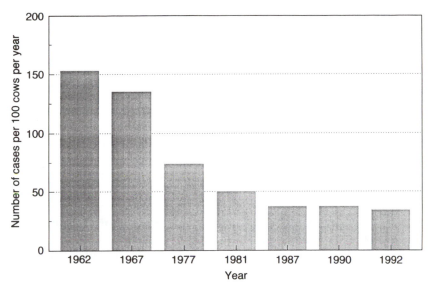

Fig. 9.2. Relative occurrence of clinical mastitis in England and Wales between 1962 and 1992 determined by various studies (Booth, 1988; Hillerton *et al.*, 1995).

but relatively little is known about infection level and it has been suggested that this has increased in Norway recently (Hamann, 1994). In the UK, parts of the USA and Australia, control was directed at reducing the level of infection as measured by the rate of clinical disease or proportion of quarters infected, and this has been successful with a reduction of at least 75% in clinical mastitis and similar success in quarters infected in the UK (Booth, 1988). This was achieved, however, without much impact on bulk milk cell count despite the vast majority of success being in reduction of contagious forms of mastitis, those most related to increasing cell count.

There is a clear need to define the target for a control programme and the perception of the problem will vary significantly between different cultures and dairy systems. Before introduction of successful control schemes, the target was to reduce infection rate from approximately 150 cases per 100 cows per year, with 30–35% quarters infected sub-clinically and bulk milk cell count averaging over 500,000 cells ml^{-1}. This still pertains in most undeveloped dairy industries. The various control methods have been so successful that in the Scandinavian countries cell count averages below 200,000 cells ml^{-1} (IDF, 1993) and in the UK mastitis levels average 35–40 cases per 100 cows per year (Fig. 9.2) and 7–12% quarters infected.

There are two main types of mastitis control schemes. In Scandinavia control has been approached by using a national, government subsidized system of supervision by mastitis laboratories monitoring bulk milk for cell count and particular pathogens. When certain bacteria, e.g. *S. agalactiae*, are detected or cell count exceeds a threshold there is intervention sampling of individual cows and introduction of therapy programmes. This system is supplemented by technical support supervising environmental and milking machine operation. These have become increasingly important. Usually there is no use of teat disinfection and administration of antibiotics is carried out only by veterinary surgeons.

Extensive epidemiological investigations in the UK led to a series of field tests in the 1960s to develop a system of hygienic milk production where various risk factors were identified and components of an integrated control programme introduced with the intention of combining preventive methods with elimination of infection. This system of local application and control of the scheme has also been tested and adopted, with minor variations, in parts of the USA and Australia. This is a form of integrated control analogous to the techniques developed for pest control. It is based on the identification of the essential components in the dynamics of infection. To allow a uniform programme for widespread application it is necessary that there is occasional redundancy or overlap in the effect of the measures, it does not rely on a single control action or need a major screening programme, and the measures are designed to enhance natural defence mechanisms.

Integrated control programmes for mastitis are targeted at changing the dynamics of infection.

Level of mastitis $\propto$ Rate of new infection − Rate of elimination
(no. of infections)

Rate of new infection $\propto$ Level of exposure $\times$ No. of susceptible quarters

Rate of elimination $\propto$ (No. of infections $\times$ Efficacy of treatment)
+ Spontaneous cure

Successful control occurs when the level of infection is held low or is decreased either by preventing new infections or eliminating existing infections. Changes may not always occur quickly but they need to be obvious to convince the farmer of the correctness of the strategy.

The dynamics are not, however, so simple in reality. They vary with the susceptibility of the individual animal which changes with age and stage of lactation and so can be season dependent. They vary with the pathogens involved and the relative importance between herds can be very considerable and also vary with time. The duration of infection may be extremely different for different pathogens. *E. coli* causes acute clinical disease but usually self-eliminates fairly quickly; it is rarely found causing sub-clinical infections. *S. agalactiae* or *S. aureus* are very persistent and the latter responds very poorly to therapy. The rate of elimination and the persistency of these pathogens are very variable. Similarly there can be large variation in the rate of new infections very much related to the identifiable risk factors including rate of teat contamination, mechanisms aiding teat penetrations and effectiveness of establishment and growth of bacteria in the mammary gland.

Dodd (1980) has given a simple description of these complex factors. The success of a control programme can be measured by the decrease in level of infection and by the speed with which this is achieved. Practical impact will require a significant effect within a year if farmers are to remain enthusiastic in application of the methods. The level of infection can be controlled quite well by lowering the rate of new infections but the speed of change very much depends on the duration of the infection and hence is related more to the rate of elimination. Unfortunately no control measures have been found so far that prevent all new infections and only culling is absolutely successful in eliminating infections. Control schemes therefore require both prevention and elimination to give optimal effect and that optimum will vary with pathogen.

Conventional Control

Prevention

Reduction of bacterial contamination of the teat
The first level of prevention is to reduce the exposure of the teat and its orifice to pathogenic bacteria. The cleanliness of the teat is a starting point but bacterial cleanliness is necessary and not only prevention of physical soiling. A physically clean teat is easier to disinfect than one covered in mud or faeces. It is also of primary importance that the teat skin should be intact. Part of mastitis prevention is the maintenance of good teat skin condition, avoidance of physical damage such as teat trampling, insect bites or other cuts and abrasions, and the prevention and elimination of infectious lesions. Damaged teat skin is readily colonized by a number of pathogens including *S. aureus* and *S. dysgalactiae* (Jackson, 1971).

Assuming teats start clean they must be kept clean. This is best achieved by management of housing and pasture where cows lie. There are many examples which show the improving effects on infection rate of providing good quality lying areas both indoors and out. Bramley and Neave (1975) demonstrated large differences in the amount of coliforms present between sand, straw and sawdust bedded cubicles. Similar effects occur with *S. uberis* which appears a more common cause of mastitis in cows bedded on straw yards (Bramley *et al.*, 1979a). These are simple examples of a very complex problem which is influenced by bedding material, temperature, humidity, season, geographical location and possible general level of mastitis in the herd.

Despite the best intentions and actions on a farm, teats will become dirty and it is then important to clean them appropriately. Washing can be achieved by a number of techniques, most of which benefit by stimulating milk let-down, of varying success in reducing bacterial numbers on the teat skin, in milk and reducing the rate of new infections. The quality of the technique is important if the only effect is not to be a more effective spread of pathogens. In the first instance the milker's hand must not be a vehicle of transfer from cow to cow and hence needs to be relatively free of contamination. This is best achieved by wearing gloves and disinfecting them between individual cows if necessary. Experimental work from a wide variety of sources has established that a detailed routine of udder preparation can be effective. Washing teats with plain water using a cloth common to all cows is of little benefit; use of separate, preferably disposable, towels, disinfectant in the water and drying of the teats after washing are essential components (Dodd and Neave, 1970). It is important how the teats are cleaned. Spread of water to a dirty udder, even if teats are dried, can lead to dirty water running down teats to contaminate the teat orifice and the teatcup liner. Good quality teat cleaning is effective in reducing the number of *S. aureus* on teat skin and in reducing the number of new infections by this pathogen (Neave, 1971). A popular alternative effective system for relatively unsoiled teats is to wash with clean water and then to dry with a disinfectant impregnated cloth (McKinnon *et al.*, 1985). Little effect of washing has been found on environmental mastitis.

In an attempt to increase the effect of pre-milking teat disinfection much higher concentrations of disinfectant than used in wash-water have been applied by dipping teats in a reportedly more active disinfectant, leaving it in contact with the skin for 30 seconds, and then wiping off immediately before teatcup attachment, to reduce the likelihood of contaminating milk. It has been claimed that significant reductions in the rate of new environmental mastitis infections can be achieved (Pankey *et al.*, 1987) although a contrary study in herds with generally much less mastitis could not produce a consistent effect distinguishable from any resulting change in attitude of the herdsman when an external interest was taken in his mastitis control (Hillerton *et al.*, 1993b).

Finally, it is essential to eliminate bacterial contamination occurring during milking and to prevent contamination for as long as possible into the inter-milking interval. Post-milking teat disinfection has been, and still is, the single most effective component of hygienic milking programmes used in the UK, USA and Australia. Teats are coated in disinfectant as soon as possible after removal of the milking. Disinfectant is applied by dipping each teat separately in a cup or by spraying disinfectant on to the teats from below. The first method has the advantage that complete coverage is fairly easy to achieve whilst spraying often coats one side of each teat only, yet might use twice as much disinfectant in creating the aerosol. It has been suggested that the quality of teat disinfection can be demonstrated by the level of infection in the herd by the minor pathogen *C. bovis* (Bramley *et al.*, 1976).

There are a large number of disinfectant formulations available commercially using a variety of active ingredients, the commonest being formulations of iodine, chlorhexidine, aldehydes, chlorine releasers and quaternary ammonium compounds. All of these products available in regulated markets have been extensively tested for efficacy and safety. They are, however, not equally active (Table 9.1). However, the biggest variables in successful use are the quality and regularity of application.

Table 9.1. Effect of different teat disinfectant formulations on the recovery of *Staphylococcus aureus* from artificially contaminated teat skin (derived from King *et al.*, 1977).

Active ingredient	Concentration	Emollient	Geometric mean bacteria recovered	
			Trial 1	Trial 2
Quaternary ammonium compound	1.5% benzalkonium 1.3% picloxydine gluconate	Triton X Citric acid		920
Sodium hypochlorite	0.1%	None	390	
Iodophor	0.5%	15% glycerine	107	
Chlorhexidine	1%	Polyvinyl pyrollidone		40
Iodophor (as above)	0.5%	None	17	
Sodium hypochlorite	4%	None	9.2	3.8

It is common for disinfectants to contain emollients; the more effective formulations appear to use glycerine, the major effect being to prevent damage to skin during adverse weather conditions. They also improve the general skin condition by speeding recovery from damage. It has been shown clearly that there are significantly fewer lesions on teats now than in the days before regular dipping in an effective disinfectant formulation was introduced (Shearn and Morgan, 1994). This reduction in the likely reservoir of contagious mastitis pathogens has probably made a significant contribution to the reduction of disease. The relative benefits of killing bacteria on skin directly, compared with improving skin condition, in reducing mastitis do not appear to have been investigated. Good skin condition may be very important because mastitis caused by lesion colonizing bacteria, i.e. *S. aureus* and *S. dysgalactiae*, appears to be well controlled by teat disinfectants compared with simple surface contaminants such as the coliforms.

Teat disinfection after milking has not been adopted universally. It has not been considered necessary in some countries and doubts have been raised on whether residues may contaminate milk. Teat disinfection is, however, recommended in problem herds in countries not advising general usage (Olsen, 1975) and in others there has been gradual adoption with time. It is not possible to determine any consistent adverse effect of disinfectants on milk. Although there is some evidence that use of iodophor disinfectants may increase milk iodine content (Galton *et al.*, 1986), probably the variability in iodine content of milk between farms is most influenced by diet (Hillerton *et al.*, 1993b).

Preventing bacteria entering the gland

Despite efforts to limit bacteria on the teat skin there are, albeit infrequent, penetrations of pathogens into the gland. These involve bacteria passing through the teat duct; there is little evidence of any other route of invasion. The teat duct possesses a number of properties defensive against penetration including a relatively efficient closure mechanism, although this may vary between cows. Higher flow rate/yielding cows, presumably with a wider duct, may fail to achieve good closure (Dodd and Neave, 1951). The keratin lining of the duct is a physical barrier to prevent bacterial invasion although contagious species appear to be able to colonize the keratin. There is some evidence that the keratin acts against some bacteria by trapping them and that they are then removed by milk flow stripping the most mature layers at the next milking (Williams and Mein, 1985). Factors that affect the integrity of the ductular lining may therefore affect the rate of infection.

The milking machine influences the rate of mastitis by providing mechanisms by which bacteria presented to the teat orifice can be 'forced' through the teat duct or can modify the teat, or the immediate intramammary tissues, sufficiently to aid bacterial growth or inhibit the defence mechanisms. Reduction of the defence mechanisms and increased colonization of the teat are chronic effects which may be achieved by too high a milking vacuum or by poor pulsation. Control of mastitis requires the use of good milking conditions (Dodd and Neave, 1970).

Two mechanisms have been proposed by which bacteria may be propelled into the teat sinus during milking. Vacuum fluctuations in the claw may lead to milk moving between teatcups. If this milk contains pathogens and moves quickly it can impact on the teat orifice and drive through the teat duct (Thiel *et al.*, 1973). Milk can be prevented from impacting on the teat end by interceptor shields (Griffin *et al.*, 1983), by use of one way valves (Griffin *et al.*, 1988) or by milking each quarter separately. These are equipment solutions to the problem. It is also possible to reduce the vacuum fluctuations, which drive the movement of milk, by use of larger vacuum pumps and balance tanks or by larger diameter pipelines and tubes. These contribute to improved vacuum stability which minimizes rapid fluctuations. Recent work, largely from the US, has described a force to drive bacteria into the teat sinus by creation of a greater vacuum inside than that in the liner – a negative pressure gradient (Zehr *et al.*, 1985). Any contribution this might make to the rate of infection and how it can be best controlled is not yet clear. There is considerable debate as to the effect of automatic cluster removers on mastitis. They are claimed to prevent over-milking which can lead to teat trauma (Peterson, 1964) and hence infection. They may also, more positively, lead to a low milk cell count although not necessarily a lower infection rate. Counter claims are made that they lead to under-milking which could lead to ideal conditions for bacteria to establish in the udder. Use of automatic cluster removers does mean that the cluster is removed after the vacuum has been released and minimizes the likelihood of impacts being created by rough cluster removal.

It is obvious that milking machine design and operation must be part of mastitis control. The machine can only make a contribution to infection rate if bacteria are present and so clean teats and a clean machine are fundamental. The contact between the cow and the milking machine occupies a very small proportion of the time the cow is in the herd but it does seem to provide a significant proportion of the opportunities for bacteria to invade the teat. Most of these invasions are associated with contagious pathogens. The evidence is that the environmental pathogens, invading directly and not by a colonization method, do so mostly in the inter-milking interval. However, when teats are contaminated with bacteria and the milking conditions are poor, allowing impacts or reverse pressure gradients, then infection may be introduced during milking. The milking process cannot be involved in invasion during the dry period when the commoner infections by *S. aureus* and *A. pyogenes* occur by growth through a keratin plug in the duct and *S. uberis* invades by an unidentified method. Risk factors to invasion suggest the major control method is to limit exposure to bacteria.

The major clinical problem in the dry period in temperate Europe and also reported in specific instances from elsewhere is the disease 'summer mastitis' which is reportedly transmitted by the muscid fly *Hydrotaea irritans* although absolute evidence is not available (Hillerton *et al.*, 1990). Good fly control, preventing this fly feeding from the teats of cows, is effective in limiting epidemic distribution of this disease. Individual cases do occur under all types of conditions and no obvious prevention other than maintaining good teat condition, especially preventing damage and exposure to the bacteria, appears effective.

Prevention of establishment of pathogens in the mammary gland
Once having gained entry to the mammary gland pathogens have to sustain their presence in order to grow and cause disease. Relatively little is known about the early fate or location of bacteria in the mammary tissue, other than that they have to avoid being milked out.

The major systems resisting establishment of bacteria in the gland are the humoral and cellular components of the immune system. Milk polymorphonuclear neutrophils are active phagocytes and their presence in milk is readily increased following bacterial invasion. Phagocytosis is normally fairly effective as antibodies to opsonize the bacteria are not normally limiting in milk. It can, however, be overcome by more rapid bacterial growth than removal and recently it has been shown that *S. uberis* can produce a toxin to inhibit phagocytosis although the neutrophils are not killed (Leigh and Field, 1994). Whilst phagocytosis may be efficient for some species, e.g. *S. aureus*, bacteria are not necessarily killed. It is clear that *S. aureus* survive within neutrophils and are even protected from antibiotics there, such that viable bacteria are released from dying neutrophils (Craven and Anderson, 1979).

The natural defence systems are complex and apparently manipulated by the bacteria. It is now proving possible to manipulate the immune system to enhance its effect. There have been many attempts to produce a mastitis vaccine but with little historical success. The majority of attempts have been to enhance the cellular defence systems, but of course by causing a mastitis to cure a mastitis. This may be only of true value when the effect of the induced defence is less than the infection, and there is evidence of this for coliform mastitis (Gonzales *et al.*, 1989). An alternative approach has been described recently where the humoral defences are stimulated, such that the proteases induced to provide essential amino acids and peptides for bacterial growth are neutralized by specific antibodies (Leigh, 1993) and hence bacterial growth is inhibited. This allows bacteria to be removed by milking out and by the natural level of phagocytic activity in the gland. It remains to be shown if this can be an effective means of preventing mastitis. It is apparent that enzyme systems specific to individual or closely related bacteria only will be targeted and so a mastitis vaccine would require to be multivalent. Initial products will be specific and require particular targeting at individual problems.

A number of antibacterial mechanisms are naturally active in milk including lactoferrin, lactoperoxidase and lysozyme but appear relatively ineffective in limiting experimental infections.

The only current, successful, method of mastitis control affecting establishment of bacteria is the prophylactic use of antibiotics in the dry period. Initial formulations were developed to prevent summer mastitis (Pearson, 1950) with later evolution of long-acting preparations (Smith *et al.*, 1967). These antibiotics are extremely effective in preventing new infections, especially early in the dry period, e.g. when most *S. uberis* infections occur (Fig. 9.3). The most effective preparations have a wide-spectrum of activity to include Gram-positive and Gram-negative bacteria and so combinations are common. The major limitation to their use is that antibiotic residues should not persist into lactation and so efficacy towards the end of a normal dry period may be low.

A number of products have become available in recent years for application to the udder at the very first indications of infection that are claimed to limit the development of clinical disease. Some of these are homeopathic methods for which, so far, benefit has not been demonstrated in controlled trials (Hamann, 1993). Massage with various ointments is a traditional remedy. Many claims are made that this 'treatment' avoids more severe clinical signs and the need for antibiotic treatment but objective studies are limited. The mechanism of action is supposed to be simple stimulation of blood supply.

Mastitis requires invasion of the mammary gland by pathogens and successful establishment. Many of the infections will be insubstantial because of rapid natural elimination of the infection. Natural 'cure', at least of clinical disease if not of the bacterial infection, is not infrequent. Probably most cures of coliform infections are entirely natural as diagnosis rarely involves identification of the pathogen and most antibiotic preparations are of limited efficacy against Gram-negative bacteria. Both the effectiveness and timing of the various mechanisms of elimination of infection are important in influencing the dynamics of mastitis.

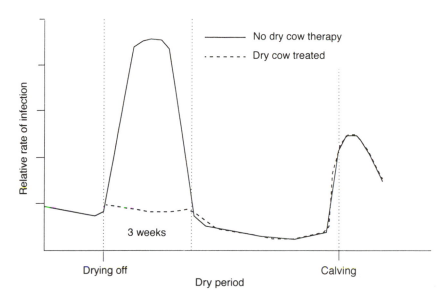

Fig. 9.3. Timing and relative effect of dry cow therapy on new infections in the dry period (Hillerton *et al.*, 1993a).

Elimination

There are at present three proven methods of elimination of disease relevant to mastitis control. These are natural cure, use of chemical therapy and culling. Natural cure, by successful deployment of the defensive systems, leads to spontaneous recovery. This may include parasite limitation if they exhaust nutrient supplies. Antibiotic therapy is used during lactation to achieve clinical cure of disease or at drying off to eliminate chronic or sub-clinical infection. The effectiveness of lactational therapy varies with the pathogen and the antibiotic. Clinical cure rates may be high but bacteriological cure is much less likely. Usually it will not be relevant for coliform infections as the antibiotics in common use are not generally effective against such bacteria. Bacteriological cure of streptococcal infections may be high but *S. aureus* infection is not easily eliminated, often in only 30% of cases during lactation (Table 9.2). The effectiveness of antibiotic therapy is limited by the problems in delivering an optimal amount of drug to the site of infection and sustaining an adequate concentration for a long enough period either to achieve kill if the active ingredient is bactericidal or to change the dynamics of defence if a bacteriostat is used. The major limitations are in the relative ineffectiveness of the drugs *in vivo*; most bacteria are usually sensitive to the drugs, but they may be poorly dispersed from the site of delivery through infected tissue; and avoidance mechanisms such as the intracellular survival of staphylococci. This is shown by the usual higher success in eliminating sub-clinical mastitis (Griffin *et al.*, 1982), when

Table 9.2. Bacteriological cure rates (% of quarters) for clinical mastitis in lactation using sodium cloxacillin (Orbenin QR) and for infections at drying off using benzathine cloxacillin (Orbenin Dry Cow), from Hillerton *et al.* (1995). The values for effect against Gram-negative bacteria are spurious as the antibiotic has no effect on such organisms. They indicate spontaneous cure.

	Pathogen				
	Strep. dysgalactiae	*Strep. uberis*	*Staphylococcus aureus*	Gram-negative bacteria	Overall
Lactation	81	65	33	71	60
Dry period	100	84	59	50	77

presentation of the drug is easier as there is much less damaged tissue to be penetrated.

Significantly greater cure of infection is achieved by application of dry cow therapy – infusion of intramammary antibiotics in a long-acting formulation into each quarter of each cow immediately after the last milking of the lactation. This may double the bacteriological cure rate for infections by *S. aureus*. Use of dry cow therapy has made a major impact into the reduction of the level of infection, especially by *S. agalactiae* and *S. aureus*.

The only absolutely effective method of elimination is by culling of the infected animal. Culling may, under certain circumstances, be the only option available, especially when all other techniques have failed to eliminate infection or prevent continual reinfections. Culling must only be considered along with the impact on the herd structure and the economic circumstances. Usually one-quarter of a herd will be culled annually and then a number of reasons will apply in selection of animals. Whilst culling at the maximum rate possible for reasons of mastitis has been shown to reduce the level of contagious infection by about 50% in some three years and can be used to achieve eradication of *S. agalactiae*, there will be a major impact on herd structure and production (Natzke and Everett, 1975). Culling has to be strategic.

Integrated Mastitis Control Schemes

The dynamic processes of mastitis have been described and targets for intervention identified. Most methods are specific to one part of the process and usually restricted in effect to particular forms of mastitis. To control such a disease complex as mastitis requires an integrated programme of prevention and elimination of infection and two major, successful strategies have been described, although individual elements of each are in common use. Where only individual elements are used mastitis control may be poor, e.g. simply

Table 9.3. Five-point mastitis control plan developed by the National Institute for Research in Dairying and the Central Veterinary Laboratory based on field trials of hygienic milking practices, with later additions based on fundamental research (Dodd and Neave, 1970).

Original plan

1. Hygienic preparation of teats for milking and disinfection of all teats after milking
2. Treatment of all cases of clinical mastitis and accurate recording of the occurrence
3. Use of dry cow antibiotic preparations on all quarters of all cows intended to recalve into the herd
4. Culling of cows with persistent mastitis based on records
5. Maintenance of the milking machine by frequent servicing

Additions

6. Maintenance of the best possible teat skin condition
7. Reduction of exposure to bacteria by management of the environment
8. Prevention of transfer of bacteria between cows via the milking machine
9. Prevention of milking conditions likely to aid bacteria penetration of the teat duct

---

using lactational antibiotic therapy may reduce clinical signs but is ineffi-cient in eliminating infection and ineffective in preventing new infections. It leads to abuse of the technique and further problems may occur, including development of resistance to antibiotics and contamination of milk supplies with residues.

The five point control plan (Table 9.3) developed in the UK in the 1960s (Dodd and Neave, 1970) was based on attacking the key areas in the dynamic processes of mastitis and the individual components of the plan proven by field testing. Its success has been well reported. It still required additional work, especially improvements in milking machine technology, and residual problems remain. In particular not all farms seem to have benefited equally and in general environmental mastitis is not well controlled. The plan, however, depends on the motivation, education and financial commitment of the herdsman and farmer to achieve good mastitis control. Recent studies have indicated that close supervision of the application of the components of the plan, especially teat disinfection, recording and treatment of all cases of mastitis and universal application of dry cow therapy, can lead to an even greater reduction of mastitis (Hillerton *et al.*, 1995). The control scheme is only as good as the use made of it. Its failings may be addressed by targeting specific additional benefits and providing incentives for more dedication to mastitis control. Unfortunately good intentions are insufficient. When the Milk Marketing Boards in the UK started payment schemes for better quality milk the improvements were immediate and obvious (Booth, 1993), suggesting that the techniques are not limiting, that education is not limiting but that the correct motivation may be essential.

Future Control

Application of conventional mastitis control measures has resulted in an 80% reduction in the rate of clinical disease compared with that known prior to introduction of control measures and in industries not having access to control plans (Booth, 1988). This does not mean that control schemes can be relaxed readily. *S. agalactiae* infections have been eradicated from particular herds but can be re-introduced. *S. aureus* and *S. dysgalactiae* infections may be at a very low rate but this suppression is only an indicator of the quality of the mastitis control. Failures in the application of the plan can result in a rapid increase in the number of mastitic infections from which it may take a long time to recover. It will always be necessary to prepare teats properly for milking, to seek to prevent bacterial invasion of the mammary gland and to eliminate infections that do occur.

There may be advantages in some relaxation of control in herds with low rates of mastitis: selective use of dry cow therapy has been suggested and is frequently used. It has been shown, however, that this can lead to higher rates of clinical mastitis (Robinson *et al.*, 1983) in the next lactation and that dry cow prophylaxis is undermined (Schukken *et al.*, 1993). Good hygiene will be increasingly necessary to help to maintain, and improve, milk quality. It is more likely that more hygienic systems will be required, perhaps including better environmental control. This will be to some extent a marrying of the UK and Scandinavian approaches to mastitis control.

More improvements in mastitis control will be possible by targeting further risk factors to infection. Additional improvement in the milking machine action to reduce physiopathological effects on teat tissue (Hamann, 1985) will be necessary and are possible. These may include interactive milking systems and customization of milking conditions for individual cows (Lind *et al.*, 1994).

It is possible that the era of chemical control of mastitis by use of antibiotics is starting to close, although there appear to be few obvious replacements for disinfectants. Promising advances have been made and shown to be effective in vaccination to reduce the severity of coliform mastitis (Gonzales *et al.*, 1989) and to prevent *S. uberis* mastitis (Leigh, 1993), so addressing the major residual problems from conventional control. In due course multivalent vaccines may be possible. It has been suggested that a transgenic approach to the control of *S. aureus* mastitis by expressing a unique antibacterial protein in milk might be possible (Williamson *et al.*, 1994). However, successful developers of such systems will face a plethora of problems.

Modern techniques in quantitative epidemiology can allow identification of additional risk factors to infection, factors that may change in response to conventional control and vary with problems in individual herds and particular variants, e.g. national and ecological, of the industry. One risk factor worthy of study is to identify why particular cows are susceptible to mastitis. Various factors related to production are known but markers to identify the 50% of cows that suffer mastitis in their lifetime would allow better and more cost effective targeting of control.

There is a need to be more successful in the identification of disease especially earlier before full clinical signs are apparent and therapy so difficult. It appears that after 50 years of research the microprocessor may allow the calculations of probability that changes in yield, temperature and milk electrical conductivity can be diagnostic of mastitis before visible changes in milk. Other sensors could detect other milk chemical changes and indeed identify the particular bacteria causing the changes. Such detection systems may lead to earlier and more specific therapy, improving the elimination of infection.

Despite many possible advances, of which vaccines are the most exciting, there will still be a need for a hygiene-based mastitis control programme. The best vaccine technology may only be effective and economically beneficial in partial control of a small number of pathogens, thus still leaving plenty of infection by the remainder of the 137 possible pathogens so far reported. Epidemiological patterns suggest that, given the most effective vaccine systems imaginable and the best quality mastitis prevention, the rate of clinical disease will not drop below 5 cases per 100 cows per year.

References

Beck, H.S., Wise, W.S. and Dodd, F.H. (1992) Cost benefit analysis of bovine mastitis in the UK. *Journal of Dairy Research* 59, 449–460.

Booth, J.M. (1988) Progress in controlling mastitis in England and Wales. *Veterinary Record* 122, 299–302.

Booth, J.M. (1993) Mastitis in the 1990's. *Cattle Practice* 1, 125–131.

Bramley, A.J. and Neave, F.K. (1975) Studies on the control of coliform mastitis in dairy cows. *British Veterinary Journal* 131, 160–169.

Bramley, A.J., Kingwill, R.G., Griffin, T.K. and Simpkin, D.L. (1976) Prevalence of *Corynebacterium bovis* in bovine milk samples. *Veterinary Record* 99, 275.

Bramley, A.J., King, J.S. and Higgs, T.M. (1979a) The isolation of *Streptococcus uberis* from cows in two dairy herds. *British Veterinary Journal* 135, 262–270.

Bramley, A.J., King, J.S., Higgs, T.M. and Neave, F.K. (1979b) Colonization of the bovine teat duct following inoculation with *Staphylococcus aureus* and *Escherichia coli*. *British Veterinary Journal* 135, 149–162.

Bramley, A.J., Godinho, K.S. and Grindal, R.J. (1981) Evidence of penetration of the bovine teat duct by *Escherichia coli* in the interval between milkings. *Journal of Dairy Research* 48, 379–386.

Craven, N. and Anderson, J.C. (1979) The location of *Staphylococcal aureus* in experimental chronic mastitis in the mouse and the effect on the action of sodium cloxacillin. *British Journal of Experimental Pathology* 60, 453–459.

Dodd, F.H. (1980) Mastitis control. In: Bramley, A.J., Dodd, F.H. and Griffin, T.K. (eds) *Mastitis Control and Herd Management*. NIRD/HRI Technical Bulletin No. 4, Reading, pp. 11–23.

Dodd, F.H. and Neave, F.K. (1951) Machine milking rate and mastitis. *Journal of Dairy Research* 18, 240–245.

Dodd, F.H. and Neave, F.K. (1970) Mastitis control. *Report of the National Institute for Research in Dairying*, pp. 21–60.

Frost, A.J., Wanasinghe, D.D. and Woolcock, J.B. (1977) Some factors affecting selective adherence of microorganisms in the bovine mammary gland. *Infection and Immunity* 15, 245–253.

Galton, D.M., Petersson, L.G. and Erb, H.N. (1986) Milk iodine residues in herds practising iodophor pre-milking teat disinfection. *Journal of Dairy Science* 69, 267–271.

Gonzales, R.N., Cullor, J.S., Jasper, D.E., Farver, T.B., Bushnell, R.B. and Oliver, M.N. (1989) Prevention of clinical coliform mastitis in dairy cows by a mutant *Escherichia coli* vaccine. *Canadian Journal of Veterinary Research* 53, 301–305.

Griffin, T.K., Dodd, F.H. and Bramley, A.J. (1982) Antibiotic therapy in the control of mastitis. *Proceedings of the British Cattle Veterinary Association*, pp. 137–152.

Griffin, T.K., Williams, R.L., Grindal, R.J., Neave, F.K. and Westgarth, D.R. (1983) Use of deflector shields to reduce intramammary infection by preventing impacts on the teat ends of cows during machine milking. *Journal of Dairy Research* 50, 397–404.

Griffin, T.K., Grindal, R.J. and Bramley, A.J. (1988) A multivalved milking machine cluster to control intramammary infection in dry cows. *Journal of Dairy Research* 55, 155–169.

Hamann, J. (1985) Infection rate as affected by teat tissue reactions due to conventional and non-conventional milking systems. *Kieler Milchwirtschaftliche Forschungsberichte* 37, 426–430.

Hamann, J. (1993) Homeopathic treatment of bovine mastitis. *International Dairy Federation Mastitis Newsletter* 134, 10–12.

Hamann, J. (1994) Trends in yield and health parameters – an attempt at a retrospective analysis. In: Lind, O. and Svennersten, K. (eds) *Prospects for Future Dairying: a Challenge for Science and Industry*. Alfa Laval Agri AB, Tumba, Sweden, pp. 57–61.

Hillerton, J.E. and Winter, A. (1992) The effects of frequent milking on udder physiology and health. In: Ipema, A.H., Lippus, A.C., Metz, J.H.M. and Rossing, W. (eds) *Prospects for Automatic Milking*. Pudoc, Wageningen, pp. 201–212.

Hillerton, J.E., Bramley, A.J. and Thomas, G. (1990) The role of *Hydrotaea irritans* in the transmission of summer mastitis. *British Veterinary Journal* 146, 147–156.

Hillerton, J.E., Shearn, M.F.H. and Dodd, F.H. (1993a) Clinical mastitis and stage of lactation. *Proceedings of the British Mastitis Conference*, Genus/Institute for Animal Health/Ciba Agriculture, Stoneleigh, pp. 94–95.

Hillerton, J.E., Shearn, M.F.H., Teverson, R.M., Langridge, S. and Booth, J.M. (1993b) Effect of pre-milking teat dipping on clinical mastitis on dairy farms in England. *Journal of Dairy Research* 60, 31–41.

Hillerton, J.E., Bramley, A.J., Staker, R.T. and McKinnon, C.H. (1995) Patterns of intramammary infections and clinical mastitis over a 5 year period in a closely monitored herd applying mastitis control measures. *Journal of Dairy Research* 62, 39–50.

IDF (1987) Environmental influences on bovine mastitis. *International Dairy Federation*, Bulletin No. 217, pp. 37.

IDF (1993) *Mastitis Newsletter* No. 18, International Dairy Federation, Brussels.

Jackson, E.R. (1971) An outbreak of teat sores in a commercial dairy herd possibly associated with milking machine faults. *Veterinary Record* 87, 2–5.

King, J.S., Neave, F.K. and Westgarth, D.R. (1977) Disinfection properties of some bovine teat dips. *Journal of Dairy Research* 44, 47–55.

Leigh, J.A. (1993) Progress in the understanding of the pathogenesis of *Streptococcus uberis* for the lactating bovine mammary gland. *33rd Annual Meeting National Mastitis Council*, Orlando, Florida, pp. 96–103.

Leigh, J.A. and Field, T.R. (1994) *Streptococcus uberis* resists phagocytosis despite the presence of bound immunoglobulin. In: Pathogenic Streptococci: Present and future. *Proceedings of the XII Lancefield Symposium on Streptococci and Streptococcal Diseases.* Lancer Publications, St. Petersburg, pp. 498–499.

Lind, O., Gisel-Ekdahl, H. and Svennersten, K. (1994) Technical aspects and demands on the milking equipment. In: Lind, O. and Svennersten, K. (eds) *Prospects for Future Dairying: a Challenge for Science and Industry.* Alfa Laval Agri AB, Tumba, Sweden, pp. 107–116.

McKinnon, C.H., Higgs, T.M. and Bramley, A.J. (1985) An examination of teat drying with disinfectant impregnated cloths on the bacteriological quality of milk and on the transfer of *Streptococcus agalactiae* before milking. *Journal of Dairy Research* 52, 355–359.

Morris, R.S. (1975) Criteria for the design and evaluation of bovine mastitis control systems. *International Dairy Federation Bulletin, Document* 85, 356–409.

Natzke, R.P. and Everett, R.W. (1975) The elimination of mastitis by culling. *International Dairy Federation Bulletin, Document* 85, 303–310.

Neave, F.K. (1971) The control of mastitis by hygiene. In: Dodd, F.H. and Jackson E.R. (eds) *The Control of Bovine Mastitis.* NIRD, Reading, pp. 55–71.

Newbould, F.H.S. and Neave, F.K. (1965) The recovery of small numbers of *Staphylococcus aureus* infused into the bovine teat cistern. *Journal of Dairy Research* 32, 157–162.

Olsen, S.J. (1975) A mastitis control system based upon extensive use of mastitis laboratories. *International Dairy Federation Bulletin, Document* 85, 410–421.

Pankey, J.W., Drechsler, P.A. and Wildman, E.E. (1991) Mastitis prevalence in primigravid heifers at parturition. *Journal of Dairy Science* 74, 1550–1552.

Pankey, J.W., Wildman, E.E., Drechsler, P.A. and Hogan, J.S. (1987) Field trial evaluation of premilking teat disinfection. *Journal of Dairy Science* 70, 867–872.

Pearson, J.K.L. (1950) The use of penicillin in the prevention of *C. pyogenes* infection of the non-lactating udder. *Veterinary Record* 62, 166–168.

Peterson, K.L. (1964) Mammary tissue injury resulting from improper machine milking. *American Journal of Veterinary Research* 25, 1002–1009.

Plastridge, W.N. (1958) Bovine mastitis: a review. *Journal of Dairy Science* 41, 1141–1181.

Robinson, T.C., Jackson, E.R. and Marr, A. (1983) Within herd comparison of teat dipping and dry cow therapy with only selective dry cow therapy in six herds. *Veterinary Record* 112, 315–319.

Schukken, Y.H., Vanvliet, J., Vandeger, D. and Grommers, F.J. (1993) A randomized blind trial on dry cow antibiotic infusion in a low somatic cell count herd. *Journal of Dairy Science* 76, 2925–2930.

Shearn, M.F.H. and Morgan, J.H. (1994) Teat skin lesions on the dairy cow. *Proceedings of the British Mastitis Conference*, Genus/Institute for Animal Health/Ciba Agriculture, Stoneleigh, pp. 17–24.

Smith, A., Westgarth, D.R., Jones, M.R., Neave, F.K., Dodd, F.H. and Brander, G.C. (1967) Methods of reducing the incidence of udder infection in dry cows. *Veterinary Record* 81, 504–510.

Stableforth, A.W., Hulse, E.C., Wilson, C.D., Chodkowski, A. and Stuart, P. (1949) Herd eradication of *Str. agalactiae* by simultaneous treatment of all cows with 5 doses of 100,000 units of penicillin at daily intervals and disinfection. *Veterinary Record* 61, 357–362.

Thiel, C.C., Cousins, C.L., Westgarth, D.R. and Neave, F.K. (1973) The influence of some physical characteristics of the milking machine on the rate of new mastitis infections. *Journal of Dairy Research* 40, 117–129.

Thomas, L.H., Haider, W., Hill, A.W. and Cook, R.S (1995) Pathologic findings of experimentally induced *S. uberis* infection in the mammary gland of cows. *American Journal of Veterinary Research* 55, 1723–1728.

Watts, J.L. (1988) Etiological agents of bovine mastitis. *Veterinary Microbiology* 16, 41–66.

Westgarth, D.R. (1971) Interpretation of bulk cell count data. In: Dodd, F.H. and Jackson, E.R. (eds) *The Control of Bovine Mastitis*. NIRD, Reading, pp. 105–111.

Wilesmith, J.W., Francis, P.G. and Wilson, C.D. (1986) Incidence of mastitis in a cohort of British herds. *Veterinary Record* 118, 199–204.

Williams, D.M.D. and Mein, G.A. (1985) The role of machine milking in the invasion of mastitis organisms and implications for maintaining low infection rates. *Kieler Milchwirtschaftliche Forschungberichte* 37, 415–425.

Williamson, C.M., Bramley, A.J. and Lax, A.J. (1994) Expression of the lysostaphin gene of *Staphylococcus simulans* in a eukaryotic system. *Applied and Environmental Microbiology* 60, 771–776.

Wilson, C.D. (1963) The microbiology of bovine mastitis in Great Britain. *Bulletin, Office International des Épizooties* 60, 533–551.

Wilson, C.D. and Kingwill, R.G. (1975) A practical mastitis control routine. *International Dairy Federation Bulletin, Document* 85, 422–438.

Zehr, P.D., Galton, D.M., Scott, N.R. and Czarniecki, C.S. (1985) Measurement of reverse pressure gradients across the streak canal during machine milking. *Journal of Dairy Science* 68 (Suppl 1), p. 200.

Controlling Lameness in Dairy Cows

10

P.R. Greenough

Department of Veterinary Anaesthesiology, Radiology and Surgery, Western College of Veterinary Medicine, University of Saskatchewan, Saskatoon, Saskatchewan S7N 0W0, Canada

Introduction

Lameness is a clinical sign of many conditions, the control of each being a matter for individual consideration. In the majority of cases the abberations of gait observed result from pain and therefore have serious implications in respect to animal welfare. In a minority of instances such as in the case of neurological disorders there seems to be little or no pain involved.

The economic importance of lameness has been well documented. Financial losses occur as the result of reduced milk production, loss of body condition, reduced fertility, culling, veterinary costs and medication, as well as time devoted to nursing an animal by the dairyman. It has been estimated (Esslemont, 1990) that a single case of sole ulcer (pododermatitis circumscripta) could cost £130–180 and one case of interdigital disease might cost £55–100. In a later study (Esslemont and Spincer, 1993) the average cost of an incident of lameness was placed at £75. Earlier estimates indicate that financial annual losses from lameness may amount to £1000 per 100 cows (Whitaker *et al.*, 1983) and the loss to the dairy industry of the UK could be £15 million each year.

Reduced fertility makes up a significant proportion of these losses (Collick *et al.*, 1989). Lame cows were found in this study to take 14 days longer to conceive than did normal animals. Amongst animals affected with sole ulcers, conception might be delayed for 40 days. Lame cows have been shown to get other diseases more frequently (usually indirectly), such as mastitis. Figures from milk-recording organizations in West Germany indicate that culling due to claw and leg problems nearly doubled from about 4–5% to 7–9% over a ten-year period starting in 1983 (Distl, 1994).

The incidence of disorders of the locomotory system has been the subject of a number of surveys. The annual incidence of lameness has been estimated by a number of workers as 5–30% (Politiek *et al.*, 1986) and 21% (Reurink and Van Arendonk, 1987). Philipot *et al.* (1990) found the incidence of lameness to be 8.2% but 25% of the animals examined had claw lesions.

The Multifactorial Concept

There are a number of predisposing factors that either separately or in conjunction with one another will influence the severity and the herd morbidity of lameness. Laminitis, for example, was referred to as a multifactorial disease by Mortensen and Hesselholt (1986), and it was they who earlier (1982) had conceived the possibility of developing an epidemiological approach to investigating the disease. Starting in 1990, an extensive epidemiological study of lameness was undertaken at the University of Liverpool (Clarkson *et al.*, 1993; Ward, 1994b) during which many of the factors that predispose to lameness were evaluated. It has become an essential premise that if herd lameness is to be controlled an effective method or technique has to be developed to collect and analyse relevant data (Greenough and Vermunt, 1994). The ideal epidemiological methodology would evaluate potential stress in the animal as well as the degree of exposure or risk to the lameness-producing disease.

Genetic considerations

Russell *et al.* (1986) demonstrated that the daughters of some bulls were more likely to suffer from lameness of digital origin than those of other sires. Heritability estimates for a 'single eye-scored claw angle' averages about 0.10 (McDaniel, 1994). McDaniel also indicates that animals with steep claw angles (50–60°) have greater longevity. The most common claw traits were discussed by the EAAP Working Group 'Claw Quality in Cattle' (Politiek *et al.*, 1986; Distl *et al.*, 1990). These traits consisted of an evaluation of the claw shape, the quality of claw horn and features of the inner structure of the claw. Several studies demonstrated that these traits had sufficiently high additive genetic variation to achieve genetic improvement. Claw measurements are significantly correlated genetically and phenotypically to the prevalence of claw disease, longevity and lifetime performance (Nielsen and Smedegaard, 1984; Reurink and Van Arendonk, 1987; Rogers and McDaniel, 1989; Rogers *et al.*, 1989; Baumgartner and Distl, 1990).

Because the angle of the joints of the limb are difficult to measure their significance has been studied less intensively than have been the characteristics of the claw. Variations in posture contribute to the difficulty of making accurate evaluations. Scores or even actual measurements of individual cows often show large changes when they are observed after the cow moves a few steps (Te Plate and McDaniel, 1990). Since the advent of photogrammetric methods for measuring hock angle (Greenough, 1987), more precise measurements have been possible. In one study, Vermunt (1994a) found that the range of the hock angle was from 154.3–177.4° and that there is a decrease in the angle with age. McDaniel (1994) states that the 'mildly straight leg' can be associated with survivability. However, it must be appreciated that the very straight limb is correlated to a high incidence of joint disorders (Bailey, 1985). McDaniel (1994) also points out that 'rear leg rear view' scoring is extremely valuable in assessing overall limb conformation.

Strategies for improving leg and claw quality are being developed. Distl (1994) states:

> Important parameters for claw and leg quality can only be identified when traits used in breeding work are closely related to claw health, longevity, lifetime performance and functional efficiency of the animal. This definition implies that claw and leg quality cannot be recorded by just one trait. The traits necessary seem to be more complex and may be of different importance in dependence of the exposure to environmental effects. Particularly, claw shape is a result of the interaction between individual factors and environment. Genetic components may respond differently to specific environments and in each specific environment other genetic components may play the prominent role.

In the late 1970s the Nordic countries introduced a system whereby claw and leg traits were given an economic rating which was included when the total merit index was being calculated.

Research during the past decade has established a rationale for contributing to the control of lameness through improving claw and limb quality. Still further work is needed to establish claw and limb traits as useful parameters for the epidemiological investigation of herd lameness.

Stockmanship

The study on lameness in dairy cows conducted by the University of Liverpool considered the role of stockmanship in foot lameness in UK dairy cattle (Ward, 1994a). How important is the farmer as a cause? They found that the amount of lameness was closely related to his/her knowledge, training and awareness. The necessity of providing short courses for dairy farmers and milkers is obvious. However, as a corollary to this problem, the education of the veterinary surgeon should also be taken into account. In some countries the knowledge of the veterinarian is limited to the treatment of foot rot (interdigital phlegmon) and some simple semi-surgical procedures. In other countries (Italy and Spain) veterinary practices specializing in digital disorders exist. From another perspective, undergraduate veterinary education hardly ever reflects the economic importance of lameness in the curriculum. Instruction on the anatomy of the digit is often completely absent from the curriculum as may be detailed instruction on such topics as bovine laminitis.

Functional hoof trimming

The term 'functional hoof trimming' implies that the 'Dutch method' of hoof care, which was originated by Toussaint-Raven (1989), is being applied. Correctly performed hoof trimming is considered to be beneficial (Manson and Leaver, 1988). However, undue stress can be counterproductive to milk production (Stanek *et al.*, 1994) as can poor trimming technique. The use of a tipping table and older type of equipment can cause the trimming period to be extended to as much as 30 minutes. The modern Danish or Dutch claw trimming crushes, combined with contemporary hydraulic hoof cutters and metal bladed, electric angle grinders, can permit the procedure to be completed in as little as 7 minutes with minimum of distress to the cow. The conclusion reached by the Liverpool workers seems to be highly appropriate (Clarkson *et al.*, 1993): 'Foot-trimming can be beneficial, but not always. It would seem that correct training in the correct technique is essential.'

Cow comfort (etho-pathology of lameness)

Significant improvement in milk production per cow has taken place during the past ten years. During the same period there has been a tendency for size of the herd to increase, concrete to take the place of pasture, and cubicles to be favoured over straw yards or tie stalls. The economics of contemporary dairy farming demands intensive management. In some cases this demand creates stresses in the cows, caused by a conflict between their normal behaviour and the environment in which they must exist.

At least two factors can link housing/behaviour to a higher than normal incidence of lameness. Cermák (1990) hypothesizes that if cows lie longer in cubicles, their exposure to slurry deposited in passageways will be subsequently shorter. This in turn would reduce the environmental challenge to the foot as well as reducing the likelihood of falls on slippery concrete surfaces. He goes on to point out that there is a forward space demand (0.7–1.1 m for a 600 kg Friesian dairy cow) as she lunges forward to rise. Cermák (1990) extrapolates this in Table 10.1 to recommended cubicle lengths.

The cubicle partition should be of 'space sharing' design and provide three zones of free space for the head, ribcage and pelvic area. The bottom division rail should be set at from 34–40 cm and the top rail at from 111–117 cm from the floor. The width of the cubicle should be from 115–122 cm.

The base and the bedding of the cubicle has a profound effect on the lying time of the cow. The more resilient and soft the lying surface, the longer the cow will rest. Cows will lie for as many as 14 hours in the most comfortable cubicles. The number of cubicles available per animal may be important, particularly for heifers (Leonard *et al.*, 1994), some of which are sensitive to social confrontation when first introduced into the milking herd (Greenough

Table 10.1. Relationship between chest girth, body weight, diagonal body length and the cubicle length (Čermák 1990)

Cow body (kg)	Chest girth (m)	Body length (m)	Cubicle length (m)
375	1.68	1.36	2.00
425	1.75	1.41	2.04
475	1.81	1.46	2.08
525	1.87	1.50	2.12
575	1.93	1.54	2.16
625	1.98	1.58	2.20
675	2.04	1.62	2.24
725	2.09	1.65	2.28
775	2.14	1.68	2.30
825	2.18	1.72	2.33

and Vermunt, 1990). Heifers should be introduced into the milking herd in groups and care should be taken to ensure that each is properly trained to use a cubicle. Heifers that stand for long periods tend to have a greater preponderance of haemorrhages in the sole of their claws. Haemorrhages are both an indication of bruising and the presence of laminitic changes.

The design of the loose housing system is important. Potter and Broom (1987, 1990) point out that space is used very competitively between rows of cubicles, around drinking troughs in milking collecting yards and at entries and exits. Space available in these 'strategic sites' must be generous if a cow is to have sufficient personal space for flight to accommodate aggressive encounters between the various animals in the social hierarchy. It is desirable that the width of passageways between the cubicles and a feeding bunk should be 3–3.5 m. The loafing or exercise area should be calculated at not less than 3.3 m² per cow (Sainsbury and Sainsbury, 1979; Zeeb, 1987). Computerized feeding devices often result in the cows spending prolonged periods waiting for access to the equipment.

The floor surfaces over which cows walk have also received a great deal of attention. Slurry creates an environment for the claw that softens the horn, which allows in turn the rate of wear to increase and enhances the risk of damage. The concrete surfaces may be extremely smooth and this state increases the risk of slippage.

There has been much study of the interaction between housing and behaviour. Most authors believe that adverse conditions result in an increased incidence of lameness. Some of this lameness is the direct result of injury. There is also a belief that animals that suffer stress are more prone to digital disease than those that are not stressed. If this is true it is still not possible to deal with the problem objectively. If the incidence of lameness in a herd is high, it should be assumed that the environment is interacting adversely with the behaviour of the animals. Careful observation of the behaviour of the animals, *vis-à-vis* aggressive behaviour and/or amount of time spent resting can provide a useful indicator of the importance of the environmental factor.

It is difficult to convince a farmer to make extensive physical alterations to facilities within an existing housing shell. However, the following recommendations should be considered in order of priority.

1. Ensure that 25% more cubicles are available than there are cows. Maximize the comfort aspect of the cubicles. The average lying time should be between 8 and 10 hours.
2. Ensure that there is a 25% greater allowance for feeding space than is needed to accommodate the number of cows being fed. Observe if the behaviour of dominant cows is detrimental to the submissive animals, and, if this is the case, investigate the possibility of ameliorating the situation.
3. Resurface a smooth concrete floor with either a roughened surface or improve the existing one by cutting grooves 10 mm wide and 40 mm apart in a quadrilateral pattern (Dumelow and Albutt, 1990).
4. If a computerized feeder is being used, ensure that it is not causing confrontational problems or depriving cows of resting time by forcing them to wait for considerable periods. If it is, discontinue its use.
5. Make sure that slurry is not permitted to accumulate.
6. Investigate the stocking rate and attempt to recommend measures that would make exercise space available at the rate of $4\,m^2$ per cow.

Cattle tracks

In countries such as New Zealand, housing is not a factor in the aetiology of lameness. The condition and use of tracks has been identified as being a contributory cause of herd lameness (Chesterton, 1989; Chesterton *et al.*, 1989; Clackson and Ward, 1991). Several factors are involved. Firstly, crowding or the concentration of animals moving on the track at one time is important. Unfavourable conditions are created if the track is narrow. The use of dogs and motor bikes to hasten the animals along tends to make them bunch up. When a cow is walking it is important for her to see where she is placing her feet in order to avoid traumatic materials. The second factor is the condition of the track. It may be very stony or abrasive. Tracks may become extremely muddy during certain seasons of the year and this provides ideal conditions for the transmission of organisms that cause interdigital and digital diseases. Controlling roadway-induced lameness involves improving the quality of the track and correcting poor handling of the animals by the stockmen.

Foot baths

Foot baths provide a traditional technique aimed at reducing the reservoirs of organisms on the interdigital skin. In recent years the installation of permanent foot baths is being discontinued in favour of portable equipment, usually fabricated from fibreglass. Trials have been conducted to evaluate the efficacy of different solutions, and formalin or formalin and copper sulphate solutions are the most effective (Serieys, 1982). Other chemical agents such as iodides or cresols fail rapidly due to the high levels of organic matter present in the washing fluid. Formalin at a concentration of 5% is considered to be effective if the ambient temperature is more than 13°C. Formalin foot baths are effective in reducing the incidence of interdigital dermatitis. Reports concerning the use of formalin for the control and treatment of digital dermatitis are extremely contradictory.

It should be noted that formalin is poorly biodegradable and that there are reports of milk taint when the product is used. Foot baths should always be placed at the exit of the milking parlour. Cleansing the digits by running the animals through a clear water bath water prior to entering the parlour not only reduces the bacterial burden on the skin but extends the life of a medicated foot bath by minimizing contamination of the bath with organic matter (Blowey, 1994).

Since digital dermatitis has become a major problem various antibiotic solutions have been used (oxytetracycline-HCl, $<8\,\mathrm{g}\,\mathrm{l}^{-1}$; Lincospectin 100, Upjohn Ltd $0.062\,\mathrm{g}\,\mathrm{l}^{-1}$). The antimicrobial activity and concentration of these products reduced significantly after a herd had used the bath possibly due to absorption in faeces and soil particles (Keulen *et al.*, 1992.)

A minimal solution foot bath is now marketed (Ward, 1994c) that has a soft foam base lying beneath a waterproof membrane. When a cow steps into the bath the fluid moves to bathe her feet. The bath needs only 10–15 litres of fluid compared with 125–200 litres in a traditional foot bath. Fluid is used at the rate of about 4 litres for every 25 cows (Ward, 1994b).

Management of replacement animals

The corium of the claws of cattle aged between 8 and 13 months are more susceptible to nutritional and management stress than is the case with older animals (Greenough *et al.*, 1990; Greenough and Vermunt, 1991).

Solear haemorrhages are a consistent finding in the claws of animals affected with sub-clinical laminitis. Beef steer calves fed high levels of energy have more haemorrhages in the soles of their claws than those fed lower energy levels. Heifers that increased in weight at rates greater than 750 g per day showed more haemorrhages in the soles of their claws than those that increased in weight less rapidly. However, there is yet no objective evidence that establishes a link between solear haemorrhage in the young animal and claw disease in later years. From a circumstantial perspective, in the dairy herds in which laminitis is recognized as a major problem a significant number of heifers are found to have haemorrhages in the soles of their claws.

The following recommendations should be considered:

1. Monitor the growth rate of heifers (particularly small framed animals) between the age of 3 and 15 months. Attempts to increase daily weight gain mainly through the use of carbohydrates is definitely contraindicated. It has been suggested that feeding high levels of carbohydrate depressed growth hormone production and increases the percentage of intercellular mammary fat deposits.
2. Allow heifers to adjust to reduced exercise and walking on a concrete surface prior to introducing them to the dry-cow unit. An acclimatization period of 8 weeks would be appropriate.
3. Avoid introducing single heifers into the dry-cow unit to minimize bullying. Ensure that each heifer is trained in the use of the cubicle. If they are slow learners, halter them and tie them in the cubicle for several hours each day.
4. Ensure that there is a choice of cubicles and adequate feeding space in order to minimize confrontation with mature cows (Cheli and Mortellaro, 1974).
5. Lightly trim the claws prior to the entry of the heifers into the dry-cow unit.

Control of Diseases of the Digital Skin

Digital dermatitis

In California alone this disease is costing the dairy industry \$1.5–5.0 m annually (Read, 1994a,b). It was first reported in Italy in 1974 by Cheli and Mortelaro and has since been reported as an increasing world problem. The etiology of the disease is not known although spirochaetes (*Borrelia* spp.) have been isolated consistently from lesions. It is thought that the condition has a multi-factorial aetiology and that bacteria and/or viruses may also be implicated (Zemljic, 1994).

Good results have been obtained from using foot baths containing 125 g of lincomycin (Lincospectin 100, Upjohn Ltd) per 200 litres of water or $6\text{--}8\,\mathrm{g\,l^{-1}}$ of oxytetracycline (Blowey, 1994). If a herd fails to respond to routine antibiotic foot bathing it is recommended that the heel region of every animal is sprayed with an antibiotic solution. Shearer and Elliot (1994) recommend that a solution of $25\,\mathrm{mg\,ml^{-1}}$ of oxytetracycline in 20% glycerine and deionized water be sprayed on the heels of the cows daily for five days.

Interdigital dermatitis

Many authors have implicated *Dichelobacter (Bacteroides) nodosus* as the cause of this condition (Thorley *et al.*, 1977; Laing and Egerton, 1978; Espinasse *et al.*, 1984). As a simple interdigital skin infection it appears not to produce lameness and is therefore of little significance. However, there are suggestions that interdigital dermatitis or the causal organism may be involved in the pathogenesis of digital dermatitis. What is more important, *D. nodosus* is keratolytic and is possibly implicated as a contributory agent causing heel erosion. This condition, in its most advanced form, will cause lameness.

Interdigital dermatitis is most severe in the presence of unhygienic conditions which are likely to occur in most operations during the winter months. Control measures must include reducing the presence of slurry. In the northern hemisphere the routine use of a foot bath medicated with formalin should commence on a weekly basis in October. More frequent use of the bath may be required later in the winter if the incidence of the condition increases. Good results have also been obtained with lincomycin (Lincospectin 100, Upjohn Ltd) at the rate of 125 g per 200 ml water.

Interdigital phlegmon (Foot rot, foul in the foot, interdigital necrobaccilosis)

Foot rot is caused by *Fusobacterium necrophorum* in the presence of other organisms. Usually the organisms enter through the damaged interdigital skin. Recently a condition referred to colloquially as 'super foul' has been described. Although similar in almost every respect to phlegmon the characteristic of 'super foul' is the rapid onset and severity of lesions without the usual initial signs of damage to the interdigital skin (Blowey, 1994).

This disease usually responds rapidly to treatment by injections of antibiotics. Penicillin is widely used but care must be taken to ensure that adequate dosages are employed. In the past the dosage recommended by the manufacturers was inadequate and frequently associated with treatment failure. Revision of the dosage recommended on the product label may occur in the near future.

Controlling foot rot is extremely difficult. Vaccines are available but they usually only provide immunity for a few months. Some reduction in incidence may be expected if regular foot bathing is practised. However, once a case has become 'open' the discharge of the causal organisms into the environment causes contamination of those areas in which the traffic of cattle is heaviest. The reduction of slurry is extremely important. Good drainage around drinking and feeding areas is important. The isolation of cows (or the use of a protective boot) during the early infectious stages of the disease is strongly recommended. Very early, adequate treatment is essential. The use of oral inorganic iodides is prohibited in some countries but where it is used it has proved helpful in some herds.

Interdigital hyperplasia (corns, fibroma)

Cîrlan (1982) found that there is a hereditary component to the condition which determines the transmission as an inconsistent autosomal dominant pattern. He found also that there was a correlation to the butterfat yield of the dam as well as the body weight and the season.

Chronic irritation from the causal organisms of interdigital dermatitis probably is more important than is genetic predisposition as a cause of this condition in dairy cattle.

Control of interdigital hyperplasia should be directed towards the elimination of seriously affected animals as well as the control of interdigital dermatitis.

Control of Diseases of the Claw Capsule

Fissures

Vertical fissures
These are small cracks that first appear at the horn skin junction and involve the coronary band. Referred to as 'sand cracks', they are a major problem of beef cattle in some areas but are relatively uncommon in dairy cows. The aetiology of the disease is unknown.

Horizontal fissures
Acute horizontal fissures or thimbles are encountered sporadically and result from a complete cessation of horn growth for an undetermined period. As horn wall growth pushes the fissure distally it can be compared to a hang nail in humans and causes considerable pain. This condition can be caused by any acute febrile disease.

Less acute horizontal fissures or hardship grooves may result from a sudden stress such as a nutritional insult that might cause laminitis. Usually horizontal fissures are not a herd problem, but if they become so the investigator should assume that a laminitis-like situation exists.

Claw deformities

Overgrowth
Overgrowth of the claws of cattle can result from insufficient exercise on hard surfaces. Lack of wear may also occur in hot dry climates where the claw horn

becomes extremely hard. Young animals that are growing rapidly will also produce claw horn at a more rapid rate than is normal. Regular claw trimming is recommended but only if performed by a skilled operator. Regular removal of old horn probably accelerates the growth of new sound horn.

Corkscrew claw

This condition affects about 3% of all Friesian cattle. It develops from strain of the abaxial collateral ligament of the lateral hind claw. The resulting ligament repair causes a build-up of bony exostosis which presses and stimulates the horn-generating tissues of the coronary corium. Once started the condition is irreversible. Anecdotally, most workers consider this to be a hereditable condition and animals with this defect should not be used for breeding replacement animals. Unfortunately the condition does not appear in a cow until she is over four years of age. Bulls rarely have the advanced condition but some have lateral hind claws more concave than the medial claw. It is wise not to use such animals for breeding purpose unless their ancestry has been carefully scrutinized for the defect.

Slipper foot

At one time this was believed to be a specific condition. Today it is considered to be synonymous with chronic laminitis. Reducing laminitis in a herd will reduce the incidence of this condition.

Control of Laminitis and Diseases Associated with Laminitis

The importance of laminitis in dairy cattle has attracted the attention of many reseachers (see review by Vermunt and Greenough, 1990; Vermunt, 1994b). The term laminitis is applied to a condition that causes impairment of the circulation to the horn-producing tissues. The pathogenesis of the disease is not certain. The acute form of the disease in cattle usually results from a sudden, often accidental, consumption of concentrate high in energy. The environment of the rumen changes dramatically, an acute acidosis follows, and a vasoactive anaemia results which interferes with the circulation of the corium. Because the occurrence of acute laminitis results from cattle having unlimited (and accidental) access to concentrates or grain, recommendations for control depend on improved supervision of the animals.

A phenomenon referred to as 'sub-clinical laminitis' was first described by Peterse in 1979. The onset of the disease is insidious and signs of the condition can not be observed until the condition is well established. As with the acute form of the condition, impairment of the circulation to the corium occurs, which results in subtle changes in the quality of the sole horn. As the texture of the horn softens, it becomes more prone to wear and damage. Diseases such as pododermatitis circumscripta (sole ulcer), white line disease, double sole and possibly heel erosion become more prevalent in herds in which sub-clinical laminitis has become established. Haemorrhage of the sole is considered to be a very significant characteristic sign of sub-clinical laminitis. Sole haemorrhages are used by many workers to evaluate the severity of laminitis in a herd or group of cattle.

Discussion of the control of the specific diseases associated with sub-clinical laminitis will not be included in this review because control of the sub-clinical laminitis should reduce the incidence of these conditions.

In 1982 Mortensen and Hesselholt introduced the concept that sub-clinical laminitis had a multifactorial aetiology. Nutritional mismanagement still remained the major causative factor but they recognized that other factors made an animal more susceptible to sub-clinical laminitis. The disease is essentially one of intensive management of animals in high production herds. Control will therefore be considered from both the nutritional perspective and other factors.

Nutrition

Carbohydrate

Sudden increases in carbohydrate intake or the continued intake of high levels of carbohydrate leads to a disturbance in the microbial environment of the rumen (Peterse et al., 1986). It is universally accepted by research workers that the resulting acidosis is associated with the release of vasoactive toxins which adversely affect the microvasculature of the digital corium.

It is dangerous to make generalized statements regarding the management of carbohydrates because circumstances vary greatly from one farm to the next. For example, if silage is being used, it would be wise to have this forage analysed at regular intervals. Farmers can and do underestimate the nutritional content of some samples. It may then be necessary to make adjustments to the fibre, carbohydrate and protein content of the concentrate ration. When concentrate and roughage are fed separately it is important to confirm that the level of roughage consumed is greater than 30% of the dry matter intake. The intake of large quantities of carbohydrate-rich concentrates only twice each day should be avoided because this may reduce the intake of roughage.

Some cows may be more sensitive to high levels of carbohydrate than others. The dairymen should be instructed to observe the feet while the cows are being milked. Cows that are unable to accept the amount of carbohydrate consumed may show a slight pinkness around the coronary band and dew claws. This is a transitory phenomenon but one that is quite significant. If most newly calved cows show this sign, it is an indication that too much concentrate is being consumed (and wasted).

Steaming up (lead feeding). Heavy concentrate feeding prior to calving can increase an animal's predisposition to laminitis. However, suddenly increasing the amount of carbohydrate fed after calving will also increase the possibility of laminitis occurring. It is recommended that the pre-calving feeding programme should be based on the animal's body condition. After calving the cow must be introduced gradually to a ration if it is formulated to be high in energy. Peterse (1982) recommended that until the concentrate offered reaches 8 kg per day the daily offering should not be increased by more than 1 kg per day. Once the cow is consuming 8 kg of concentrate per day, further daily increases should be limited to 0.5 kg per day.

Buffers. Some producers add sodium bicarbonate to the ration at 1% of the dry matter. Including more buffer may reduce the palatability of the ration. Although not a buffer, providing rock salt licks will increase salivation in cows and the saliva increases the rumen pH.

Barley. Several authors (Nilsson, 1963; Maclean, 1966, 1970; Weaver, 1971; Little and Kay, 1979) associate feeding barley with an increase in the incidence of laminitis. It is probably wise to avoid the use of barley in dairy cow rations.

Protein

The literature is controversial on the role of protein in the pathogenesis of laminitis. In some cases feeding protein at levels in excess of 18% is associated with laminitis (Manson and Leaver, 1988; Bargai *et al.*, 1992). In other instances no such relationship could be established (Greenough *et al.*, 1990). There is no evidence that any particular source of protein is more dangerous than any other. However, protein-rich grass has been associated with the occurrence of laminitis (Vermunt, 1992) but it is unclear to what extent allergic reactions to protein exist. It should also be borne in mind that grass that is growing extremely rapidly tends to be low in fibre.

Roughage

The level and quality of fibre intake is considered to be a critical factor in the aetiology of laminitis (Peterse, 1982). Peterse suggests that the fibre in hay is superior to that in fast-growing pasture grass. He also recommends that the level of roughage should never be allowed to drop below 30% of the dry matter intake. Heavy manuring of pasture or the generous use of nitrate fertilizers can sometimes cause problems. If the weather is warm and wet, pasture will grow rapidly and during the period immediately prior to seeding have a high nitrate content. Nitrate is converted to nitrite in the rumen. Nitrite can reach toxic levels and because of its vasoactive properties may contribute to laminitis in cattle at pasture (Vermunt, 1992). Although definitive recommendations can not be offered regarding pasture management, farmers should be made aware of the potential risks.

If chopped, forage should not have a particle size of less than 2.5 cm (Vermunt, 1990).

Other factors

Exercise

Exercise is extremely important for the health of claws. Movement causes blood to circulate freely through the vasculature of the digit. Lack of movement allows blood to pool in the digit and the tissues have less opportunity to be oxygenated. Any managemental change that reduces the opportunity for an animal to walk freely is detrimental. This situation can occur when young animals are taken from pasture and placed in relatively confined spaces. It can also occur if heifers stand for prolonged periods when they are introduced to the dry herd. Aggressive social interaction with dominant cows together with unfamiliarity with the cubicle system causes this to happen. In some cases this may be the first time that the animal has walked on concrete and it is also a time at which the diet changes. The cumulative effect may stress the animal.

Rearing replacement heifers

Over the past decade interest has focused on the finding that laminitis commonly affects dairy heifers (Peterse and Van Vuuren, 1984; Moser and Divers, 1987; Bradley *et al.*, 1989; Colam-Ainsworth *et al.*, 1989; Vermunt, 1990; Greenough and Vermunt, 1991; Bargai *et al.*, 1992; Frankena *et al.*, 1992; Leonard *et al.*, 1992, 1994). The fact that so many workers have identified laminitis-like problems in young dairy heifers is probably the most significant finding of recent years. Too high a rate of growth during puberty, the stress of social interaction, sudden changes in diet, reduced exercise and negative reaction to a new and hostile environment are all factors that cumulatively predispose heifers to the occurrence of laminitis. There is evidence that once damage to the vasculature of the corium has occurred the animal will become increasingly sensitive to future insults.

Discussion

The subject of controlling lameness is a vast topic. It is a problem that is likely to increase in the future if the trend towards more intensive management continues. Excellent research models have been developed and ample evidence has been forthcoming to point the direction for future research.

An extremely important lameness problem that requires much more study is that of digital dermatitis. The occurrence of the disease is increasing extremely rapidly and significant economic losses are being reported.

Another profitable direction for future study is to provide guidelines for the genetic selection of dairy cattle in order to increase resistance to digital disease and improve longevity. The dairy industry is still relatively unaware of the considerable importance of claw size, shape and quality.

Heifer management is emerging as one of the most important and urgent matters requiring attention. Pressure from the industry to bring animals into milk at an earlier age has to be considered in relationship to possible reduction in the longevity of the animal. Animal behaviour in reaction to the inadequacy of modern facilities needs further study.

Finally, the most important factor: veterinary education and the transfer of knowledge to dairy farmers has failed lamentably to keep up with the excellent information that has been made available from research over the past few years. One exception and a good example for the future is the work of the *Centre d'Ecopathologie*, Villeurbanne, France, which has directed significant resources towards the better understanding of lameness problems by farmers and veterinarians.

References

Bailey, J.V. (1985) Bovine arthritides; classification, diagnosis, prognosis and treatment. In: The Veterinary Clinics of North America: *Food Animal Practice* 1, 39–51.

Bargai, U., Shamir, A., Lubin, A. and Bogin, E. (1992) Winter outbreaks of laminitis in calves: aetiology and laboratory, radiological and pathological findings. *Veterinary Record* 131, 411–414.

Baumgartner, C. and Distl, O. (1990) Genetic and phenotypic relationships of claw disorders and claw measurements in first lactating German Simmental cows with stayability, milk production and fertility traits. *Proceedings of the VIth International Symposium on Disorders of the Ruminant Digit, Liverpool, UK*, pp. 199–218. British Cattle Veterinary Association, University of Liverpool, UK.

Blowey, R.W. (1994) Studies on the pathogenesis and control of digital dermatitis. *Proceedings of the VIIIth International Symposium on Disorders of the Ruminant Digit, Banff, Canada*, pp. 168–173. Continuing Veterinary Education Section, University of Saskatchewan, Saskatoon, Saskatchewan, Canada.

Bradley, H.K., Shannon, D. and Neilson, D.R. (1989) Subclinical laminitis in dairy heifers. *Veterinary Record* 125, 177–179.

Čermák, J. (1990) Notes on welfare of dairy cows with reference to spatial and comfort aspects of design of cubicles. *Proceedings of the VIth International Symposium on Diseases of the Ruminant Digit, Liverpool, UK*, pp. 85–90. British Cattle Veterinary Association, University of Liverpool, UK.

Cheli, R. and Mortellaro, C.M. (1974) La dermatite digitale del bovino. Cited in Mortellaro C.M. (1994) Digital Dermatitis. *Proceedings of the VIIth International Symposium on Diseases of the Ruminant Digit, Banff, Canada*, pp. 137–141. Continuing Veterinary Education Section, University of Saskatchewan, Saskatoon, Saskatchewan, Canada.

Chesterton, R.N. (1989) Examination and control of lameness in dairy herds. *New Zealand Veterinary Journal* 37, 133–134.

Chesterton, R.N., Pfeiffer, D.U., Morris, R.S. and Tanner, C.M. (1989) Environmental and behavioural factors affecting the prevalence of foot lameness in New Zealand dairy herds. *New Zealand Veterinary Journal* 37, 135–142.

Cîrlan, M. (1982) Hyperplasia interdigitalis in bulls. A genetic and epizootiologic 6-year study. *Proceedings of the IVth International Symposium on Disorders of the Ruminant Digit, Paris, France*, pp. 23–24. Sociéte Française de Buiatrie, 94704 Maisons-Alfort, CEDEX.

Clackson, D.A. and Ward, W.R. (1991) Farm tracks, stockman's herding and lameness in dairy cattle. *Veterinary Record* 129, 511–512.

Clarkson, M.J., Downham, D.Y., Faull, W.B., Hughes, J.W., Manson, F.J., Merritt, J.B., Murry, R.D., Russell, W.B., Sutherst, J.E. and Ward, W.R. (1993) *An Epidemiological Study to Determine the Risk Factors of Lameness in Dairy Cows.* (Ref: CSA 1379). Final report, University of Liverpool, UK.

Colam-Ainsworth, P., Lunn, G.A., Thomas, R.C. and Eddy, R.G. (1989) Behaviour of cows in cubicles and its possible relationship with laminitis in replacement dairy heifers. *Veterinary Record* 125, 573–575.

Collick, D.W., Ward, W.R., and Dobson, H. (1989) Associations between types of lameness and fertility. *Veterinary Record* 125, 103–106.

Distl, O. (1994) Genetic improvement of claw and leg traits. *Proceedings of the VIIIth International Symposium on Disorders of the Ruminant Digit, Banff, Canada*, pp. 124–135. Continuing Veterinary Education Section, University of Saskatchewan, Saskatoon, Saskatchewan, Canada.

Distl, O., Koorn, D.S., McDaniel, B.T., Peterse, D., Politiek, R.D. and Reurink, A. (1990) Claw traits in cattle breeding programs: Report of the EAAP working group 'Claw Quality in Cattle'. *Livestock Production Science* 25, 1–13.

Dumelow, J. and Albutt, R. (1990) The effects of floor design on skid resistance in dairy cattle buildings. *Proceedings of the VIth International Symposium on Diseases of the Ruminant Digit, Liverpool*, pp. 130–134. British Cattle Veterinary Association, University of Liverpool, UK.

Espinasse, J., Savey, M., Thorley, C.M., Toussaint Raven, E. and Weaver, A.D. (1984) *Colour Atlas on Disorders of Cattle and Sheep Digit – International Terminology.* Maisons-Alfort: Editions du Point Vétérinaire, Paris, France.

Esslemont, R.J and Spincer, I. (1993) *The Incidence and Costs of Diseases in Dairy Herds.* DAISY Report No. 2, Department of Agriculture, University of Reading, UK.

Esslemont, R.J. (1990) The costs of lameness in dairy herds. *Proceedings of the VIth International Symposium on Diseases of the Ruminant Digit, Liverpool, UK*, pp. 237–251. British Cattle Veterinary Association, University of Liverpool, UK.

Frankena, K., Van Keulen, K.A.S., Noordhuizen, J.P., Noordhuizen-Stassen, E.N., Gundelach, J., de Jong, D-J. and Saedt, I. (1992) A cross-sectional study into prevalence and risk indicators of digital haemorrhages in female dairy calves. *Preventive Veterinary Medicine* 14, 1–12.

Greenough, P.R. (1987) A method for measuring conformation of cattle from a 35 mm transparency with a digitizing pad and computer. *Proceedings of the XXIIIth World Veterinary Congress Montreal, Canada*, p. 243. XXIII World Veterinary Congress, PO Box 1117, Succursale Desjardins, Montréal, Québec, Canada.

Greenough, P.R. and Vermunt, J.J. (1990) Evaluation of subclinical laminitis and associated lesions in dairy cattle. *Proceedings of the VIth International Symposium Disorders of the Ruminant Digit, Liverpool, UK*, pp. 45–154. British Cattle Veterinary Association, University of Liverpool, UK.

Greenough, P.R. and Vermunt, J.J. (1991) Evaluation of subclinical laminitis in a dairy herd and observations on associated nutritional and management factors. *Veterinary Record* 128, 11–17.

Greenough, P.R. and Vermunt, J.J. (1994) In search of an epidemiologic approach to investigating bovine lameness problems. *Proceedings of the VIIIth International Symposium on Disorders of the Ruminant Digit, Banff, Canada*, pp. 186–196. Continuing Veterinary Education Section, University of Saskatchewan, Saskatoon, Saskatchewan, Canada.

Greenough, P.R., Vermunt, J.J., McKinnon, J.J., Fathy, F.A., Berg, P.A. and Cohen, R.H. (1990) Laminitis-like changes in the claws of feedlot cattle. *Canadian Veterinary Journal* 31, 202–208.

Keulen Van, K.A.S., Spaans, J. and Pijpers, A. (1992) The antimicrobial activity of two tetracyclines and lincomycin in walk-through footbaths for dairy cattle: A longitudinal study. *Abstracts of the VIIth International Symposium on Disorders of the Ruminant Digit, Rebilt, Denmark 16*. K.W. Mortensen, Albertslund, Denmark, Abstr. 16.

Laing, E.A. and Egerton, J.R. (1978) The occurrence, prevalence and transmission of *Bacteroides nodosus* infection in cattle. *Research in Veterinary Science* 24, 300–304.

Leonard, L., O'Connell, J. and O'Farrell, K. (1992) The effect of cubicle design on behaviour and foot lesions in a group of in-calf heifers: Preliminary findings. *Abstracts of the VIIth International Symposium on Disorders of the Ruminant Digit, Rebilt, Denmark*. K.W. Mortensen, Albertslund, Denmark, Abstr. 29.

Leonard, F., O'Connell, J. and O'Farrell, K. (1994) Effect of overcrowded housing conditions on foot lesion development in first-calved Friesian heifers. *Proceedings of the VIIIth International Symposium on Disorders of the Ruminant Digit, Banff, Canada*, pp. 299–300. Continuing Veterinary Education Section, University of Saskatchewan, Saskatoon, Saskatchewan, Canada.

Little, W. and Kay, R.M. (1979) The effects of rapid rearing and early calving on the subsequent performance of dairy heifers. *Animal Production* 29, 131–142.

Maclean, C.W. (1966) Observations on laminitis in intensive beef units. *Veterinary Record* 78, 223–231.

Maclean, C.W. (1970) A post-mortem X-ray study of laminitis in barley beef animals. *Veterinary Record* 86, pp. 457–462.

McDaniel, B.T. (1994) Feet and leg traits of dairy cattle. *Proceedings of the VIIIth International Symposium on Disorders of the Ruminant Digit, Banff, Canada*, pp. 102–109. Continuing Veterinary Education Section, University of Saskatchewan, Saskatoon, Saskatchewan, Canada.

Manson, F.J. and Leaver, J.D. (1988) The influence of dietary protein intake and of hoof trimming on lameness in dairy cattle. *Animal Production* 47, 191–199.

Mortensen, K. and Hesselholt, M. (1982) Laminitis in Danish dairy cattle – an epidemiological approach. *Proceedings of the IVth International Symposium on Disorders of the Ruminant Digit, Paris, France*, pp. 31–32. Sociéte Française de Buiatrie, 94704 Maisons-Alfort, CEDEX.

Mortensen, K. and Hesselholt, M. (1986) The effects of high concentrate diet on the digital health of dairy cows. *Proceedings of the International Production Congress, Belfast, Northern Ireland*.

Moser, E.A. and Divers, T.J. (1987) Laminitis and decreased milk production in first-lactation cows improperly fed a dairy ration. *Journal of the American Veterinary Medical Association* 190, 1575–1576.

Nielsen, E. and Smedegaard, H.H. (1984) Disease in legs and hooves with Black and White dairy cattle in Denmark. *568 Report National Institute Animal Science, Copenhagen, Denmark*.

Nilsson, S.A. (1963) Clinical, morphological and experimental studies of laminitis in cattle. *Acta Veterinaria Scandinavica (Suppl)* 4, 304 pp.

Peterse, D.J. (1979) Nutrition as a possible factor in the pathogenesis of ulcers of the sole in cattle. *Tijdschrift voor Diergeneeskunde* 104, 966–970.

Peterse, D.J. (1982) Prevention of laminitis in Dutch dairy herds. *Proceedings of the IVth International Symposium on Disorders of the Ruminant Digit, Paris, France.* K.W. Mortensen, Albertslund, Denmark.

Peterse, D.J. and Van Vuuren, A.M. (1984) The influence of a slow or rapid concentrate increase on the incidence of foot lesions in freshly calved heifers. *Proceedings of the EAAP Congress, the Hague, The Netherlands.*

Peterse, D.J., Van Vuuren, A.M. and Ossent, P. (1986) The effects of daily concentrate increase on the incidence of sole lesions in cattle. *Proceedings of the Vth International Symposium on Disorders of the Ruminant Digit, Dublin, Ireland,* pp. 39–46. AD Weaver, Charlton Mecrell Court, Somerton, UK.

Philipot, J.M., Pluvinage, P., Cimarosti, I. and Luquet, F. (1990) On indicators of laminitis and heel erosion in dairy cattle: research on observation of digital lesions in the course of an ecopathological survey. *Proceedings of the VIth International Symposium on Disorders of the Ruminant Digit, Liverpool, UK,* pp. 184–198. British Cattle Veterinary Association, University of Liverpool, UK.

Politiek, M.J, Distl, O. and Fjeldaas, T. (1986) Importance of claw quality in cattle: review and recommendations to achieve genetic improvement. Report of the EAAP working group on 'claw quality in cattle'. *Livestock Production Science* 115, 133–152.

Potter, M.J. and Broom, D.M. (1987) The behaviour and welfare of cows in relation to cubicle house design. In: Wierenga, H.K. and Peterse D.J. (eds) *Cattle Housing Systems, Lameness and Behaviour,* pp. 159–165. Martinus Nijhoff Publishers, Boston.

Potter, M.J. and Broom, D.M. (1990) Behaviour and welfare aspects of cattle lameness in relation to the building design. *Proceedings of the VIth International Symposium on Disorders of the Ruminant Digit, Liverpool, UK,* pp. 80–84. British Cattle Veterinary Association, University of Liverpool, UK.

Read, D.H. and Walker, R.L. (1994a) Papillomatous digital dermatitis of dairy cattle in California: clinical characteristics. *Proceedings of the VIIIth International Symposium on Disorders of the Ruminant Digit, Banff, Canada.* p. 159. Continuing Veterinary Education Section, University of Saskatchewan, Saskatoon, Saskatchewan, Canada.

Read, D.H. and Walker, R.L. (1994b) Papillomatous digital dermatitis of dairy cattle in California: Pathologic findings. *Proceedings of the VIIIth International Symposium on Disorders of the Ruminant Digit, Banff, Canada,* p. 156. Continuing Veterinary Education Section, University of Saskatchewan, Saskatoon, Saskatchewan, Canada.

Reurink, A. and Van Arendonk, J.A.M. (1987) Relationships of claw disorders and claw measurements with efficiency of production in dairy cattle. *Proceedings of the 38th Annual meeting EAAP Lisbon, Portugal.*

Rogers, G.W. and McDaniel, B.T. (1989) The usefulness of selection for yield and functional type traits. *Journal of Dairy Science* 72, 187–193.

Rogers, G.W., McDaniel, B.T., Dentine, M.R. and Funk, D.A. (1989) Genetic correlations between survival and linear type traits measured in first lactation. *Journal of Dairy Science* 72, 523–527.

Russell, A.M., Bloor, A.P. and Davies, D.C. (1986) The influence of sire on lameness in cows. *Proceedings of the Vth International Symposium on Disorders of the Ruminant Digit, Dublin, Ireland,* pp. 92–99. AD Weaver, Charlton Meckrell Court, Somerton, UK.

Sainsbury, D. and Sainsbury, P. (1979) *Livestock Health and Housing,* 2nd edn Bailliere Tindall, London, UK, pp. 251–262.

Serieys, F. (1982) Comparison of eight disinfectants for cattle footbaths. *Proceedings of the IVth International Symposium on Disorders of the Ruminant Digit, Paris, France.* Sociéte Française de Buiatrie, 94704 Maisons-Alfort, CEDEX.

Shearer, J.K. and Elliot, J.B. (1994) Preliminary observations on the application of cowslips as adjunct to treatment of lameness in dairy cows. *Proceedings of the VIIIth International Symposium on Disorders of the Ruminant Digit, Banff, Canada,* p. 71. Continuing Veterinary Education Section, University of Saskatchewan, Saskatoon, Saskatchewan, Canada.

Stanek, Ch., Thonhauser, M.-M. and Schroder, G. (1994) Does the claw trimming procedure affect milk yield and milk quality factors? *Proceedings of the VIIIth International Symposium on Disorders of the Ruminant Digit, Banff, Canada,* pp. 306–317. Continuing Veterinary Education Section, University of Saskatchewan, Saskatoon, Saskatchewan, Canada.

Te Plate, H.A.M. and McDaniel, B.T. (1990) Description and evaluation of measuring rear legs side and rear view using photogrammetry in Holstein Friesians. *Report to Holstein Association (unpublished).* Cited in McDaniel, B.T. (1994) Feet and leg traits of dairy cattle. *Proceedings of the VIIIth International Symposium on Disorders of the Ruminant Digit, Banff, Canada,* pp. 102–109. Continuing Veterinary Education Section, University of Saskatchewan, Saskatoon, Saskatchewan, Canada.

Thorley, C.M., Calder, H.A.M. and Harrison, W.J. (1977) Recognition in Great Britain of *Bacteroides nodosus* in foot erosions of cattle. *Veterinary Record* 100, 137.

Toussaint-Raven, E. (1989) *Cattle Foot Care and Claw Trimming.* Farming Press Books, Ipswich, UK, 59 pp.

Vermunt, J.J. (1990) Lesions and structural characteristics of dairy heifers in two management systems. M.V.Sc. thesis, University of Saskatchewan, Saskatoon, Canada.

Vermunt, J.J. (1992) 'Subclinical' laminitis in dairy cattle. *New Zealand Veterinary Journal* 40, pp. 133–138.

Vermunt, J.J. (1994a) Hock conformation in dairy heifers in two management systems. *Proceedings of the VIIIth International Symposium on Disorders of the Ruminant Digit, Banff, Canada,* pp. 122–123. Continuing Veterinary Education Section, University of Saskatchewan, Saskatoon, Saskatchewan, Canada.

Vermunt, J.J. (1994b) Predisposing causes of laminitis. *Proceedings of the VIIIth International Symposium on Disorders of the Ruminant Digit, Banff, Canada,* pp. 236–258. Continuing Veterinary Education Section, University of Saskatchewan, Saskatoon, Saskatchewan, Canada.

Vermunt, J.J. and Greenough, P.R. (1990) Observations on management and nutrition in a herd of Holstein dairy cows affected by subclinical laminitis. *Proceedings of the VIth International Symposium on Disorders of Ruminant Digit, Liverpool, UK,* pp. 22–30. British Cattle Veterinary Association, University of Liverpool, UK.

Ward, W.R. (1994a) The role of stockmanship in foot lameness in UK dairy cattle. *Proceedings of the VIIIth International Symposium on Disorders of the Ruminant Digit,*

Banff, Canada, p. 301. Continuing Veterinary Education Section, University of Saskatchewan, Saskatoon, Saskatchewan, Canada.

Ward, W.R. (1994b) Recent studies on the epidemiology of lameness. *Proceedings of the VIIIth International Symposium on Disorders of the Ruminant Digit, Banff, Canada*, pp. 197–203. Continuing Veterinary Education Section, University of Saskatchewan, Saskatoon, Saskatchewan, Canada.

Ward, W.R. (1994c) The minimal solution footbath – an aid to treatment of digital dermatitis. *Proceedings of the VIIIth International Symposium on Disorders of the Ruminant Digit, Banff, Canada*, pp. 184–185. Continuing Veterinary Education Section, University of Saskatchewan, Saskatoon, Saskatchewan, Canada.

Weaver, A.D. (1971) Solar penetration in cattle: its complications and economic loss in one herd. *Veterinary Record* 89, 288–290.

Whitaker, D.A., Kelly, J.M. and Smith, E.J. (1983) Incidence of lameness in dairy cows. *Veterinary Record* 113, pp. 60–62.

Zeeb, K. (1987) The influence of the housing system on locomotory activities. In: Wierenga, H.K. and Peterse D.J. (eds) *Cattle Housing Systems, Lameness and Behaviour*. Martinus Nijhoff Publishers, Boston, pp. 101–106.

Zemljic, B. (1994) Current investigations into the cause of dermatitis digitalis in cattle. *Proceedings of the VIIIth International Symposium on Disorders of the Ruminant Digit, Banff, Canada*, pp. 164–167. Continuing Veterinary Education Section, University of Saskatchewan, Saskatoon, Saskatchewan, Canada.

Amelioration of Heat Stress in Dairy Cattle

11

J.T. Huber

Department of Animal Sciences, University of Arizona,
Tucson, Arizona 85719, USA

Introduction

High ambient temperatures (AT) often coupled with high relative humidities seriously depress milk production and reproductive efficiency of dairy cattle in many areas of the world. Heat stress will occur when the effective temperature (ET) exerted by the environment exceeds that of the cow's thermal neutral zone (TNZ), generally thought to range between 5 and 27°C (Armstrong, 1994), but depending greatly on relative humidity (RH). A temperature–humidity index (THI) has been developed to more accurately assess the potential for an environment to produce heat stress in cattle (NOAA, 1976), and is calculated by the following equation:

$$THI = td - (0.55 - 0.55\,RH)(td - 58)$$

where td is the dry bulb temperature and RH is expressed in decimal form. Johnson (1980) reported that an average daily THI higher than 72 is likely to result in decreased milk production. No clear-cut relationship between THI and propensity for cows to conceive has been reported, but decreases probably parallel those shown for decrease in milk yield. Dairy cows react to heat stress by several complicated physiological changes (Kibler and Brody, 1953), which are manifested by a number of overt reactions, including reduced feed intake, increased water intake and evaporative water losses from the body, increased body temperatures, respiration rate, metabolic rate, and maintenance requirements, as well as altered endocrine levels. High yielding cows have been shown to be more susceptible to heat stress than low yielders (West, 1994). Florida workers (Beede and Collier, 1986) proposed three methods for minimizing stressful thermal challenges in dairy cows: namely, (i) physical modification of environment; (ii) nutritional management; and (iii) genetic development of animals less sensitive to hot environments. This chapter will address physical alterations and nutritional schemes shown to help the dairy cow's ability to withstand heat stress.

Physical Modification of Environment

The four environmental factors that should be considered when physical altera- tion of the environment is undertaken and which affect the ET to which cows are exposed include: ambient temperature, relative humidity, air movement, and solar radiation (Armstrong, 1994). Several methods employed to change one or more of these factors will be discussed. The chart devised by Wiersma in 1990 (Fig. 11.1) shows the relationship between ambient temperature and relative humidity, illustrating degree of stress when cows are subjected to varying THI.

Shade

Protection from radiant heat through providing shade for cows is often suffi- cient if temperatures are not too severe (Wiersma, 1982). Types of shade vary considerably in their ability to relieve thermal stress. Shade from trees is quite effective because of the evaporative moisture released from leaves. Other shade materials often used are branches of trees, wood, cloth (such as polypropylene fabric) and corrugated steel (Armstrong, 1994). Painting of white metal shade material on upper surfaces in addition to providing some insulation (≈ 2.5 cm) will greatly enhance effectiveness of shade systems (Armstrong, 1994).

Shade space requirements vary with climate and pen conformation, but range from 3.5 to 5.6 m^2 per cow. For hot, humid conditions, the higher figure would be recommended to allow for more ventilation of cows (Buffington *et al.*, 1983). Placement and maintenance of shade is very important to allow for maximum protection during the hottest parts of the day and to avoid excessive urine and faecal accumulation in shaded areas. Cleaning of space under shades is imperative for their maximal effectiveness. When cattle are on pasture devoid of trees, portable shades can be used which are moved at 1–2-day intervals, to prevent damage to the grass under the shade and allow the cows to stay clean (Armstrong, 1994).

Cooling of cows

Various methods of cooling cows have been employed in different areas depen- ding on conditions and needs. In hot, humid climates (such as southeastern USA) considerable benefit has been derived by providing shade and a good source of ventilation because the air will generally contain sufficient moisture to

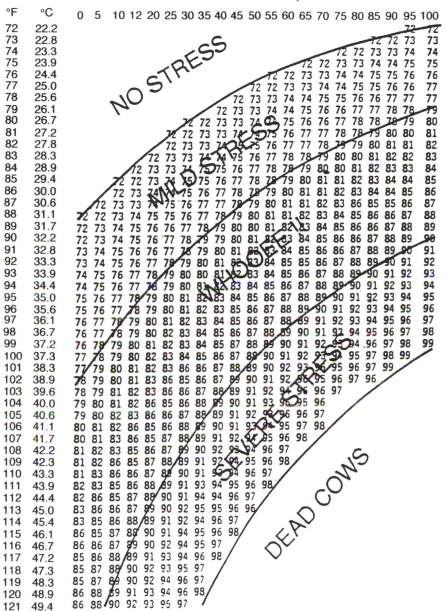

Fig. 11.1. Temperature-humidity index table for dairy producer to estimate heat stress for dairy cows. Deg = Degrees. Relative humidity expressed as percentage. (From F. Wiersma, Department of Agricultural Engineering, University of Arizona, as adapted by Armstrong, 1994).

produce a cooling effect. However, a combination of sprinkling and fans (Fig. 11.2) for shaded cows resulted in a marked increase in milk yields (12%) in a Florida study (Strickland *et al.*, 1988) compared with shade alone.

In hot, dry climates, spray systems that provide sufficient moisture to evaporatively cool cows have increased milk production and conception rates (Armstrong, 1994). They have also decreased body temperatures and respiration rates. Placement of coolers can be in holding pens, exit lanes, above feed mangers or under shades in pens.

Holding-pen cooling is very beneficial for large dairies located in arid areas such as southwestern USA. Cows might spend up to 60 minutes, two to three times daily, in a very crowded condition which would result in considerable heat stress if cooling were not provided. Overhead sprinklers (not foggers) with large fans installed in holding pens (Fig. 11.3) were tested in an Arizona trial (Wiersma and Armstrong, 1983) where maximum daily temperatures ranged between 27 and 46°C. Body temperatures of cows were decreased by 1.7°C and there were small increases in milk yields ($0.8 \, \text{kg day}^{-1}$). Estimated benefits from this cooling system were sufficient to pay for the investment in one year. Israeli scientists (Wolfenson *et al.*, 1984) reported that cows sprayed five times daily in the holding pen produced $2.4 \, \text{kg day}^{-1}$ more milk than uncooled controls.

Spraying of cows as they exit the milking parlour allows them to return to pens in a cooled condition and should encourage greater feed intake at a time when they are usually stimulated to consume larger quantities of feed than before milking. However, care should be exercised to not drench cows to the extent that the post-milking teat dip is washed off. Until now, little data is available on production and financial benefits of exit lane spraying, but it is being employed on many dairies during hot weather.

Misting of cows over feed mangers (Fig. 11.4) during intermittent periods of hot weather improved milk production and reproductive efficiency in California studies (Schultz, 1988). Increased response from feed line spraying occurs when cows are shaded or when there are natural air currents to improve the evaporative cooling effect. Careful design of such systems is imperative to prevent natural air currents from blowing the spray away from cows so that it is ineffectual. Also, the accumulation of excessive water on feed or the feeding area of cows can be problematic. Both Missouri (Igono *et al.*, 1987) and Florida (Bucklin *et al.*, 1989) workers reported little benefit from misting systems in hot, humid environments if they were not coupled with use of ventilating fans.

Pen evaporative cooling under shades has become a standard practice on many dairies located in hot dry climates in several areas of the world and is probably the most effective cooling method. It is estimated that over 50% of the dairies in the state of Arizona employ pen cooling during the hot summer months (Stott *et al.*, 1972; Armstrong and Wiersma, 1986; Ryan *et al.*, 1992). Even in areas of higher humidity, the daytime RH is often low enough for

Fig. 11.2. Fan and sprinkler system for cooling cows in an open pen. Courtesy of D.V. Armstrong.

this method of evaporative cooling to be beneficial (Brown *et al.*, 1974; Taylor *et al.*, 1986).

One system extensively used releases fine particles of mist from pressure jets about 1 m above the cows' bodies. The mist is agitated by a powerful circular fan which results in cooling the air temperature surrounding the cows by about 5°C. More importantly, the mist settles on the body surface of cows and, in conjunction with the air movement from the fan, a powerful evaporative cooling effect is exerted, reducing body temperatures of cows by as much as 2°C. A weighted curtain installed on the side of the prevailing wind automatically drops to prevent diffusion of the mist away from the cows. Inactivation of the cooler or high wind currents result in raising of the curtain. Figure 11.5 shows a 'Korral Kool' system in operation in a commercial dairy farm in Saudi Arabia.

A second and less costly system of evaporative cooling injects water mist at lower pressures into an air stream generated from normal fans suspended below shade roofs. This design has been developed and successfully used in a number of arid and semiarid regions. Without the drop curtain (as described previously), cooling is not as well controlled, but on still, hot days heat stress of cows can be greatly alleviated. Table 11.1 lists data from a number of studies comparing evaporative cooling (EC), spray plus fans, and uncooled systems. Both methods of pen cooling resulted in substantial increases in milk yields over uncooled controls, but the EC system was superior to spray plus fans for cows cooled during all stages of lactation and the dry period.

Fig. 11.3. Cooling of cows in holding pen with a mister system. Courtesy of D.V. Armstrong.

Fig. 11.4. A manger mister system for combatting periods of intermittent heat stress. Courtesy of D.V. Armstrong.

Fig. 11.5. Cooling of cows in a shaded pen. This system is commonly used in hot, arid climates of the world. Courtesy of D.V. Armstrong.

In addition to increased milk production, several of the pen cooling studies have reported considerable improvement in reproductive efficiencies in evaporatively cooled cows (Armstrong *et al.*, 1985; Smith *et al.*, 1993). Personal observation with pen evaporative cooling of University of Arizona cows during the hot weather has shown dramatic improvements in conception rates for several years in cows receiving cooling ($\approx$40% conceived at first insemination) compared with those housed with only shade ($\approx$10%).

Sprinklers with fans over feed mangers and additional fans in the cubicle area effectively cool cows in free-stall barns where excess water on floor surfaces is manageable and of no particular problem. Such systems provide direct evaporative cooling from body surfaces rather than relying mostly on convection cooling of cows (Strickland *et al.*, 1988).

Summary

Marked alleviation of the negative effects of hot summer temperatures in dairy cows can be realized through physical modification of the environment. In hot, humid environments, shades coupled with ventilation are quite beneficial. Amelioration of heat stress in hot, arid climates requires spray or misting systems in addition to ventilation to give an evaporative cooling effect. Such

Table 11.1. Estimated milk production response for different types of cooling systems under cattle shades compared with no cooling, in semiarid climates with an average daytime humidity less than 30% *.

Daily maximum temperature	Increase in milk production (kg milk per cow^{-1} d^{-1})	
	Evaporative cooling	Spray and fan
High production, ($>$38.5 kg d^{-1})		
$>$40.5°C	7.5	4.0
35–40°C	6.0	3.2
$<$34.5°C	5.3	2.8
Medium production, (29.5–38.5 kg d^{-1})		
$>$40.5°C	6.4	3.5
35–40°C	5.2	2.8
$<$34.5°C	4.5	2.5
Low production, ($<$29.5 kg d^{-1})		
$>$40.5°C	5.6	3.2
35–40°C	4.5	2.6
$<$34.5°C	3.9	2.3
*Dry cows***		
$>$40.5°C	2.0	1.4
35–40°C	1.4	0.9
$<$34.5°C	0.9	0.6

* Based upon research projects in southwestern United States, Mexico, and Saudi Arabia.
From Armstrong (1994).
** Milk production in first 120 d of subsequent lactation.

coolers can be constructed in holding pens, exit lanes, over feed mangers or in regular pens where cows are housed. Increased milk yield and reproductive efficiencies result in rapid amortization of costs of establishment and operation of the cooling systems.

Role of Nutrients in Combating Heat Stress

As mentioned, heat stress severely limits production and reproduction of dairy cows. There have been relatively few studies that have investigated dietary alterations which might allow high producing cows to better cope with hot environmental temperatures. The remainder of this chapter will deal with nutritional considerations to alleviate detrimental effects of heat stress in dairy cows.

Water

Water is the most important nutrient for all animals including dairy cattle. Its protective role in alleviating heat stress through evaporative cooling of the body is well known. This effect occurs principally through increased activity of sweat glands and respiratory organs. Loss of an estimated 20% of body water is generally fatal. Water needs by lactating dairy cows represent a larger percentage of their body weight than in other animals because of large amounts secreted in milk and the need for disposing of body heat generated by the rumen fermentation of fibrous feeds. Special water needs as related to amelioration of heat stress in lactating dairy cows will be discussed.

Water requirements of lactating cows are directly related to ambient temperatures as indicated in the following equation developed by Murphy *et al.* (1983):

$$\text{Water intake (kg day}^{-1}) = 15.99[0.271] + 1.6 \times \text{dry matter intake (DMI)}$$
$$(\text{kg day}^{-1})$$
$$+ 0.9[0.157] \times \text{milk yield (kg day}^{-1}) +$$
$$0.05[0.023] \times \text{Na intake (g day}^{-1}) +$$
$$1.20[0.106] \times \text{weekly mean minimum temperature}$$
$$[0.106] \times \text{weekly mean minimum temperature (°C).}$$

Weekly mean maximum temperature and weekly mean temperature were also significantly correlated with water intake to approximately the same extent, so any one of these temperature factors could have been used in the prediction equation. The other factors listed in this equation, milk yield, dry matter and Na intakes, undoubtedly affect water requirements, but other nutritional factors that also influence water intakes disproportionately to their contribution to the overall dry matter portion of the diet are concentrations of fibre and rumen degradable protein (RDP). As fibre content of diet increases, salivary flow (primarily water) also increases. Moreover, fibrous foods produce more heat of fermentation in the rumen and require larger amounts of moisture in faecal elimination than do concentrates. Pertaining to the relationship of water intake and rumen degradable protein, Higginbotham *et al.* (1989b) demonstrated that cows fed diets of 65% RDP (of the total CP) in hot climatic conditions drank about 15% more water than those fed diets containing about 60% RDP.

Using the Murphy prediction equation (Murphy *et al.*, 1983), which assumed a moisture content of 38% for the diet, Beede (1993) calculated that intake of drinking water would change 0.9 kg kg^{-1} change in milk yield, 1.5 kg kg^{-1} change in DMI, 0.05 kg g^{-1} change in Na intake and 1.2 kg per 1°C change in ambient temperature. Hence, a change from 15 to 40°C in minimum daily environmental temperature would result in an increase in drinking water intake of only about 30 l day^{-1}. This estimate is probably low, considering increases of about 100% (60–70 l day^{-1}) in water consumption observed in the field when such large changes in temperature occur, particularly in hot, dry climates (Beede, 1993). A possible reason for the discrepancies might be that RH, diet moisture and other factors mentioned later, but not taken into account in the equation of Murphy *et al.* (1983).

When estimating water needs of cows, diet moisture content should be considered. An air dried ration (90% DM) will furnish only about 2% of the total water requirement, whereas diets composed largely of silage, pasture or freshly cut herbage might contain 50% DM and provide 25 l of water or 15–20% of the cow's needs. Another water source is metabolic water, which might represent 15–20% of that required by the cow. This water undergoes some change with diet composition (higher with fat than carbohydrate), but its variability in cows is not sufficient to be considered in the equation for predicting water intakes.

Comparing cows maintained at 20 vs. 30°C, McDowell and Weldy (1967) showed that the higher temperature resulted in a 58% increase in total body water losses and 176% higher sweat losses. Respiratory and urinary losses were increased by 54 and 26%, and faecal losses decreased by 25%, probably due to decreased DMI. Beede (1993) suggests that little is actually known pertaining to water requirements of lactating cows during heat stress. However, factors such as rate of feed intake, physical form of the diet, physiological state and breed of the animal, as well as water quality, accessibility and temperature, are likely to affect water intake in hot weather (NRC, 1981).

Several studies have been conducted on temperature of drinking water for cows during hot weather. Data from three reports by Texas workers (Milam *et al.*, 1986; Stermer *et al.*, 1986; Wilks *et al.*, 1990) showed some advantage in milk yield of providing chilled (10°C) compared to warm water (26°C), but it was probably uneconomical. Comparisons by Florida researchers (NRC, 1974; Challis *et al.*, 1987) of chilled (11–15°C) to well water (24–27°C) showed no significant advantage of the colder water in milk yield. Therefore, the maintenance of well water at temperatures below 30°C through shading of water lines and drinking tanks, or other measures, provides water of sufficiently cool temperatures to not warrant chilling (Beede, 1993).

Of critical importance is the provision of high quality, clean water, devoid of toxic substances, undesirable odours and tastes. Good water should be sufficiently low in chemicals (such as NO_3, Fe, Na, SO_4 or Fl), which limit water intake or are injurious to health of cows. Physiochemical properties such as pH, total dissolved solids (TDS), hardness, and total dissolved oxygen affect water intake and should be maintained within their desirable range (NRC, 1974).

A dramatic example of decreased milk yield in cows drinking high saline water during hot weather in Saudi Arabia was reported by Challis *et al.* (1987). They found that desalination of the drinking water, which had previously contained 4400 ppm total dissolved salts (TDS) (of which 2400 ppm was

Table 11.2. Relative changes in maintenance and dry matter (DM) requirements for 600 kg cows producing 27.5 kg milk of 3.7% fat at various ambient temperatures along with estimates of actual intakes of DM and water. (Adapted from McDowell, 1974.)

Temperature[1] (°C)	Maintenance requirements (% of req. at 10°C)	Dry matter needed[2] (kg d⁻¹)	Dry matter intake[3] (kg d⁻¹)	Milk yield (kg d⁻¹)	Water intake (l d⁻¹)
−16	151	21.3	20.4	20.0	49.1
−10	126	19.8	19.8	25.0	56.1
0	110	18.8	18.8	27.0	61.9
10	100	18.2	18.2	27.5	64.6
20	100	18.2	18.2	27.0	65.4
25	104	18.4	17.7	25.0	71.2
30	111	18.9	16.9	23.0	76.2
35	120	19.4	16.7	17.2	116.1
40	132	20.2	10.5	12.0	102.6

[1] Values for 25°C and higher temperatures are for days with at least 6 h exceeding the temperature class but not more than 12 h.
[2] Estimated DM intake requirements for maintenance and milk production.
[3] Estimates of intakes of DM and water and milk yield on water-free choice and *ad libitum* feeding of a ration of 60% hay and corn silage with 40% concentrates.

sulphates, improved milk production by 30% (27 vs. 35 kg d⁻¹), as well as water intake (20%) and grain intake (32%).

In conclusion, *ad libitum* provision of cool (<30°C), clean, potable drinking water to lactating cows is essential for maintenance of high milk yields during hot weather. There are numerous factors that affect water intake, the most important of which are dry matter intake and ambient temperatures. The equation of Murphy *et al.* (1983) may be useful for planning purposes, but probably does not account for the full impact of heat stress on water consumption. Generally, hot, dry climates compared with cool climates, will cause a doubling of water intake by cows.

Feed intake

Perhaps the most detrimental event leading to decreased milk production in heat-stressed cows is a reduction in feed intake, which occurs concurrently with increased body temperatures and respiration rates (McDowell *et al.*, 1976). Maintenance requirements of lactating dairy cows increase by about 30% if ambient temperatures are raised from 25 to 40°C for 6 h d⁻¹ (Table 11.2), and DMI decreases to about 55% of that eaten by cows in the thermal neutral zone (TNZ) (4–25°C). The depressed intake, coupled with other detrimental effects, causes milk yields to drop to less than 50% of that produced by cows in the TNZ. A general increase in water consumption is expected up to about 35°C, but further increases in ambient temperatures over extended periods will decrease water intake due to inactivity of cows and lowered feed intake.

Mean and maximum daily temperatures often have a variable effect on feed intake and consequent milk production, depending on relative humidity (RH) and amount of time cows are exposed to stressful temperatures (McDowell, 1974). At moderately high temperatures, decreases in milk production might be magnified by high RH and/or poor acclimatization of cows to heat stress conditions. A change in eating patterns from day to night feeding has been associated with hot days and cool nights and is one way cows acclimatize to hot temperatures.

Cooling systems for modification of the environment to alleviate heat stress exert their effect largely through increased intakes. Corral and holding pen cooling systems markedly increase time that lactating cows spend at the feed manger. A typical observation has been that after cows are cooled in holding pens before milking and in exit lanes after milking, they eat more feed for longer periods when returned to corrals, than do uncooled cows. Moreover, cows which are pen cooled during hot weather approach the feed manger more frequently than those that are not cooled. As mentioned previously, misting of cows at the feed manger during sporadic periods of high ambient temperatures increases feed intake and consequent milk production.

Energy concentration of diets

Body heat production, rectal temperatures, and respiration rates are increased for high forage compared with moderate or high concentrate diets. The greater heat increment has been associated with higher acetate production in the rumen of cows fed high forage diets (Tyrell *et al.*, 1979). Often cows voluntarily limit their forage consumption during hot weather, even to the extent of drastically shifting acetate to propionate ratios and lowering butterfat content of milk. Addition of buffers (about 1% $NaHCO_3$, K_2CO_3 or $KHCO_3$ and 0.6% MgO) partially alleviates milk fat depression and helps cows maintain a healthier rumen fermentation during periods of heat stress when forage consumption is less than 1% of body weight (Coppock and West, 1987). Moreover, less heat is produced in fermentation of high quality compared with low quality forages (due to differences in their fibre content), but a minimum fibre of about 20% ADF or 28% NDF is recommended for maintenance of good rumen function.

Providing supplemental fat for lactating cows is a common dietary practice on many dairies. The possibility that added fat alleviates heat stress by providing non-fermentative energy (McDowell *et al.*, 1969; Tyrell *et al.*, 1979) to cows is of interest. In a study by Moody *et al.* (1967), supplementation of 10% fat (from soyabean oil or hydrogenated vegetable fat) to diets of thermally stressed cows did not increase milk yield. However, such high percentages of ruminally unprotected fat, particularly from soyabean oil, would probably depress fibre digestion if included in diets for high yielding dairy cows (Palmquist, 1987).

More recently, added fat (from whole oilseeds and/or commercial fat sources) has been fed to heat-stressed cows in commercial herds. Ruminally-inert fats, such as calcium soaps of fatty acids, prilled fatty acids, or moderately saturated triglycerides (tallow), which minimize fatty acid inhibition of rumen microorganisms (Palmquist, 1987), have been added at 2–3% of DM to alfalfa-based diets containing about 10% whole cottonseed. The supplemental fat increased milk yields by about 2 kg d^{-1} (Huber *et al.*, 1993). However, total fat content of dietary DM should not exceed 6–7% (Huber *et al.*, 1993; Palmquist *et al.*, 1987).

One study at the University of Arizona (Chan *et al.*, 1992) in heat-stressed cows subjected to maximum daily temperatures of over 40°C, with mean rectal temperatures of 39.6°C and respiration rates of 85 breaths min^{-1}, showed increased milk yield (by 1.2 kg) and milk fat (2 g kg^{-1}) from diets containing alfalfa hay, 10% whole cottonseed, and 2.5% supplemental fat as prilled fatty acids (PFA) compared with unsupplemented diets. A second study in Arizona (Chan *et al.*, 1993) with similar diets conducted during hot summer temperatures (mean rectal temperature of cows was 39.3°C and respiration rate was 83 breaths min^{-1}) showed only small increases in milk yields (0.6 kg) when evaporatively cooled (EC) or uncooled cows were fed 2.5% PFA compared with controls which received no added fat. However, EC tended to improve milk yield (by 1.6 kg), regardless of fat supplementation. Some, but not all studies (Chan *et al.*, 1992, 1993; Huber *et al.*, 1993) suggest that response from added fat is less for cows suffering heat stress than for those which are not (Table 11.3). The reason for the lesser response to added fat in heat-stressed cows is not obvious.

The hypothesis that added fat reduces heat of fermentation in cows by providing non-rumen fermentable energy during hot summer weather apparently was erroneous, because neither rectal temperatures nor respiration rates decreased in cows fed supplemental fat compared with those of cows fed no added fat (Chan *et al.*, 1992, 1993).

Skaar *et al.* (1989) showed that fat supplementation tended to increase milk yield of cows that calved during the warm, but not the cool, season; however, maximum temperatures during the warm months did not exceed 35°C, and amount of added fat was double (5%) that of the Arizona studies. Knapp and Grummer (1991) also fed 5% prilled fatty acids to cows housed in environmental chambers in a Latin-square design with 15-day periods, and showed a 1.1 kg d^{-1} non-significant increase in milk yield for cows in the

Table 11.3. Response to added fat in hot or moderate temperatures (Huber *et al.*, 1994).

Ambient temperature	DMI (kg d⁻¹)		NE$_L$ intake[1] (Mcal d⁻¹)		Milk yield (kg d⁻¹)		Milk protein (%)	
	Control	Added fat[2]	Control	Added fat[2]	Control	Added fat[2]	Control	Added fat[2]
Hot	22.9	24.0	37.3	41.0	32.9	34.1[a]	3.14	3.10
	25.8	24.7	42.3	42.7	30.3	30.9	2.95	2.95
Moderate	24.2	25.3	40.6	45.3	31.6	34.2[a]	3.13	3.01[b]
	27.2	26.8	44.6	46.4	32.6	34.3[a]	3.03	3.08

[1] Estimated from NRC (1989).
[2] EB, Energy Booster 100 (Milk Specialties Co., Dundee, IL).
[a] Probability that treatment with added fat = control <0.10.
[b] Probability that treatment with added fat = control <0.05.

cool environment fed added fat compared with non-fat controls; but only a $0.3 \, \text{kg} \, \text{d}^{-1}$ increase due to added fat was noted for cows in the warm environment. If there is a lower response to fat during heat stress, it might be attributed to reduced metabolic rate and a lesser response to most nutritional improvements in heat-stressed cows.

Further study is needed to delineate clearly the effects of fat supplementation in heat stressed cows. Such studies should allow for adequate adaptation of cows to hot weather and should establish a response gradient to varying amounts of added fat during heat stress.

Heat stress and protein nutrition

Cattle under heat stress often have negative N balances because of reduced intakes; hence, less protein is available for productive functions if dietary protein concentrations are not increased (Huber et al., 1994).

In a hot environment (19 and 31°C minimum and maximum daily temperatures, respectively), Hassan and Roussel (1975) observed increased feed intake, milk yield, and milk protein yield of Holstein cows fed 21% CP compared with cows fed 14% CP. Even the 14% CP diet exceeded NRC (1989) requirements for protein by 23%, but the increased milk was attributed to a higher feed intake elicited by 21% CP.

Total CP content of diets fed to dairy cows in many regions of the US, particularly in the southwest, where hot summer temperatures prevail, is commonly as high as 19–20% and often exceeds NRC (1989) recommendations by 25–30%, even though rumen-undegradable protein (RUP) content generally would meet that suggested by NRC (1989). The reduced energy consumption and increased maintenance requirement during heat stress often results in considerably more protein being metabolized to meet energy needs of cows in heat stress than in cows maintained at moderate temperatures (Beede and Collier, 1986). When excess amino acids (AAs) are provided, they are deaminated at an energy cost of synthesizing and excreting the amino group as urea, estimated to be $7.3 \, \text{kcal} \, \text{g}^{-1}$ of urea synthesized (Tyrell et al., 1970). Resultant increases in blood ammonia can be injurious to cow health (Visek, 1984), and reproductive efficiency (Chalupa and Ferguson, 1989).

Protein concentration and degradability

Three separate trials involving 60 cows were conducted during the hot summer months in Tucson (Higginbotham et al., 1989b). Diets containing high (18.5% CP) or medium protein (16.1% CP) concentrations and two percentages of RDP (65 vs. 59% of total CP) as calculated from NRC (1989) were compared. During the trials, heat stress conditions were confirmed because mean daily maxima for THI, calculated according to Johnson and Vanjonack (1976), were over 80.

During the late spring in Provo, Utah, diets of similar protein concentrations and degradabilities were tested at moderate ambient conditions (Higginbotham *et al.*, 1989a).

Table 11.4 summarizes these studies. In the hot environment, milk yields averaged 3.1 kg d^{-1} less for cows fed the high protein, 65% RDP diet than the other diets. This depression in milk production was equivalent to about 8.4 MJ (2.1 Mcal) of NE_L (NRC, 1989). The energy expenditure for formation of urea in the kidney (assumed to be 31 KJ g^{-1} of urinary N; Tyrell *et al.*, 1970) from the additional N consumed on the high than medium protein diets would equal about 25% of the difference in milk energy. Oldham (1984) calculated that the energy cost for excretion of 100 g of excess N in urine was 4.21 MJ. The lower DMI of cows fed high rather than medium protein equalled about 12.2 MJ of NE_L and might be an additional explanation for the lower milk yields on the 18.5% CP, 65% RDP diet. The reason for the reduction in consumption on the high compared with the medium protein diet is not clear.

Danfaer *et al.* (1980) reported that milk energy output was reduced by 4.6 MJ d^{-1} of NE_L when dietary CP increased from 19 to 23% of DM. The calculated net energy cost used for synthesizing extra urea, primarily in the liver, and for excreting excess N accounted for the loss in milk (Oldham, 1984). In the Arizona study (Higginbotham *et al.*, 1989b), the HP diets contained 24 g CP kg^{-1} DM more than the medium protein diets. Hence, excess protein would explain a large part of the reduced milk yields observed for the high protein, high RDP diet.

The high protein diets increased ruminal NH_3 and blood urea N, but diets with more RDP increased blood urea N and water intake. In some studies (Hassan and Roussel, 1975), increased blood urea N has been implicated in reduced reproductive performance, which might be accentuated during heat stress. Rectal temperatures and respiration rates were not affected by dietary protein. However, these measurements were higher in the Arizona (Higginbotham *et al.*, 1989b) than in the Utah study (Higginbotham *et al.*, 1989a). Rumen VFA, blood glucose, thyroid hormones, and cortisol were not different among diets, but those heat-sensitive hormones (Johnson and Vanjonack, 1976; Deresz, 1987), namely cortisol and thyroid hormones, were much lower for cows in Arizona (Higginbotham *et al.*, 1989b) than for cows in Utah (Higginbotham *et al.*, 1989a).

At moderate temperatures (Higginbotham *et al.*, 1989a), milk yields followed a different pattern than they did in the hot environment. Yields were lowest for the medium protein treatment of higher RDP, perhaps because of insufficient availability of essential AAs. Even though total CP met NRC (1989) recommendations, intake of RUP was only 88% of the recommended amount. Milk fat was reduced by the lower RDP diets (Higginbotham *et al.*, 1989a). As previously mentioned, water intakes were increased when more RDP was fed in hot weather (Higginbotham *et al.*, 1989b).

In two subsequent trials (Taylor *et al.*, 1991), the interaction of RDP and

Table 11.4. Influence of protein concentration (PC) and rumen-degradable protein (RDP) fed to lactating cows at hot and moderate temperatures (Huber et al., 1994).

	Treatment							
	Hot environment[1]				Moderate environment[2]			
	1	2	3	4	1	2	3	4
	18.4% PC[3] 65.3% RDP[4]	18.5% PC[3] 58.3% RDP[4]	16.1% PC[3] 65.0% RDP[4]	16.1% PC[3] 60.0% RDP[4]	18.5% PC[3] 66.2% RDP[4]	18.0% PC[3] 55.1% RDP[4]	15.4% PC[3] 67.1% RDP[4]	15.0% PC[3] 57.3% RDP[4]
Milk (kg d^{-1})[5]	25.5	28.9	28.5	28.4	36.9	35.4	34.8	36.7
3.5% FCM (kg d^{-1})[6]	23.5	26.8	26.5	26.1	34.7	31.8	32.2	32.4
DMI (kg d^{-1})[7]	21.2	21.6	23.0	22.8	-	-	-	-
Milk fat (%)[8,9]	2.72	3.04	3.01	2.95	3.11	2.89	3.04	2.78
Milk protein (%)	3.04	3.04	3.13	3.11	2.89	2.94	2.92	2.96

[1] Conducted in Tucson, Arizona during summer (60 cows used in three trials).
[2] Conducted in Provo, Utah during late spring (60 cows used in one trial).
[3] Percentage of DM.
[4] Percentage of CP.
[5] Treatment 1 versus all others ($P < 0.05$) for hot environment.
[6] Interaction effect (($P < 0.06$) for moderate environment.
[7] Protein concentration effect ($P < 0.01$).
[8] Treatment 1 versus all others ($P < 0.08$) for hot environment.
[9] Protein degradability effect ($P < 0.05$) for moderate environment.

Table 11.5. Effect of protein degradability and evaporative cooling on milk yields and feed intakes (from Taylor *et al.*, 1991).[1]

	Cooled		Shaded		Effects (*P<*)		
	HD	LD	HD	LD	D	C	D × C
Trial 1[2]							
Milk yield (kg d^{-1})	31.9	29.9	30.0	29.3	0.04	0.05	0.31
3.5% FCM yield (kg d^{-1})	25.7	26.1	24.3	22.9	0.57	0.02	0.34
DMI (kg d^{-1})	21.3	21.1	21.8	22.7	0.52	0.09	0.37
Milk/DMI	1.54	1.47	1.38	1.27	0.14	0.01	0.67
Trial 2[2]							
Milk yield (kg d^{-1})	26.8	32.8	29.1	29.6	0.02	0.76	0.04
3.5% FCM yield (kg d^{-1})	24.4	26.4	26.1	25.0	0.75	0.87	0.04
DMI (kg d^{-1})	23.9	21.6	22.6	23.8	0.60	0.64	0.09
Milk/DMI	1.18	1.39	1.29	1.24	0.44	0.94	0.03

[1] Covariate-adjusted means. HD, high degradability; LD, low degradability; D, degradability effect, and C, cooling effect.
[2] For Trial 1, rumen-degradable protein of HD and LD diets was 65 and 55%, respectively; Trial 2, respective values were 61 and 47%.

cooling of cows during hot summer weather was studied using a factorial arrangement of treatments. All diets averaged 18.3% CP; Trial 1 compared 64 with 55% RDP (estimated from NRC, 1989), but RDP estimates for Trial 2 were 61 and 47%. As shown in Table 11.5 (Taylor *et al.*, 1991) evaporative cooling increased milk yield in one trial, but not the other. Degradability exerted opposite effects in the trials; milk yields were higher for the 64% RDP in Trial 1 but lower for 61% RDP in Trial 2. Cooler ambient temperatures and lower THI in Trial 1 than 2 (71 vs. 77 average THI) may have negated the significant protein by heat stress interaction previously observed for milk production (Higginbotham *et al.*, 1989b). In Trial 2, conditions were more conducive to heat stress; maximum daily THI averaged 80 and 83 for cooled and shaded groups, respectively. Rectal temperatures (39.2 vs. 39.7°C) and respiration rate (70 vs. 88 breaths min^{-1}) also were lower for the cooled groups. Cows fed 47% RDP yielded more milk (3.4 kg d^{-1}) than did those fed 61% RDP, and the degradability by cooling interaction also was significant in that cooled cows responded more to lower RDP than did uncooled cows. As in the studies of Higginbotham *et al.* (1989a,b), neither rectal temperature nor respiration rate were affected by protein degradability.

Amount of RDP in excess of actual needs appears more related to the negative effect of RDP in heat stressed cows than actual percentage of RDP in diets. For example, Trials 1 and 2 in the study of Taylor *et al.* (1991) and the study of Higginbotham *et al.* (1989b) showed lowest milk yields on diets exhibiting highest RDP intakes in excess of requirements (NRC, 1989). The data suggest that overconsumption of more than about $100 \, g \, N \, d^{-1}$ as RDP will reduce milk yield during hot ambient temperatures. However, further experimentation is required to quantify such a relationship.

Protein quality

A positive response in milk yield or milk protein concentration to inclusion of an increased dietary supply of essential AAs (Chen *et al.*, 1993) or to specific AAs such as lysine (King *et al.*, 1991; Polan *et al.*, 1991; Schwab, 1992) or lysine plus methionine (Schwab, 1993), has been demonstrated. However, little is known concerning a possible interaction between protein quality and heat stress. The study of Chen *et al.* (1993) addressed this question by comparing a diet of low lysine content (using corn gluten meal as the principal protein supplement) with one containing 67% higher lysine (from a combination of soyabean, fish, and blood meals). Both diets contained about 57% RDP (NRC, 1989), and the content of other essential AAs was similar (except for arginine, which was higher for the high lysine diet). The low lysine diet averaged 19.2% CP, and 137 g lysine d^{-1}, while the high lysine diet was 18.6% CP and provided 241 g lysine d^{-1}. The protein treatments received evaporative cooling plus shade or shade only. As shown in Table 11.6 (Chen *et al.*, 1993), cows fed the high lysine diet produced 11% more milk than those fed low lysine, and milk yields of cows evaporatively cooled plus shade were about 9% higher than those of cows receiving shade only. Differences in milk yield between the high and low lysine diets were numerically greater for the evaporatively cooled plus shade ($3.8 \, kg \, d^{-1}$) group than for cows receiving shade only ($2.4 \, kg \, d^{-1}$). Moreover, cooling effects were greater for the high lysine than the low lysine diets ($3.2 \, kg \, d^{-1}$ for high lysine vs. $1.8 \, kg \, d^{-1}$ for low lysine). Cows receiving the shade only and fed high lysine produced slightly more milk than those fed low lysine with both evaporative cooling and shade, suggesting that protein quality of a diet can compensate for lack of cooling in hot weather.

For maintenance of maximum milk yields during periods of heat stress of cows fed diets typical of those in the western USA (which generally contain 18% or more CP), research suggests that RDP should not exceed 61% of total dietary CP or 100 g of $N \, d^{-1}$ more than NRC (1989) recommendations.

Minerals

Higher levels of Na and K for lactating cows during hot weather than indicated by NRC recommendations (1989) were suggested by Coppock and West (1987) and Beede (1987). Raising dietary Na from 0.18 to 0.55% of DM as $NaHCO_3$ or NaCl resulted in increased milk yields (Table 11.7; Beede, 1987).

Table 11.6. Effect of supplemental protein quality and evaporative cooling on feed intake, milk yield, composition, and efficiency of feed utilization during hot weather (from Chen et al., 1994).

	Treatment[1]					Treatment effects[2]	
	HL−EC+S	LL−EC+S	HL−S	LL−S	SEM	Protein	Cooling
DMI (kg d^{-1})	25.5	24.3	23.9	22.7	0.9	0.18	0.10
Milk (kg d^{-1})	31.9	28.1	28.7	26.3	0.9	0.01	0.02
3.5% FCM (kg d^{-1})	30.2	26.6	27.2	24.4	1.2	0.01	0.03
SCM (kg d^{-1})	28.9	25.4	25.0	24.0	1.1	0.06	0.03
FCM/DMI	1.23	1.11	1.13	1.08	0.07	0.24	0.35
Fat							
%	3.20	3.28	3.22	3.18	0.11	0.86	0.72
kg d^{-1}	1.02	0.91	0.91	0.83	0.04	0.05	0.04
Protein							
%	3.09	3.07	3.13	3.22	0.06	0.60	0.14
kg d^{-1}	0.99	0.87	0.89	0.84	0.04	0.03	0.04

[1] HL, high lys; LL, low lys; EC+S, evaporative cooling plus S; S, Shade. Covariant-adjusted least squares means. The S cows were higher than EC+S (*P* < 0.05) in mean rectal temperatures (39.1 vs. 38.6°C) and respiration rates (82 vs. 64 breaths min^{-1}). Mean maximum ambient temperature was 35.8°C, and mean temperature-humidity index was 76.6.
[2] No interactive effects were significant.

Table 11.7. Summary of increases in actual daily milk yield with increasing dietary potassium or sodium in complete diets (adapted from Beede, 1987).

	Dietary level (%) lower to higher	Milk yield % increases
Potassium experiment		
1	0.66-1.08	+4.6
	0.66-1.64	+2.6
	1.08-1.64	no change
2	1.00-1.50	+2.8
3	1.30-1.80	+4.2
4	1.07-1.51	+3.4
5	1.07-1.58	no change
6	1.14-1.58	no change
Sodium experiment		
7	0.20-0.43	+3.6
8	0.18-0.55	+9.6
	0.18-0.88	+10.8
	0.55-0.88	no change
9	0.28-0.47	+3.4
10	0.16-0.42	+9.2
11	0.24-0.62	no change

Exps 1, 2, 3, 4, 7, 8 and 9 (all studies conducted during the warm weather); 5, 6, 10 and 11 (all studies conducted during moderate or cool weather).

The addition of $NaHCO_3$ increased milk fat more than did NaCl. The need for more Na in heat-stressed cows was attributed to increased urinary secretion of Na (Beede, 1987) and the increased dietary requirement of K in heat-stressed cows was attributed to greater excretion of K in sweat. Also, less forage is eaten in hot weather, often decreasing K content of diets. Positive responses in intakes have been obtained with cows fed 1.5–1.8% K (Malonee *et al.*, 1985; Beede, 1987), compared with NRC (1989) recommendations of 0.9–1.2%.

West *et al.* (1991) demonstrated that heat-stressed lactating cows responded to increasing dietary cation–anion balance (DCAB, Na + K − Cl) from 120 to 460 meq kg^{-1} with higher dry matter intake; but response was independent of whether Na or K were used to increase DCAB. Large differences in DCAB were recommended for dry vs. lactating cows (West, 1993). Further, it was reported that diets high in cationic salts (as recommended for lactating cows in hot weather) will cause a higher incidence of milk fever if fed to dry cows. Diets higher in anionic salts (Cl and S) were recommended for dry cows. The desirable DCAB range for dry cows is −10 to −15 meq kg^{-1}. Inclusion of

S in a DCAB formula for dry cows was recommended: $[(Na + K) - (Cl + S)]$ (West, 1993).

The review by Sanchez *et al.* (1994) suggests that cows respond differently in the summer (when heat-stressed) and winter seasons to varying concentrations of several minerals as well as to DCAB. By combining data from 15 experiments with over 19,000 milk yield (MY) and DMI observations, these workers showed increased FCM and DMI during the summer, but not the winter, when dietary P was increased from 0.35 to 0.65% of diet DM. This response to P was attributed to a dramatic decrease in uptake of P by the portal-drained viscera of cows during heat stress, perhaps because of increased blood flow to peripheral tissues, thus requiring more dietary P.

This large database supported previously mentioned findings that Na and K needs were higher during summer than winter. The increased excretion of K (via sweat) and Na (via urine) were associated with decreased plasma aldosterone (El-Nouty *et al.*, 1980).

Responses to increasing dietary Cl from 0.15% to 1.15% of the diet DM were negative in the summer but not the winter. It was suggested that this Cl-by-season interaction was due to disturbances in the acid–base balance of heat-stressed cows which undergo both respiratory acidosis and alkalosis, but a definitive clarification of the complex reactions involved in such a response must await further experimentation.

Changes in MY and DMI to increasing dietary Ca and Mg were affected quadratically by season but in opposite directions. Summer data showed there was decreased lactational performance when Ca was raised from 0.5 to 0.85% of DM, but MY and DMI were increased by further increases of Ca to about 1.3% of DMI. The increases at high Ca might be related to a dietary buffer effect. Response to dietary changes in Ca and Mg were not observed during the winter season. The effects of Ca were attributed to a reduction in readily available blood Ca during heat stress caused by an interactive relationship between blood pH, proteins, HCO_3 and CO_2. Pertaining to why cows were affected differently with changing Mg during the summer than winter, poultry studies (Belay *et al.*, 1992) have shown that heat stress lowered Mg balance in birds, and the authors suggest that lactating cows are affected similarly. Changes in DCAB did not appear to affect MY or DMI differently during the summer and winter even though the study by West *et al.* (1991) suggests a slightly lower optimum for DCAB in the summer months.

In conclusion, it appears that heat stress in lactating cows affects needs for the macrominerals differently than in cool conditions. These differences are largely related to reduction in feed intake and mineral losses in sweat and urine, as well as shifts in inter-tissue water movement and acid–base balances which are compensatory mechanisms for combatting heat stress.

Dietary additives and heat stress

Various additives used in diets for lactating cows have been shown to benefit performance during hot as well as cool weather. Texas workers (Escobosa *et al.*, 1984) demonstrated improved DMI and milk production of cows fed $NaHCO_3$ during heat stress conditions, probably because of the Na response (Beede, 1987). In a study on niacin use during the summer feeding period, Muller *et al.* (1986) showed that cows fed niacin produced $0.9 \, kg \, d^{-1}$ more milk ($P < 0.05$) than controls, but response to niacin supplementation was $2.4 \, kg \, d^{-1}$ in higher yielding cows ($>34 \, kg \, milk \, d^{-1}$).

Huber *et al.* (1994) summarized effects of addition of a fungal culture (from *Aspergillus oryzae*) to diets of heat-stressed cows. In 6 of 19 treatment comparisons, rectal temperatures (RT) and respiration rates were lower for cows receiving the fungal cultures compared with controls. Studies where the fungus showed significant effects on heat stress indicators were conducted during hot summer conditions. Mean increases in milk production due to fungal addition was $1 \, kg \, d^{-1}$ with little difference between groups showing RT effects and those which did not.

Use of recombinant bovine somatotrophin (bST) in heat-stressed cows

Interaction of bST with heat stress in lactating cows was reviewed recently by West (1994). It was concluded from these data (Table 11.8) that more variation in response to bST has been shown in hot compared with cool environments. Mean response in milk production to administration of bST in hot weather (nine comparisons) was $3.9 \, kg \, d^{-1}$ with a range of $+0.8$ to $+8.5 \, kg \, d^{-1}$ (or 3–49% above controls). In the five studies which measured intakes, DMI was increased in two and decreased in two, which suggest a less consistent effect on stimulating feed consumption in hot than in cool environments.

Despite this, Elvinger *et al.* (1987) and Hutchinson *et al.* (1986), reported that daily injections of bST resulted in similar responses (25% over controls) in heat-stressed as those in cool regions. In Arizona Sullivan *et al.* (1992), using biweekly injections of bST (500 mg per 14 days), demonstrated milk yield responses in hot summer months (June, July, August) the same as in cooler months (March, April, May) (Table 11.9). The bST-treated cows tended to have higher RT than controls. Responses to bST were similar in companion studies in New York, Utah, and Missouri where heat effects were less (Hard *et al.*, 1988).

In temperature-controlled chambers, daily injections of bST under hot conditions increased milk yields (20–25%) as much as under thermal neutral

Table 11.8. Response to bST administration in hot environments (from West *et al.*, 1994).

Temperature (°C)		Relative humidity (%)		Milk yield (kg d^{-1})		DM intake (kg d^{-1})		Reference
Max.	Min.	Min.	Max.	Control	bST	Control	bST	
28.9	–	55	–	20.5	+8.0	–	–	Mohammed and Johnson, 1985
35	22	65	60	28.8	+6.1	–	–	Johnson *et al.*, 1991
35	24	65	55	21.0	+7.3	14.5	+2.7	Staples *et al.*, 1988
28.1	–	–	–	16.9[1]	+2.6	–	–	West *et al.*, 1990
34.6	22.2	100	59.8	17.5[2]	+8.5	15.2	+1.4	
				22.0[3]	+4.7	16.3	0	
				26.5[4]	+0.9	17.3	–1.3	
32.4	–	–	–	22.7	+2.1	17.8	–0.5	Zoa-Mboe *et al.*, 1989
39.2	–	–	–	18.8	+2.5	–	–	Elvinger *et al.*, 1992

[1] 3.5% FCM.
[2] Cows at low yield pretreatment.
[3] Cows at medium yield pretreatment.
[4] Cows at high yield pretreatment.

Table 11.9. Effect of bST on 3.5% fat corrected milk during moderate and hot temperature (adapted from Sullivan *et al.*, 1992).

	March	April	May	June	July	August
bST-treated (kg d^{-1})	37.7	35.8	36.1	33.7	32.1	28.0
Control (kg d^{-1})	33.5	31.6	30.8	28.6	26.0	23.5
Difference (kg)	4.2	4.2	5.3	5.1	6.1	4.5
(%)	12.5	13.3	17.2	17.8	23.5	19.2
Average max. temp. (°C)	22.4	28.4	32.5	37.2	35.5	35.9

(TN) conditions (Johnson *et al.*, 1991). Feed intakes also increased, but less in hot than TN conditions. Manalu *et al.* (1991) demonstrated that bST increased both heat production and heat dissipation, with no significant change in heat balance or rectal temperatures (Table 11.10). Calorigenic hormones (T3 and cortisol) were reduced by bST injection in heat-stressed cows. An often quoted study by Mollett *et al.* (1986) in which milk production was not improved, during an extended period of high temperatures and relative humidity, prompted the suggestion that for optimum results, dosages of bST may be lower in hot compared with cool environments (Kronfeld, 1989); however, other studies do not support this conclusion.

West (1994) suggested that high yielding cows are more sensitive to elevated temperatures, whether the high yield is attributable to greater genetic potential or technologies such as bST. This observation might explain the large variations reported for response of cows administered with bST under heat-stress conditions. If bST is used in hot summer conditions, cows should be cooled by one or more of the methods mentioned previously (shade plus fans, fans and sprinklers, or pressurized evaporative cooling in holding pens, exit lanes and pens, or manger misters). Administration of bST in severely heat-stressed cows may result in an uneconomical response and might be detrimental to animal health.

In conclusion, milk yield response from bST administration of cows in hot environments has been comparable with those of cows in moderate environments, but variation appears to be greater, possibly because of additional stress put on cows for higher production and less consistent increases in feed intake when cows are heat stressed and treated with bST. For consistent benefits from bST use in hot weather, cows should be artificially cooled by one or more of the systems mentioned earlier in this chapter.

Table 11.10. Effect of bST on heat production and heat losses in thermoneutral (TN) and hot environments.[1]

	TN		Hot		Main effect probability	
	Control	bST	Control	bST	bST	T[2]
Heat production (MJ d^{-1})	128	152	127	159	0.0001	0.4
Gross energy intake (MJ d^{-1})	351	396	272	318	0.005	0.0001
Milk energy (MJ d^{-1})	63	88	53	79	0.0001	0.03
Skin vaporization (MJ d^{-1})	31	43	47	59	0.002	0.0001
Respiratory vaporization (MJ d^{-1})	7.5	9.2	10.5	12.1	0.003	0.0001
TEHL[3] (MJ d^{-1})	38	52	57	72	0.001	0.0001
Cooling heat loss[4] (MJ d^{-1})	7.5	11.3	10.0	11.7	0.03	0.2
Rectal temperature (°C)	38.3	38.5	39.7	39.9	0.3	0.0001

[1] Adapted from data of Manalu *et al.* (1991) by West (1994).
[2] Temperature environment.
[3] Total evaporative heat loss.
[4] Heat loss due to increase in temperature (equal to body temperature) when total water consumed left the body. Calculated as total water intake times difference in temperature between drinking water and body temperature.

References

Armstrong, D.V. (1994) Heat stress interaction with shade and cooling. *Journal of Dairy Science* 77, 2044–2050.

Armstrong, D.V. and Wiersma, F. (1986) How to beat the heat in lactating dairy cows. *Proceedings of the Southwest Nutrition Conference*, University of Arizona, Tucson, USA, pp. 60–66.

Armstrong, D.V., Wiersma, F., Fuhrman, T.J., Tappan, J.M. and Cramer, S.M. (1985) Effect of evaporative cooling under a corral shade on reproduction and milk production in a hot-arid climate. *Journal of Dairy Science* 68 (Suppl. 1), 167 (Abstr.).

Beede, D.K. (1987) Dietary macrominerals and sodium bicarbonate for heat-stressed dairy cows. *Proceedings of the Southwest Nutrition and Management Conference*, Tempe, Arizona, USA, pp. 39–43.

Beede, D.K. (1993) Water nutrition and quality for dairy cattle. *Proceedings of the Western Large Herd Dairy Management Conference*, Las Vegas, Nevada, USA, pp. 194–205.

Beede, D.K. and Collier, R.J. (1986) Potential nutritional strategies for intensively managed cattle during thermal stress. *Journal of Animal Science* 62, 543–554.

Belay, T., Wiernusz, C.J. and Teeter, R.B. (1992) Mineral balance and urinary and faecal mineral excretion profile of broilers housed in thermoneutral and heat-distressed environments. *Poultry Science* 71, 1043–1047.

Brown, W.H., Fuquay, J.S., McGee, W.H. and Lyengar, S.S. (1974) Evaporative cooling for Mississippi dairy cows. *Transactions of the American Society of Agricultural Engineers* 17, 513–515.

Bucklin, R.A., Strickland, J.T., Beede, D.K. and Bray, D.R. (1989) Cow cooling pays in hot humid Florida. *Hoard's Dairyman* 134, 344.

Buffington, D.E., Collier, R.J. and Canton, G.H. (1983) Shade management systems to reduce heat stress for dairy cows. *Transactions of the American Society of Agricultural Engineers* 26, 1798–1802.

Challis, D.J., Zeinstra, J.S. and Anderson, M.J. (1987) Some effects of water quality on the performance of high yielding cows in an arid climate. *Veterinary Record* 120, 12–15.

Chalupa, W. and Ferguson, J.D. (1989) The impact of nutrition on reproduction of cows. *Proceedings of the Minnesota Nutrition Conference*, Bloomington, USA, p. 59.

Chan, S.C., Huber, J.T., Wu, Z., Chen, K.H. and Simas, J. (1992) Effect of fat supplementation and protein source on performance of dairy cows in hot environmental temperatures. *Journal of Dairy Science*, 75 (Suppl. 1), 175 (Abstr.).

Chan, S.C., Huber, J.T., Wu, Z., Simas, J., Chen, K.H., Santos, F., Rodrigues, A. and Varela, J. (1993) Effects of supplementation of fat and evaporative cooling of dairy cows subjected to hot temperatures. *Journal of Dairy Science* 76 (Suppl. 1), 184 (Abstr.).

Chen, K.H., Huber, J.T., Theurer, C.B., Armstrong, D.V., Wanderlay, R., Simas, J., Chan, S.C. and Sullivan, J. (1993) Effect of supplemental protein quality and evaporative cooling on lactation performance of Holstein cows in hot weather. *Journal of Dairy Science* 76, 819–825.

Coppock, C.E. and West. J.W. (1987) Feeding systems for relieving heat stress: minerals and vitamins. *Proceedings of the Southwest Nutrition and Management Conference*, Tempe, Arizona, USA, pp. 79–83.

Danfaer, A., Thysen, I. and Ostergaard, V. (1980) The effect of the level of dietary protein on milk production. 1. Milk yields, liveweight gain and health. Beret Statens Husdrbrugsfors., p. 492.

Deresz, F. (1987) Effect of different cooling systems on concentrations of certain hormones and free fatty acids at varying times during lactation of Holstein cows. PhD dissertation, University of Arizona, Tucson. University Microfilms International, Ann Arbor, Michigan, USA.

El-Nouty, F.D., Elbanna, I.M., Davis, T.P. and Johnson, H.D. (1980) Aldosterone and ADH response to heat and dehydration in cattle. *Journal of Applied Physiology Respiration and Environmental Exercise Physiology* 48, 249–255.

Elvinger, F., Head, H.H., Wilcox, C.J. and Natzke, R.P. (1987) Effects of administration of bovine somatotropin on lactation, milk yield and composition. *Journal of Dairy Science* 70 (Suppl. 1), 121 (Abstr.).

Escobosa, A., Coppock, C.E., Rowe, L.D., Jr., Jenkins, W.L. and Gates, C.E. (1984) Effects of dietary sodium bicarbonate and calcium chloride on physiological responses of lactating dairy cows in hot weather. *Journal of Dairy Science* 67, 574–584.

Hard, D.L., Cole, W.J., Franson, S.E., Samuels, W.A., Bauman, D.E., Erb, H.N. and Huber, J.T. (1988) Effect of long term sometribove, USAN (recombinant methionyl bovine somatotropin) treatment in a prolonged release system on milk yield, animal health and reproductive performance – pooled across four sites. *Journal of Dairy Science* 71 (Suppl. 1), 210 (Abstr.).

Hassan, A. and Roussel, J.D. (1975) Effect of protein concentration in the diet on blood composition and productivity of lactating Holstein cows under thermal stress. *Journal of Agricultural Science*, Cambridge, 85, 409–415.

Higginbotham, G.E., Huber, J.T., Wallentine, W.V., Johnston, N.P. and Andrus, D. (1989a) Influence of protein percent and degradability on performance of lactating cows during moderate temperature. *Journal of Dairy Science* 72, 1818–1823.

Higginbotham, G.E., Torabi, M. and Huber, J.T. (1989b) Influence of dietary protein concentration and degradability on performance of lactating cows during hot environmental temperature. *Journal of Dairy Science* 72, 2554–2564.

Huber, J.T., Wu, Z., Chan, S.C. and Simas, J. (1993) Feeding of fat during heat stress conditions, and combinations of fat sources. *Proceedings of the Southwest Nutrition and Management Conference*, Phoenix, AZ, Department of Animal Sciences, University of Arizona, Tucson, USA, pp. 90–99.

Huber, J.T., Higginbotham, G., Gomez-Alarcon, R.A., Taylor, R.B., Chen, K.H., Chan, S.C. and Wu, Z. (1994) Heat stress interactions with protein, supplemental fat, and fungal cultures. *Journal of Dairy Science* 77, 2080–2090.

Hutchinson, T.F., Tomlinson, J.E. and McGee, W.H. (1986) The effects of exogenous recombinant or as pituitary extracted bovine growth hormone on performance of dairy cows. *Journal of Dairy Science* 69 (Suppl. 1), 152 (Abstr.).

Igono, M.O., Johnson, H.D., Steevens, B.J., Krause, G.F. and Shanklin, M.D. (1987) Physiological productive and economic benefits of shade, spray, and fan system versus shade for Holstein cows during summer heat. *Journal of Dairy Science* 70, 1069–1079.

Johnson, H.D. (1980) Bioclimate effects on growth, reproduction and milk production. In: Johnson, H.D. (ed.) *Bioclimatology and the Adaptation of Livestock*. Elsevier, Amsterdam, The Netherlands, p. 35.

Johnson, H.D. and Vanjonack, W.J. (1976) Effects of environmental and other stressors on blood hormone patterns in lactating cows. *Journal of Dairy Science* 59, 1603–1617.

Johnson, H.D., Li, R., Manalu, W., Spencer-Johnson, K.J., Becker, B.A., Collier, R.J. and Baile, C.A. (1991) Effects of somatotropin on milk yield and physiological responses during summer farm and hot laboratory conditions. *Journal of Dairy Science* 74, 1250–1262.

Kibler, H.H. and Brody, S. (1953) Influence of humidity on heat exchange and body temperature regulation in Jersey, Holstein, Brahman, and Brown Swiss cattle. *Missouri Agricultural Experiment Station Bulletin 522*, Columbia, Missouri, USA.

King, K.J., Bergen, W.G., Sniffen, C.J., Grant, A.L., Grieve, D.B., King, V.L. and Ames, N.K. (1991) An assessment of absorbable lysine requirements in lactating cows. *Journal of Dairy Science* 74, 2530–2539.

Knapp, D.M. and Grummer, R.R. (1991) Response of lactating dairy cows to fat supplementation during heat stress. *Journal of Dairy Science* 74, 2573–2579.

Kronfeld, D.S. (1989) The continuing challenge of BST. *Large Animal Veterinarian* (Sep–Oct), 6–7.

McDowell, R.E. (1974) Effect of environment on the functional efficiency of ruminants. *Proceedings of the International Livestock Environment Symposium*, American Society of Agricultural Engineers, St. Joseph, Missouri, USA, pp. 200.

McDowell, R.E. and Weldy, J.R. (1967) Water exchange of cattle under heat stress. *Biometeorology* 2, 414.

McDowell, R.E., Moody, E.G. Van Soest, P.J., Lehmann, R.P. and Ford, G.L. (1969) Effect of heat stress on energy and water utilization of lactating cows. *Journal of Dairy Science* 52, 188–194.

McDowell, R.E., Hooven, N.W. and Camoens, J.K. (1976) Effect of climate on Holsteins in first lactation. *Journal of Dairy Science* 59, 965–973.

Mallonee, P.G., Beede, D.K., Collier, R.J. and Wilcox, C.J. (1985) Production and physiological responses of dairy cows to varying dietary potassium during heat stress. *Journal of Dairy Science* 65 (Suppl. 1), 212 (Abstr.).

Manalu, W., Johnson, H.D., Li, R., Becker, B.A. and Collier, R.J. (1991) Assessment of thermal status of somatotropin-injected Holstein cows maintained under controlled-laboratory thermoneutral, hot and cold environments. *Journal of Nutrition* 121, 2006–2019.

Milam, K.Z., Coppock, C.E., West, J.W., Lanham, J.K., Nave, D.H., Labore, J.M., Stermer, R.A. and Brasington, C.F. (1986) Effects of drinking water temperature on production responses in lactating Holstein cows in summer. *Journal of Dairy Science* 69, 1013–1019.

Mollett, T.A., DeGeeter, M.J., Belyea, R.L., Youngguist, R.A. and Lanza, G.M. (1986) Biosynthetic or pituitary extracted bovine growth hormone induced galactopoiesis in dairy cows. *Journal of Dairy Science* 69 (Suppl. 1), 118 (Abstr.).

Moody, E.G. Van Soest, P.J., McDowell, R.E. and Ford, G.L. (1967) Effect of high temperature and dietary fat on performance of lactating cows. *Journal of Dairy Science* 50, 1909–1916.

Muller, L.D., Heinrichs, A.J., Cooper, J.B. and Atkin, Y.H. (1986) Supplemental niacin for lactating cows during summer feeding. *Journal of Dairy Science* 69, 1416–1420.

Murphy, M.R., Davis, C.L. and McCoy, G.C. (1983) Factors affecting water consumption by Holstein cows in early lactation. *Journal of Dairy Science* 66, 35–38.

National Oceanic and Atmospheric Administration (1976) Livestock hot weather stress. Reg. Operations Letter C-31-76. US Department of Commerce, National Weather Service, Central Region, Kansas City, Missouri, USA.

National Research Council (1974) *Nutrients and Toxic Substances in Water for Livestock and Poultry*. National Academy of Science, Washington, DC, USA.

National Research Council (1981) *Effect of Environment on Nutrient Requirements of Domestic Animals*. National Academy Press, Washington, DC, USA.

National Research Council (1989) *Nutrient Requirements of Dairy Cattle*, 6th edn National Academy of Science, Washington, DC, USA.

Oldham, J.D. (1984) Protein–energy interrelationships in dairy cows. *Journal of Dairy Science* 67, 1090–1114.

Palmquist, D.L. (1987) Adding fat to dairy diets. *Animal Health and Nutrition*, Watt Publ. Co., Mount Morris, Illinois, USA, pp. 32–35.

Polan, C.E., Cummins, K.A., Sniffen, C.J., Muscato, T.V., Vincini, J.L., Crooker, B.A., Clark, J.H., Johnson, D.G., Otterby, D.E., Guillaume, B., Muller, L.D., Varga, G.A., Murray, R.A. and Pierce-Sandner, S.B. (1991) Responses of dairy cows to supplemental rumen-protected forms of methionine and lysine. *Journal of Dairy Science* 74, 2997–3013.

Ryan, D.P., Boland, M.P., Kopel, E., Armstrong, D.V., Munyakazi, L., Godke, G.A. and Ingraham, R.H. (1992) Evaluating two different evaporative cooling management systems for dairy cows in a hot, dry climate. *Journal of Dairy Science* 75, 1052–1059.

Sanchez, W.K., McGuire, M.A. and Beede, D.K. (1994) Macromineral nutrition by heat stress interactions in dairy cattle: review and original research. *Journal of Dairy Science* 77, 2051–2079.

Schwab, C.G. (1992) Amino acid limitation and flow to duodenum at four stages of lactation. 1. Sequence of lysine and methionine limitation. *Journal of Dairy Science* 75, 3486–3502.

Skaar, T.C., Grummer, R.R., Dentine, M.R. and Stuffacher, R.H. (1989) Seasonal effects of prepartum and postpartum fat and niacin feeding on lactation performance and lipid metabolism. *Journal of Dairy Science* 72, 2028–2038.

Schultz, T.A. (1988) California dairy corral manger mister installation. Paper No. 88–4056, American Society of Agricultural Engineers, St. Joseph, Michigan, USA.

Smith, J.F., Armstrong, D.V., Correa, A., Avendeno, L., Rubio, A. and DeVise, S.K. (1993) Effects of a spray and fan system on milk production and reproduction efficiency in a hot-arid climate. *Journal of Dairy Science* 76 (Suppl. 1), 240 (Abstr.).

Staples, C.R., Head, H.H. and Darden, D.E. (1988) Short-term administration of bovine somatotropin to lactating dairy cows in a subtropical environment. *Journal of Dairy Science* 71, 3274.

Stermer, R.A., Brasington, C.F., Coppock, C.E., Lanham, J.K. and Milam, K.Z. (1986) Effect of drinking water temperature on heat stress of dairy cows. *Journal of Dairy Science* 69, 546–551.

Stott, G.H., Wiersma, F. and Lough, O. (1972) Consider cooling possibilities: the practical aspects of cooling dairy cattle. Arizona Extension Service Reproduction P-25, University of Arizona, Tucson, USA.

Strickland, J.T., Bucklin, R.A., Nordstedt, R.A., Beede, D.K. and Bray, D.R. (1988) Sprinkling and fan evaporative cooling for dairy cattle in Florida. Paper No. 88–4042, American Society of Agricultural Engineers, St. Joseph, Michigan, USA.

Sullivan, J.L., Huber, J.T., DeNise, S.K., Hoffman, R.G., Kung, L., Franson, S.W. and Madsen, K.S. (1992) Factors affecting response of cows to biweekly injections of recombinant methionyl bovine somatotropin (sometribove). *Journal of Dairy Science* 75, 756–763.

Taylor, S.E., Buffington, D.E., Collier, R.J. and DeLorenzo, A. (1986) Evaporative cooling for dairy cattle in Florida. Paper No. 86–4022, American Society of Agricultural Engineers, St. Joseph, Michigan, USA.

Taylor, R.B., Huber, J.T., Gomez-Alarcon, R.A., Wiersma, F. and Pang, X. (1991) Influence of protein degradability and evaporative cooling on performance of dairy cows during hot environmental temperatures. *Journal of Dairy Science* 74, 243–249.

Tyrell, H.F., Moe, P.W. and Flatt, W.P. (1970) Influence of excess protein intake on energy metabolism in the dairy cows. In: Schurch, A. and Wenk, C. (eds) *Energy Metabolism of Farm Animals, Proceedings of the 5th Symposium on Energy Metabolism*, Vitznau, Switzerland. European Association of Animimal Production, Publ. No. 13. Juris Druck and Verlag, Zurich, Switzerland, p. 69.

Tyrell, H.F., Reynolds, P.J. and Moe, P.W. (1979) Effect of diet on partial efficiency of acetate use for body tissue synthesis by mature cattle. *Journal of Animal Science* 48, 598–606.

Visek, W.J. (1984) Ammonia: its effects on biological systems, metabolic hormones and reproduction. *Journal of Dairy Science* 67, 481.

West, J.W. (1993) Cation-anion balance: Its role in the nutrition of dry and lactating cows. *Proceedings of the Southwest Nutrition and Management Conference*, Phoenix, Arizone, USA pp. 113–121.

West, J.W. (1994) Interactions of energy and bovine somatotropin with heat stress. *Journal of Dairy Science* 77, 2091–2102.

West, J.W., Mullinix, B.G., Johnson, J.C., Jr, Ash, K.A. and Taylor, V.N. (1990) Effects of bovine somatotropin dry matter intake, milk yield and body temperature in Holstein and Jersey cows during heat stress. *Journal of Dairy Science* 73, 2896–2906.

West, J.W., Mullinix, B.G. and Sandifer, T.C. (1991) Changing dietary electrolyte balance for dairy cows in cool and hot environments. *Journal of Dairy Science* 74, 1662–1674.

Wiersma, F. (1982) Shades for dairy cattle. University of Arizona Extension Service, WREP 51, University of Arizona, Tucson.

Wiersma, F. and Armstrong, D.V. (1983) Cooling dairy cattle in the holding pen. Paper No. 83–4507, American Society of Agricultural Engineers, St. Joseph, Michigan, USA.

Wilks, D.L., Coppock, C.E., Lanham, J.K., Brooks, K.N., Baker, C.C., Bryson, W.L., Elmore, R.G. and Stermer, R.A. (1990) Responses of lactating Holstein cows to chilled drinking water in high ambient temperatures. *Journal of Dairy Science* 73, 1091–1099.

Wolfenson, D., Her, E., Flamenbaum, I., Folman, Y., Kaim, M. and Berman, A. (1984) Effect of cooling heat stressed cows on thermal, productive and reproductive responses. *Proceedings of the Society for the Study of Fertility*, University of Reading, England, p. 38.

Zao-Mboe, A., Head, H.H., Bachman, K.C., Baccari, F., Jr and Wilcox, C.J. (1989) Effects of bovine somatotropin on milk yield and composition, dry matter intake, and some physiological functions of Holstein cows during heat stress. *Journal of Dairy Science* 72, 907–916.

Milking and Milk Technology IV

Production and Utilization of Dairy Cow's Milk and Products with Increased Unsaturated Fatty Acids

12

R.J. Baer

Dairy Science Department, Minnesota-South Dakota Dairy Foods Research Center, South Dakota State University, Brookings, South Dakota 57007–0647, USA

Introduction

Milk fat is the most complex of all common fats and is the milk constituent with the greatest variability in its percentage. Typically, milk fat is composed of 95–98% triglycerides. The majority of fatty acids in milk fat are saturated (66%), and about 30% are monounsaturated and 4% are polyunsaturated (United States Department of Agriculture, 1976). Several factors are known to produce variation in milk fatty acid composition (Christie, 1980; Baer, 1991). This chapter will discuss the various conditions that are known to affect milk fatty acid composition, as well as recent developments directed at increasing the unsaturated fatty acid content of milk fat and the effects on milk products. Alteration of milk fatty acid composition to contain lower amounts of saturated fatty acids seems desirable, as many health professionals currently endorse diets lower in saturated fatty acids.

Factors Affecting Milk Fatty Acids

Stage of lactation

Colostral milk fat secreted during the first day of lactation contains lower levels of short-chain fatty acids, while during the next 15 days palmitic acid (16:0) levels increase (Sneft and Klobasa, 1970). There is general agreement that short-chain fatty acids, with the possible exception of butyric acid (4:0), increase for the first 8–10 weeks of lactation, while palmitic acid (16:0) remains unchanged and stearic (18:0) and oleic (18:1) acids decrease. After 10 weeks of lactation, changes in fatty acid composition tend to be relatively minor (Christie, 1980).

Season of year

The effects of seasonal variation on milk fatty acid composition has been extensively studied (Boatman et al., 1965; Hall, 1970). Milk fat produced in summer months generally contains less palmitic acid (16:0) and higher levels of stearic (18:0) and oleic (18:1) acids than milk fat produced in winter months. It has been proposed that seasonal changes in milk fatty acid composition are due to dietary effects (Jensen and Clark, 1988). Dietary changes in spring and summer months to fresh pasture grasses that contain more linoleic acid (18:2) probably have the greatest influence on seasonal differences in milk fatty acid composition (Christie, 1980). The seasonal effect on milk fatty acid composition may occur only in cows grazed during the summer, which is becoming less common in many areas of the United States.

Genotype

There is limited information on the effect of dairy cattle breed on milk fatty acid composition. Jersey milk fat contains more caproic (6:0), caprylic (8:0), capric (10:0), lauric (12:0), myristic (14:0) and stearic (18:0) acids; and lower levels for oleic acid (18:1) compared with milk fat from Holsteins (Palmquist and Beaulieu, 1992). There is also considerable variation within breeds. The effect of breed on milk fatty acid composition deserves further study, particularly with the advent of genetic engineering.

In a study of Ayrshire twin cows, Edwards et al. (1973) showed that proportions of different fatty acids were subject to a high degree of genetic control. However, they concluded that due to the small size of genetic variances for individual fatty acids, manipulation would be a long-term undertaking. The molar proportion of fatty acid components of milk fat has a moderate heritability of 0.3 with a 0.05–0.02 coefficient of variation, thus genetic response is possible (Gibson, 1991). Gradual genetic improvement of cows to produce milk fat with altered fatty acid composition would produce little or no return to the breeder. Gibson (1991) summarized that changing milk fatty acid composition is difficult with conventional dairy cattle breeding programmes, therefore compositional changes would occur slowly.

Transgenic dairy cows may provide a futuristic route for altering milk fatty acid composition (Gibson, 1991). Transgenic cows would require substantial research and development efforts and major compositional alterations would be necessary to be cost-effective (Hoeschele, 1990). Successful transgenic improvements would most likely require at least several decades to affect the entire dairy cattle population. It is more likely that individual herds will produce milk for the manufacture of novel or niche products for a specialized marketplace. There is uncertainty as to the contribution that transgenic dairy cattle will make in altering milk fatty acid composition (Gibson, 1991).

Low milk fat syndrome

The low milk fat syndrome occurs when cows fed diets low in roughage produce milk containing less fat (Chilson and Sommer, 1953). Milk fat contents from cows with low milk fat syndrome may be as low as 10–20 g kg^{-1} milk. Besides changing the fat content, the level of short-chain fatty acids decreases, and oleic acid (18:1) increases (Christie, 1980).

Bovine somatotrophin

Treatment of dairy cows with bovine somatotrophin increases milk production (Eppard *et al.*, 1985). In several studies (Bitman *et al.*, 1984; Baer *et al.*, 1989), reductions in short- and medium-chain fatty acids and increases in long-chain fatty acids were reported when cows were administered bovine somatotrophin. Milk fat from treated cows was more unsaturated, due to higher amounts of oleic acid (18:1). Increases in the amount of oleic acid (18:1) are probably due to mobilization of fatty acids from adipose tissue (Grummer, 1991). In another study (Lynch *et al.*, 1988), milk fatty acid composition was not altered.

During the early lactation period (about the first 2 months) dairy cows are typically in a negative energy balance. Milk fatty acid composition changed most during the first 2 months of bovine somatotrophin treatment, therefore most changes seem to be related to the energy status of the cow (Eppard *et al.*, 1985; Baer *et al.*, 1989) and probably not to a direct effect of bovine somatotrophin. Future research is needed to establish a possible association between bovine somatotrophin and milk fatty acid composition in high milk producing cows.

Dietary fat supplementation

The inclusion of fat in dairy rations has been reviewed by Palmquist and Jenkins (1980). Dietary fat can be safely increased to 5–7% in rations. The limit for feeding added fat is 8–10%, particularly for lipids high in polyunsaturated fatty acids because rumen microflora and fibre digestibility can be adversely affected (Henderson, 1973; Kowalczyk *et al.*, 1977). Slight reductions in milk protein percentages may occur when feeding added fat (Palmquist and Jenkins, 1980; DePeters and Cant, 1992). This would reduce cheese yields and future

research is needed to address this problem. Milk production is increased 3–8%
by feeding added fat.

All short-chain (4:0 to 10:0) and half of the medium-chain (12:0 to 17:0)
fatty acids in milk fat are synthesized from acetate and β-hydroxybutyrate in
the mammary gland epithelial cells. The other half of medium-chain and almost
all long-chain (18:0 and longer) fatty acids are derived from blood plasma fatty
acids of dietary origin or from mobilization of body fat stores (Eppard *et al.*,
1985). Manipulation of milk fatty acid composition by dietary changes appears
to offer the most potential for producing milk and milk products with increased
unsaturated fatty acids.

Protected lipids

Australian researchers (Scott *et al.*, 1970, 1971) have developed a method for
protecting polyunsaturated lipids from hydrogenation to saturated lipids in the
rumen. Oils that were high in polyunsaturated fatty acids were coated with a
protein and treated with formaldehyde which protected the oil from hydrogenation by rumen microorganisms. Dairy cows fed the protected lipids produced
milk with elevated linoleic acid (18:2). Linoleic acid (18:2) increased from its
usual content of 2–3% of milk fatty acids up to 30–35%. There have been
numerous studies involving the use of protected lipids and oils in dairy cattle
rations. Fat sources that have been fed include canola (Astrup *et al.*, 1980; Ashes
et al., 1992), coconut oil (Anderson, 1974), safflower (Cook and Scott, 1972;
Bitman *et al.*, 1973; Edmondson *et al.*, 1974; Astrup *et al.*, 1979), sunflower
seeds (Badings *et al.*, 1976; Tamminga *et al.*, 1976) and tallow (Dunkley *et al.*,
1977; Kronfeld *et al.*, 1980). Development of encapsulating methods that utilize
substances approvable by regulatory agencies would be advantageous.

Unprotected lipids

Unprotected fat sources fed to dairy cows include coconut oil (Anderson, 1974),
cottonseed oil (Steele and Moore, 1968), marine oil (Astrup *et al.*, 1981),
oleic acid (Selner and Schultz, 1980), menhaden oil (Hagemeister *et al.*, 1988),
rapeseed (Murphy *et al.*, 1990), safflower seeds (Palmquist and Conrad, 1980;
Stegeman *et al.*, 1992b), soyabean oil (Larson and Schultz, 1970; Steele *et al.*,
1971a,b; Macleod and Wood, 1972; Banks *et al.*, 1980), soyabeans (Casper *et al.*,
1990) and sunflower seeds (McGuffey and Schingoethe, 1982; Rafalowski and
Park, 1982; Casper *et al.*, 1988; Stegeman *et al.*, 1992b). The fat in oilseeds is
somewhat protected due to the seed coat compared with oils. There is general
agreement that when cows are fed fats high in long-chain unsaturated fatty
acids, the milk fat contained increased amounts of stearic (18:0) and oleic (18:1)
acids, whereas the amounts of short- and medium-chain fatty acids decreased.

Fatty Acid Alteration – Effect on Milk and Products

Filled dairy products

Filled milk products are produced when milk fat is replaced with another fat or oil. Although filled milk is not a new concept (Henderson, 1971), these products still appear in the marketplace (Reiter, 1994). The future of filled milk products with added unsaturated vegetable oils to reduce the saturated fatty acid content remains in doubt, as widespread consumption has not materialized.

Protected lipids

There has been more research in the past on the production of milk and dairy products from cows fed protected lipids compared with unprotected lipids. Recently this trend is reversing and today more research is being conducted with the feeding of unprotected supplemental fat. In milk fat, the feeding of protected lipids reduces the amounts of lauric (12:0), myristic (14:0) and palmitic (16:0) acids, while stearic (18:0), oleic (18:1) and linoleic (18:2) acids increase. The greatest change is the increase in linoleic acid (18:2) (Table 12.1). Some researchers report that milk high in polyunsaturated fatty acids was more susceptible to oxidation (Edmondson *et al.*, 1974; Goering *et al.*, 1976). Milk that contains more than 20% of the fatty acid content as linoleic acid (18:2) generally has an oxidized off-flavour (Edmondson *et al.*, 1974). The supplementation of vitamin E in the cows' diet or direct addition to milk partially prevents development of oxidized off-flavours.

The manufacture of butter from milk produced by cows fed protected lipids has been researched (Buchanan and Rogers, 1973; Kieseker *et al.*, 1974; Badings *et al.*, 1976; Wong *et al.*, 1982; Cadden *et al.*, 1984a; Sporns *et al.*, 1984; Urquhart *et al.*, 1984). The amount of fatty acids with a high melting point (for example, palmitic acid) (Table 12.2) decreases in concentration (Table 12.1) and butter produced was softer and more spreadable at refrigeration temperatures, but became oily faster than normal butter at higher temperatures (Wood *et al.*, 1975). This is due to the change in the ratio of saturated to unsaturated fatty acids. Milk and butter flavour is not affected by prolonged feeding of protected canola and no increase in oxidation has been observed (Cadden *et al.*, 1984b).

The production of Gouda cheese with 104–117 g linoleic acid (18:2) per kg fatty acid resulted in a bland flavour and was sometimes oxidized after four months of ageing (Badings *et al.*, 1976). Cheese texture at 4°C was slightly mealy, but acceptable; however, at 14°C the texture was very soft and mealy.

Milk from cows fed a protected safflower and oil-casein fat source produced Cheddar cheese that contained 300 g linoleic acid (18:2) per kg fatty acid (Wong *et al.*, 1973). Body and texture defects were reported and flavour scores decreased as the amount of linoleic acid (18:2) increased in the cheeses.

Table 12.1. Fatty acid composition of normal milk fat and milk fat from cows fed added lipid.

	g per 100 g in			
Fatty acid[1]	Normal[2]	Normal[3]	Unprotected lipid[4]	Protected lipid[5]
4:0	3.5	3.3	2.5	3.9
6:0	1.9	2.3	1.3	2.0
8:0	1.3	1.2	0.7	0.8
10:0	2.5	2.8	1.6	1.4
12:0	2.9	3.4	2.1	1.3
14:0	10.7	11.4	8.7	4.0
14:1	–	2.6	1.9	0.5
16:0	27.9	29.5	20.6	10.7
16:1	2.5	3.4	3.5	0.3
18:0	12.7	9.8	12.8	13.3
18:1	26.6	27.4	36.8	30.2
18:2	2.5	2.8	3.5	28.0
18:3	1.6	–	–	1.6
Other	3.4	–	4.1	2.0
Saturated	63.4	63.8	50.3	37.4
Unsaturated	33.2	36.2	45.6	60.6

–, not reported.

[1] Expressed as number of carbons : number of double bonds.
[2] United States Department of Agriculture (1976).
[3] Barbano (1990).
[4] Middaugh et al. (1988).
[5] Wood et al. (1975).

Cheddar cheese that contained 15% of the fatty acids as linoleic acid (18:2) was the most acceptable. As the cheeses aged, off-flavours were less detectable.

Unprotected lipids

Current research indicates that feeding cows unprotected lipids appears to offer the most promise in altering milk fatty acid composition, while still being able to produce a marketable product. Feeding cows protected lipids provides the best potential to alter milk fatty acid composition; however, formaldehyde treated protected lipids are not approved by the United States Food and Drug Administration. Ample research has been reported on the feeding of unprotected lipids to cows to increase the energy content of rations. Feeding cows different fat sources can alter milk fatty acid composition (Fig. 12.1).

Table 12.2. Milk fatty acid melting points and effect of milk fatty acids on human blood cholesterol levels.

Fatty acid[1]	Melting point (°C)[2]	Cholesterol effect	Effect references
4:0	−8	None	Bonanome and Grundy (1988)
6:0	10	None	Hashim *et al.* (1960)
8:0	15	None	Grundy (1990)
10:0	31	None	Grundy (1990)
12:0	48	May raise	Hegsted *et al.* (1965)
14:0	58	Raises	Hegsted *et al.* (1965)
16:0	64	Raises	Hegsted *et al.* (1965)
18:0	70	Lowers	Bonanome and Grundy (1988)
18:1 (cis)	13	Lowers	Grundy (1986)
18:1 (trans)	44	Raises	Mensink and Katan (1990)
18:2	−5	Lowers	Mattson and Grundy (1985)

[1] Expressed as number of carbons : number of double bonds.
[2] Atherton and Newlander (1977); DeMan (1980).

Oil seeds fed that are higher in polyunsaturated fatty acids tend to produce milk fat with higher amounts of unsaturated fatty acids and lower amounts of saturated fatty acids, compared with the feeding of more saturated fat sources. Diets high in unprotected lipids may cause a depression in milk protein (Palmquist and Jenkins, 1980). A milk fat depression may also be observed when feeding unprotected lipids (Middaugh *et al.*, 1988), especially highly unsaturated lipids. This may be due to reduced ruminal fibre digestion and/or formation of trans isomers of unsaturated fatty acids which may interfere with mammary fatty acids synthesis or triglyceride synthesis. The feeding of protected or unprotected lipids to alter milk fatty acid composition has the greatest effect on dairy products by changing the melting properties of milk fat (Palmquist *et al.*, 1993).

Few studies have evaluated the effect of feeding unprotected lipids on the sensory properties of milk, and studies evaluating dairy products from this milk are even more limited. In one study milk and butter was produced from cows fed high oleic and regular sunflower seeds (Middaugh *et al.*, 1988). The butters were softer, more unsaturated and had acceptable flavour. The milk fat was lower in amounts of short- and medium-chain fatty acids and higher in concentrations of long-chain fatty acids (Table 12.1). Conclusions were that this research has application for production of fluid milk and other dairy products with lowered saturated fatty acid content.

Milk from cows fed full fat rapeseed with the seed coat broken was manufactured into butter that contained 40% of the fatty acids as oleic acid (18:1) (Murphy, 1992). This 'mono-butter' was reported by 70% of 135 households in Dublin, Ireland to have superior spreadability characteristics compared with

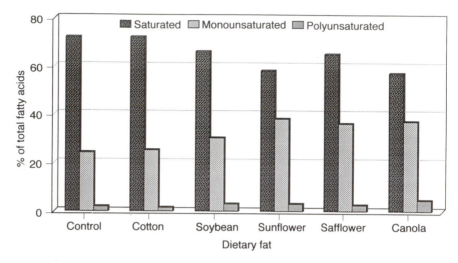

Fig. 12.1. Saturated, monounsaturated and polyunsaturated fatty acid content of milk fat from cows fed unprotected highly unsaturated dietary fat. From Casper *et al.*, 1988; Middaugh *et al.*, 1988; Mohamed *et al.*, 1988; Casper *et al.*, 1990; Khorasani *et al.*, 1991; Stegeman *et al.*, 1991; and Lightfield *et al.*, 1993.

their product normally purchased (Cowan and McIntyre, 1991). The butter was softer and more spreadable due to an increase in oleic acid (18:1) and decrease in palmitic acid (16:0) (Murphy and Connolly, 1992). The 'mono-butter' tasted much better or better according to 48% of consumers, and 36% of consumers indicated the product was similar to their usual product purchased (Cowan and McIntyre, 1991). Market potential with this product is being evaluated with additional test marketing in Germany. The possible production of mono-cheese and mono-yogurt has also been considered by the authors.

The effects of feeding sunflower seeds or safflower seeds in combination with bovine somatotrophin treatment on milk fatty acid composition have been evaluated (Stegeman, 1992a,b). Milk from cows fed the added fat and treated with bovine somatotrophin were not susceptible to oxidized off-flavours, although they were higher in unsaturated fatty acids. Milk fat percentages measured by the Mojonnier fat extraction and mid-infrared spectroscopic methods were 2.99, 2.97; 2.73, 2.56; and 2.86, 2.74 for the control diet, sunflower seed diet with bovine somatotrophin treatment and safflower seed diet with bovine somatotrophin treatment, respectively (Stegeman *et al.*, 1991). The mid-infrared spectroscopic method underestimated the fat content in the milk higher in unsaturated fatty acids. For each 1% increase in the unsaturated fatty acid content of milk fat, there was a 0.01% increase in the difference of percent milk fat between the mid-infrared spectroscopic method and Mojonnier fat extraction method (Stegeman *et al.*, 1991). The mid-infrared spectroscopic method measures fat by quantifying ester bonds, and fats which contain higher concentrations of long-chain fatty acids have more mass per unit ester bond (Dunkley *et al.*, 1977). This method uses a B filter (3.4–3.5 μm) which accounts for differences in fatty acid chain length. Milk fat high in unsaturated fatty acids may be underestimated because differences in unsaturation are not accounted for which can alter molecular weight. Wiesen *et al.* (1990) reported no differences between the mid-infrared spectroscopic method and the Babcock method when long-chain fatty acid composition was altered by feeding rapeseed screenings.

Franke *et al.* (1977) advised special calibration of mid-infrared spectroscopic instruments when measuring milk fat from cows fed high fat diets. With more dairy producers feeding highly unsaturated fat sources, there may be discrepancies in milk fat measurement with mid-infrared spectroscopic instruments. Further research is necessary to address this potential problem.

Cheddar cheese manufactured from milk produced by cows fed a sunflower seed or extruded soyabean diets has been researched by Lightfield *et al.* (1993). Milks and cheeses contained 48% more unsaturated fatty acids. The extruded soyabeans were preferred over the sunflower seeds because they are readily available to more dairy producers and they did not reduce milk fat contents as much as the sunflower seeds. The high unsaturated fatty acid cheeses did not exhibit a 'weak' body, in contrast to previous research (Middaugh *et al.*, 1988; Murphy, 1992; Stegeman *et al.*, 1992a) which showed that butter containing increased unsaturated fatty acids was softer. Milk from the extruded soyabean diet was more susceptible to oxidation when 0.1 mg copper kg^{-1} milk was added but no oxidation was reported in any of the cheeses. The high unsaturated fatty acid cheeses had acceptable flavour, body and texture characteristics. High unsaturated fatty acid reduced fat Cheddar cheese (19.3–19.7% fat) has been made from milk from cows fed extruded soyabeans (Lentsch *et al.*, 1994a). These cheeses had acceptable flavour, manufacturing, and storage characteristics. The cheeses exhibited improved rheological characteristics, however, these differences were not detected by a sensory panel (Lentsch *et al.*, 1994b).

The results of many unprotected lipid studies (Rafalowski and Park, 1982; Middaugh *et al.*, 1988) indicate that the amount of short-chain fatty acids decreases. This may be of concern since the short-chain fatty acids which are abundant in milk fat, particularly butyric acid (4:0), is necessary for flavour development in some cheeses and fermented dairy products (Barbano and Lynch, 1987).

Human Health Implications

Different dietary fatty acids are known to decrease, increase, or have no effect on human serum cholesterol levels (Grundy, 1990) (Table 12.2). Hegsted *et al.*

(1965) reported that myristic (14:0) and palmitic (16:0) acids have a hyper-cholesterolaemic effect on human serum cholesterol levels (Table 12.2). Previous studies (McGuffey and Schingoethe, 1982; Middaugh *et al.*, 1988; Murphy, 1992) on feeding highly unsaturated unprotected lipids indicated that the resulting milk fat was lower in these two fatty acids. The milk fat also contained more stearic (18:0) and oleic (18:1) acids which are known to reduce human serum cholesterol levels. These changes have positive ramifications for human health, because dairy products supply 25% of the saturated fatty acids while supplying only 15% of total fat in the typical United States diet (O'Donnell, 1993).

Dairy products typically contain low levels of *trans* isomers of fatty acids (O'Donnell, 1993). *Trans* fatty acids have been implicated in raising human serum cholesterol levels. There is concern due to the high concentrations of *trans* fatty acids in margarine (Mensink and Katan, 1990). Casper *et al.* (1988) reported 8.1% *trans*-oleic acid (18:1) in high unsaturated milk fat, compared with 0.7% in the control. Research is needed to minimize production of *trans* fatty acids when altering milk fatty acid composition.

The Ideal Milk Fat

At a recent Wisconsin Milk Marketing Board Milk Fat Round Table discussion, the general opinion for an ideal nutritional fat, based on current dietary recommendations, was up to 8% saturated fatty acids, greater than or equal to 82% monounsaturated fatty acids and less than 10% polyunsaturated fatty acids, including omega-3 and omega-6 fatty acids (O'Donnell, 1989). There is limited research on increasing the omega-3 fatty acids in milk fat. No published reports were found on the production of dairy products higher in omega-3 fatty acids. Normal milk fat contains 64–70% saturated fatty acids, 4% polyunsaturated fatty acids and 26–32% monounsaturated fatty acids, which differs considerably from the ideal fat. From a nutritional standpoint the most promising research to date is in the area of changing milk fatty acid composition by feeding highly unsaturated fat sources. However, more research is necessary to find methods to further increase the unsaturated fatty acid composition of milk fat. This may be accomplished by combining factors that alter milk fatty acid composition and/or new technologies. Dairy producers need incentives to produce milk fat with increased levels of unsaturated fatty acids. Separation and transportation of high unsaturated fatty acid milks and development of consumer demand and marketing are practical considerations in producing these products (Palmquist *et al.*, 1993). Though production of the ideal milk fat is unlikely, changes in feeding practices could make milk fat consumption more nutritionally desirable.

References

Anderson, M. (1974) Milk fat globule membrane composition and dietary change: supplements of coconut oil fed in two physical forms. *Journal of Dairy Science* 57, 399–404.

Ashes, J.R., St. Vincent Welch, P., Gulati, S.K., Scott, T.W., Brown, G.H. and Blakeley, S. (1992) Manipulation of the fatty acid composition of milk by feeding protected canola seeds. *Journal of Dairy Science* 75, 1090–1096.

Astrup, H.N., Vik-Mo, L., Lindstad, P. and Ekern, A. (1979) Casein protected oil supplement fed at a low level to milk cows. *Milchwissenschaft* 34, 290–291.

Astrup, H.N., Baevre, L., Vik-Mo, L. and Ekern, A. (1980) Effect on milk lipolysis of restricted feeding with and without supplementation with protected rape seed oil. *Journal of Dairy Research* 47, 287–294.

Astrup, H.N., Fauske, O. and Baevre, L. (1981) Feeding of various hydrogenated marine oil products to the dairy cow. *Acta Agriculturae Scandinavica* 31, 75–80.

Atherton, H.V. and Newlander, J.A. (1977) *Chemistry and Testing of Dairy Products*, 4th edn. AVI Publishing Company, Inc., Westport, Connecticut, p. 6.

Badings, H.T., Tamminga, S. and Schaap, J.E. (1976) Production of milk with a high content of polyunsaturated fatty acids. 2. Fatty acid composition of milk in relation to the quality of pasteurized milk, butter and cheese. *Netherlands Milk and Dairy Journal* 30, 118–131.

Baer, R.J. (1991) Alteration of the fatty acid content of milk fat. *Journal of Food Protection* 54, 383–386.

Baer, R.J., Tieszen, K.M., Schingoethe, D.J., Casper, D.P., Eisenbeisz, W.A., Shaver, R.D. and Cleale, R.M. (1989) Composition and flavor of milk produced by cows injected with recombinant bovine somatotrophin. *Journal of Dairy Science* 72, 1424–1434.

Banks, W., Clapperton, J.L. and Kelly, M.E. (1980) Effect of oil-enriched diets on the milk yield and composition, and on the composition and physical properties of the milk fat, of dairy cows receiving a basal ration of grass silage. *Journal of Dairy Research* 47, 277–285.

Barbano, D.M. (1990) Seasonal and regional variation in milk composition in the U.S. *Proceedings of Cornell Nutrition Conference of Feed Manufacturers*, Cornell University, Ithaca, New York.

Barbano, D.M. and Lynch, J. (1987) Milk from BST-treated cows – its composition and manufacturing properties. *National Invitational Workshop on Bovine Somatotrophin*, St. Louis, Missouri, pp. 67–73.

Bitman, J., Dryden, L.P., Goering, H.K., Wrenn, T.R., Yoncoskie, R.A. and Edmondson, L.F. (1973) Efficiency of transfer of polyunsaturated fats into milk. *Journal of the American Oil Chemists' Society* 50, 93–98.

Bitman, J., Wood, D.L., Tyrrell, H.F., Bauman, D.E., Peel, C.J., Brown, A.C.G. and Reynolds, P.J. (1984) Blood and milk lipid responses induced by growth hormone administration in lactating cows. *Journal of Dairy Science* 67, 2873–2880.

Boatman, C., Hotchkiss, D.K. and Hammond, E.G. (1965) Effect of season and stage of lactation on certain polyunsaturated fatty acids of milk fat. *Journal of Dairy Science* 48, 34–37.

Bonanome, A. and Grundy, S.M. (1988) Effect of dietary stearic acid on plasma cholesterol and lipoprotein levels. *The New England Journal of Medicine* 318, 1244–1248.

Buchanan, R.A. and Rogers, W.P. (1973) Manufacture of butter high in linoleic acid. *The Australian Journal of Dairy Technology* 28, 175–78.

Cadden, A.M., Urquhart, A. and Jelen, P. (1984a) Evaluation of milk and butter from commercial dairy herds fed canola-based protected lipid feed supplement. *Journal of Dairy Science* 67, 2041–2044.

Cadden, A.M., Urquhart, A. and Jelen, P. (1984b) Storage stability of canola-based protected lipid feed supplement and its effect on characteristics of milk and butter. *Journal of Dairy Science* 67, 1414–1420.

Casper, D.P., Schingoethe, D.J., Middaugh, R.P. and Baer, R.J. (1988) Lactational responses of dairy cows to diets containing regular and high oleic acid sunflower seeds. *Journal of Dairy Science* 71, 1267–1274.

Casper, D.P., Schingoethe, D.J. and Eisenbeisz, W.A. (1990) Response of early lactation cows to diets that vary in ruminal degradability of carbohydrates and amount of fat. *Journal of Dairy Science* 73, 425–444.

Chilson, W.H. and Sommer, H.H. (1953) A study of milk fat from cows on special roughage diets. *Journal of Dairy Science* 36, 561 (Abstract).

Christie, W.W. (1980) The effects of diet and other factors on the lipid composition of ruminant tissues and milk. *Progress in Lipid Research* 17, 245–277.

Cook, L.J. and Scott, T.W. (1972) Formaldehyde-treated casein-safflower oil supplement for dairy cows II. Effect on the fatty-acid composition of plasma and milk lipids. *Journal of Dairy Research* 39, 211–218.

Cowan, C. and McIntyre, B. (1991) Consumers' views on monounsaturated butter. *Farm and Food* 1, 6–7.

DeMan, J.M. (1980) *Principles of Food Chemistry*. AVI Publishing Company, Westport, Connecticut, p. 78.

DePeters, E.J., and Cant, J.P. (1992) Nutritional factors influencing the nitrogen composition of bovine milk: a review. *Journal of Dairy Science* 75, 2043–2070.

Dunkley, W.L., Smith, N.E. and Franke, A.A. (1977) Effects of feeding protected tallow on composition of milk and milk fat. *Journal of Dairy Science* 60, 1863–1869.

Edmondson, L.F., Yoncoskie, R.A., Rainey, N.H. and Douglas, F.W., Jr. (1974) Feeding encapsulated oils to increase the polyunsaturation in milk and meat fat. *Journal of the American Oil Chemists' Society* 51, 72–76.

Edwards, R.A., King, J.W.B. and Yousef, I.M. (1973) A note on the genetic variation in the fatty acid composition of cow milk. *Animal Production* 16, 307–310.

Eppard, P.J., Bauman, D.E., Bitman, J., Wood, D.L., Akers, R.M. and House, W.A. (1985) Effect of dose of bovine growth hormone on milk composition: alpha-lactalbumin, fatty acids, and mineral elements. *Journal of Dairy Science* 68, 3047–3054.

Franke, A.A., Dunkley, W.L. and Smith, N.E. (1977) Influence of feeding protected tallow on accuracy of analyses by infrared milk analyzer. *Journal of Dairy Science* 60, 1870–1874.

Gibson, J.P. (1991) The potential for genetic change in milk fat composition. *Journal of Dairy Science* 74, 3258–3266.

Goering, H.K., Gordon, C.H., Wrenn, T.R., Bitman, J., King, R.L. and Douglas, F.W., Jr. (1976) Effect of feeding protected safflower oil on yield, composition, flavor, and oxidative stability of milk. *Journal of Dairy Science* 59, 416–425.

Grummer, R.R. (1991) Effect of feed on the composition of milk fat. *Journal of Dairy Science* 74, 3244–3257.

Grundy, S.M. (1986) Comparison of monounsaturated fatty acids and carbohydrates for lowering plasma cholesterol. *The New England Journal of Medicine* 314, 745–748.

Grundy, S.M. (1990) Trans monounsaturated fatty acids and serum cholesterol levels. *The New England Journal of Medicine* 323, 480–481.

Hagemeister, H., Precht, D. and Barth, C.A. (1988) Studies on transfer of omega-3 fatty acids into bovine milk fat. *Milchwissenschaft* 43, 153–158.

Hall, A.J. (1970) Seasonal and regional variations in the fatty acid composition of milk fat. *Dairy Industry* 35, 20–24.

Hashim, S.A., Arteaga, A. and van Itallie, T.B. (1960) Effect of a saturated medium-chain triglyceride on serum-lipids in man. *Lancet* 1, 1105–1108.

Hegsted, D.M., McGandy, R.B., Myers, M.L. and Stare, F.J. (1965) Quantitative effects of dietary fat on serum human cholesterol. *American Journal of Clinical Nutrition* 17, 281–295.

Henderson, C. (1973) The effects of fatty acids on pure cultures of rumen bacteria. *Journal of Agricultural Science* 81, 107–112.

Henderson, J.L. (1971) *The Fluid Milk Industry*, 3rd edn. The AVI Publishing Company, Inc., Westport, Connecticut, p. 455.

Hoeschele, I. (1990) Potential gain from insertion of major genes into dairy cattle. *Journal of Dairy Science* 73, 2601–2618.

Jensen, R.G. and Clark, R.W. (1988) Lipid composition and properties. In: Wong, N.P. (ed.) *Fundamentals of Dairy Chemistry*, 3rd edn. Van Nostrand Reinhold Company, New York, pp. 171–213.

Khorasani, G.R., Robinson, P.H., De Boer, G. and Kennelly, J.J. (1991) Influence of canola fat on yield, fat percentage, fatty acid profile, and nitrogen fractions in Holstein milk. *Journal of Dairy Science* 74, 1904–1911.

Kieseker, F.G., Hammond, L.A. and Zadow, J.G. (1974) Commercial scale manufacture of dairy products from milk containing high levels of linoleic acid. *The Australian Journal of Dairy Technology* 29, 51–53.

Kowalczyk, J., Orskov, E.R., Robinson, J.J. and Stewart, C.S. (1977) Effect of fat supplementation on voluntary food intake and rumen metabolism in sheep. *British Journal of Nutrition* 37, 251–257.

Kronfeld, D.S., Donoghue, S., Naylor, J.M., Johnson, K. and Bradley, C.A. (1980) Metabolic effects of feeding protected tallow to dairy cows. *Journal of Dairy Science* 63, 545–552.

Larson, S.A. and Schultz, L.H. (1970) Effects of soyabeans compared to soyabean oil and meal in the ration of dairy cows. *Journal of Dairy Science* 53, 1233–1240.

Lentsch, M.R., Baer, R.J., Schingoethe, D.J., Madison, R.J. and Kasperson, K.M. (1994a) Characteristics of milk and reduced fat Cheddar cheese from cows fed unsaturated dietary fat and niacin. *Journal of Dairy Science* 77 (Suppl. 1), 22 (Abstract).

Lentsch, M.R., Baer, R.J., Schingoethe, D.J., Madison, R.J. and Kasperson, K.M. (1994b) Texture profile analysis of reduced fat Cheddar cheese high in unsaturated fatty acids. *Journal of Dairy Science* 77 (Suppl. 1), 26 (Abstract).

Lightfield, K.D., Baer, R.J., Schingoethe, D.J., Kasperson, K.M. and Brouk, M.J. (1993) Composition and flavor of milk and Cheddar cheese higher in unsaturated fatty acids. *Journal of Dairy Science* 76, 1221–1232.

Lynch, J.M., Barbano, D.M., Bauman, D.E. and Hartnell, G.F. (1988) Influence of somatribove (recombinant methionyl bovine somatotrophin) on the protein and fatty acid composition of milk. *Journal of Dairy Science* 71 (Suppl. 1), 100 (Abstract).

McGuffey, R.K. and Schingoethe, D.J. (1982) Whole sunflower seeds for high producing dairy cows. *Journal of Dairy Science* 65, 1479–1483.

Macleod, G.K. and Wood, A.S. (1972) Influence of amount and degree of saturation of dietary fat on yield and quality of milk. *Journal of Dairy Science* 55, 439–445.

Mattson, F.H. and Grundy, S.M. (1985) Comparison effects of dietary saturated, monounsaturated, and polyunsaturated fatty acids on plasma lipids and lipoproteins in man. *Journal of Lipid Research* 26, 194–202.

Mensink, R.P. and Katan, M.B. (1990) Effect of dietary trans fatty acids on high-density and low-density lipoprotein cholesterol levels in healthy subjects. *The New England Journal of Medicine* 323, 439–445.

Middaugh, R.P., Baer, R.J., Casper, D.P., Schingoethe, D.J. and Seas, S.W. (1988) Characteristics of milk and butter from cows fed sunflower seeds. *Journal of Dairy Science* 71, 3179–3187.

Mohamed, O.E., Satter, L.D., Grummer, R.R. and Ehle, F.R. (1988) Influence of dietary cottonseed and soyabean on milk production and composition. *Journal of Dairy Science* 71, 2677–2688.

Murphy, J.J. (1992) Altering the composition of milk fat by dietary means. *Irish Grassland and Animal Production Journal* 25, 3–7.

Murphy, J.J. and Connolly, J.F. (1992) Manipulating the diet of the cow to produce a softer milk fat high in monounsaturated fatty acids. *Irish Journal of Agricultural and Food Research* 31, 92 (Abstract).

Murphy, J.J., McNeill, G.P., Connolly, J.F. and Gleeson, P.A. (1990) Effect on cow performance and milk fat composition of including full fat soyabeans and rapeseeds in the concentrate mixture for lactating dairy cows. *Journal of Dairy Research* 57, 295–306.

O'Donnell, J.A. (1989) Milk fat technologies and markets: a summary of the Wisconsin milk marketing board 1988 milk fat round table. *Journal of Dairy Science* 72, 3109–3115.

O'Donnell, J.A. (1993) Future of milk fat modification by production or processing: integration of nutrition, food science, and animal science. *Journal of Dairy Science* 76, 1797–1801.

Palmquist, D.L. and Beaulieu, A.D. (1992) Differences between Jersey and Holstein cows in milk fat composition. *Journal of Dairy Science* 75 (Suppl. 1), 292 (Abstract).

Palmquist, D.L. and Conrad, H.R. (1980) High fat ration for dairy cows. Tallow and hydrolyzed blended fat at two intakes. *Journal of Dairy Science* 63, 391–395.

Palmquist, D.L. and Jenkins, T.C. (1980) Fat in lactation rations: review. *Journal of Dairy Science* 63, 1–14.

Palmquist, D.L., Beaulieu, A.D. and Barbano, D.M. (1993) Feed and animal factors influencing milk fat composition. *Journal of Dairy Science* 76, 1753–1771.

Rafalowski, W. and Park, C.S. (1982) Whole sunflower seed as a fat supplement for lactating cows. *Journal of Dairy Science* 65, 1484–1492.

Reiter, J. (1994) New product review. *Dairy Foods* 95 (2), 31.

Scott, T.W., Cook, L.J., Ferguson, K.A., McDonald, I.W., Buchanan, R.A., and Hills, G.L. (1970) Production of polyunsaturated milk fat in domestic ruminants. *Australian Journal of Science* 32, 291–293.

Scott, T.W., Cook, L.J. and Mills, S.C. (1971) Protection of dietary polyunsaturated fatty acids against microbial hydrogenation in ruminants. *Journal of the American Oil Chemists' Society* 48, 358–364.

Selner, D.R. and Schultz, L.H. (1980) Effects of feeding oleic acid or hydrogenated vegetable oils to lactating cows. *Journal of Dairy Science* 63, 1235–1241.

Sneft, B. and Klobasa, F. (1970) Studies on the fatty acid composition of cow's colostrum. *Milchwissenschaft* 25, 391–394.

Sporns, P., Rebolledo, J., Cadden, A.M. and Jelen, P. (1984) Compositional changes in fatty acids of butter caused by feeding canola-based protected lipid feed supplement. *Milchwissenschaft* 39, 330–332.

Steele, W. and Moore, J.H. (1968) The effects of dietary tallow and cottonseed oil on milk fat secretion in the cow. *Journal of Dairy Research* 35, 223–235.

Steele, W., Noble, R.C. and Moore, J.H. (1971a) The effects of 2 methods of incorporation soyabean oil into the diet on milk yield and composition in the cow. *Journal of Dairy Research* 38, 43–48.

Steele, W., Noble, R.C. and Moore, J.H. (1971b) The effects of dietary soyabean oil on milk-fat composition in the cow. *Journal of Dairy Research* 38, 49–56.

Stegeman, G.A., Baer, R.J., Schingoethe, D.J. and Casper, D.P. (1991) Influence of milk fat higher in unsaturated fatty acids on the accuracy of milk fat analyses by the mid-infrared spectroscopic method. *Journal of Food Protection* 54, 890–893.

Stegeman, G.A., Baer, R.J., Schingoethe, D.J. and Casper, D.P. (1992a) Composition and flavor of milk and butter from cows fed unsaturated dietary fat and receiving bovine somatotrophin. *Journal of Dairy Science* 75, 962–970.

Stegeman, G.A., Casper, D.P., Schingoethe, D.J. and Baer, R.J. (1992b) Lactational responses of dairy cows fed unsaturated dietary fat and receiving bovine somatotrophin. *Journal of Dairy Science* 75, 1936–1945.

Tamminga, S., Steg-Beers, A., van Hoven, W. and Badings, H.T. (1976) Production of milk with a high content of polyunsaturated fatty acids. 1. Experiments in relation to the efficiency of production. *Netherlands Milk and Dairy Journal* 30, 106–117.

United States Department of Agriculture (1976) *Composition of Foods; Dairy and Egg Products Raw Processed Prepared*. Agriculture Handbook Number 8-1. United States Government Printing Office, Washington, D.C.

Urquhart, A., Cadden, A.M. and Jelen, P. (1984) Quality of milk and butter related to canola-based protected lipid feed supplement. *Milchwissenschaft* 39, 1–6.

Wiesen, B., Kincaid, R.L. and Hillers, J.K. (1990) The use of rapeseed screenings in diets for lactating cows and subsequent effects on milk yield and composition. *Journal of Dairy Science* 73, 3555–3562.

Wong, N.P., Walter, H.E., Vestal, J.H., LaCroix, D.E. and Alford, J.A. (1973) Cheddar cheese with increased polyunsaturated fatty acids. *Journal of Dairy Science* 56, 1271–1275.

Wong, W., Jelen, P. and DeMan, J.M. (1982) Softening of butter related to feeding low doses of protected-tallow supplement. *Journal of Dairy Science* 65, 1632–1638.

Wood, F.W., Murphy, M.F. and Dunkley, W.L. (1975) Influence of elevated polyunsaturated fatty acids on processing and physical properties of butter. *Journal of Dairy Science* 58, 839–845.

Robotic Milking of Dairy Cows

13

A. Kuipers[1] and W. Rossing[2]

[1]Research Station for Cattle, Sheep and Horse Husbandry (PR), Runderweg 6, 8219 PK Lelystad, The Netherlands; [2]Institute for Agricultural and Environmental Engineering (IMAG-DLO), Mansholtlaan 10–12, 6708 PA Wageningen, The Netherlands

Introduction

The goal of automatic milking is to make the continuous presence of the dairyman during milking unnecessary. The entry of cows into the milking unit, the attachment of the teat cups as well as the removal of the teat cups are automated. Realization of automatic milking in combination with management-by-exception programmes will be a very important achievement in the automation of the dairy farm. It will have a large impact on the management of the dairy farm, the cow and on the farm family as part of society.

However, several technical, economical and social uncertainties still remain. Some experience is being obtained with milking robots on research farms in various countries. In this chapter, most information is gathered with two types of milking robots in use on the experimental farms De Waiboerhoeve in Lelystad and De Vijf Roeden in Duiven, The Netherlands. About 25 milking robots are also placed on practical farms in The Netherlands. The techniques of automatic milking systems in development in various countries, the effects of automatic milking on the cow, as well as the economics and introduction aspects are analysed.

History of Milking

The first attempts to mechanize the milking of dairy cows were made in the early 19th century. The earliest devices were metal tubes (cannulae) inserted into the teat canal to allow milk in the teat and udder sinuses to flow out under the forces of gravity and intramammary pressure.

Some of the machines invented in the second half of the 19th century sought to mimic hand milking by applying pressure to the outside of the teat, rather than by applying a vacuum. Some were extremely crude, but there were many amazing examples of mechanical ingenuity using plates, bars, rollers or belts, operated manually through mechanical, hydraulic or pneumatic transmission (Bramley *et al.*, 1992).

Machines using a vacuum were developed in the middle of the 19th century. Some of the machines invented in this period had individual teat cups. The vacuum applied to the teat was continuous.

In the late 19th century, double action machines based on the principle of alternating vacuum (the forerunners of all modern milking machines) were developed. These were improved in the early years of this century, but the double action principle remained fundamentally unchanged. Although nearly all the machines still transferred milk into a bucket or can, various types of milking parlours were designed to enable larger herds to be milked.

The mechanization of dairying continued, and in the 1960s the equipment for automatic teat cup removal became standard. This equipment not only saved labour but also improved the milking process. However, milking still required a large part of the dairy farmer's total efforts.

The latest innovation in the milking process to have major implications for dairy farming is automatic milking. It makes manual intervention by the dairy farmer during milking redundant, thereby saving labour and enabling the daily farm work to be separated from the milking of the cows.

Automatic Milking Systems

Equipment for the automatic attachment of milking units has been developed in a number of countries (Hogewerf *et al.*, 1992; Marchal *et al.*, 1992; Schön *et al.*, 1992; Street *et al.*, 1992). All these systems have the same aims: the complete automation of the milking process and efficient milk production that takes human and animal welfare into account. However, the principles and techniques employed differ.

In The Netherlands three milking robots are currently being developed. A commercial company, Prolion (Vijfhuizen, The Netherlands), has produced a special container (Fig. 13.1a) in which all the necessary equipment is housed (Bottema, 1992; Hogewerf *et al.*, 1992). All the teat cups are on a milk rack and they are attached from the side, after two sensor units have located the teats. The first unit (two ultrasonic sensors with diverging fields) has to find the front right teat as a reference point. The second unit, an ultrasonic sensor called the fine sensor, moves up and down and has a rotating field. It measures the distance between the teats and the reference teats. If the cow moves in the stall after the teat positions have been located, the arm moves with the reference teat. After all the teats have been located, the attachment of all four teat cups starts. The teat cups are positioned under the teats one by one and moved upwards. Prolion manufactures installations with one, two, three or four stalls. In situations with several stalls, the robot arm with sensors moves along a rail next to the stalls. Each stall has an arm, which remains under the animal during milking after the teat cups have been attached. At the end of milking the arm is activated and removes the teat cups. The cleaning of the teats is carried out in the teat cups. In late 1994 the Prolion automatic milking system was in use in The Netherlands on three experimental farms and on 15 commercial farms.

Another system, developed by Gascoigne Melotte (Emmeloord, The Netherlands), attaches the teat cups from behind the cow, between her hind legs (Fig. 13.1b; van der Linden and Lubberink, 1992). The coordinates of the teats are stored in a computer database. Before the teat cups are attached they are correctly positioned. Then the arm bearing the four teat cups approaches from the rear and attaches the cups. After the cups have been attached the database is updated with new teat coordinate data. A brush cleans the teats and the udder before the cups are attached. This robot is being experimentally tested in The Netherlands on one experimental farm.

Lely Industries (Maasland, The Netherlands) has built a milking robot which functions in a similar way to the Prolion unit (Fig. 13.1c). It too attaches the teat cups from the side of the animal. The arm moves under the animal and then moves upwards until a laser sensor locates the top of a teat. It then moves towards the rear of the animal until the two hind teats are located. The teat cups are positioned under the teats individually and moved upwards. When they have been attached to the hind teats, the arm moves back and searches for the front teats. After it has located them, the cups are attached. Before the cups are attached, the teats are cleaned by a system using rollers and soft towels. During milking, the cleaning unit is cleaned and disinfected. This system is currently in use on ten commercial farms in The Netherlands.

The automatic milking system developed by the French Institute of Agricultural and Environmental Engineering Research and currently being evaluated on an experimental farm has four arms, one for each teat cup. The arms for the front teats approach from the sides, and those for the rear teats come up from the floor (Marchal *et al.*, 1992). Teat position is established by a global sensor and a local sensor. The three-dimensional vision system is based on the triangulation principle and uses a CCD camera and laser. The local sensor is a network of infrared-emitting devices and photo transistors around each teat cup.

A British system developed by Silsoe Research Institute uses a pneumatic robot to do the attaching (Street *et al.*, 1992). The position of the animal is detected by sensors which are pressed gently against the animal's flanks and back. A matrix of eight infrared light beams is arranged across the top of the end effector to detect the teat and allow the robot's position to be corrected until the cup is centred on the teat. The four teat cups are attached one by one by the robot arm. This robot is being tested on an experimental farm.

A system being developed by Düvelsdorf (Ottersberg-Posthausen, Germany)

Fig. 13.1. Examples of milking robots. (a) Prolion; (b) Gascoigne Melotte.

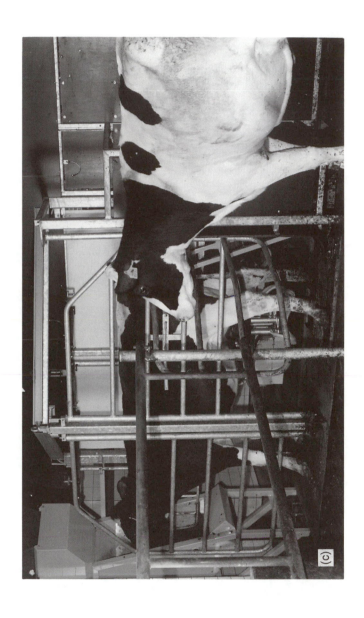

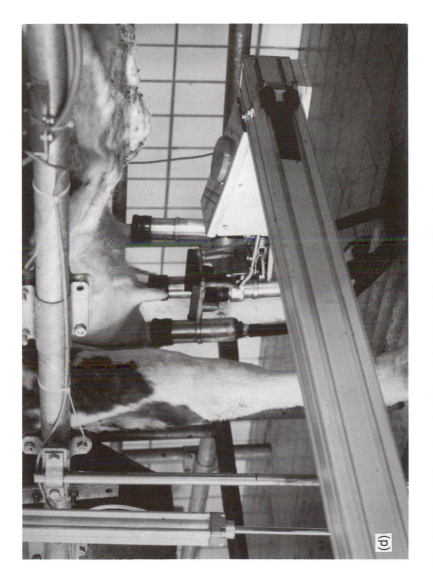

Fig. 13.1. (contd) Examples of milking robots. (c) Lely Astronaut; (d) Düvelsdorf.

uses a frame in which an arm is moved by electric motors (Fig. 13.1d; Dück, 1992). Data on the positions of the teats are stored in the computer database. The arm brings the teat cups into these positions, and the exact position is established by an ultrasonic sensor and light barriers. The robot arm then attaches the four teat cups one by one. This robot is being tested on a commercial farm in Germany.

The Federal Research Centre for Agriculture in Germany has also developed and tested an automatic milking system (Artmann, 1992). This robot has three linear axles and one rotary axle, all driven by asynchronous motors. The clusters are attached one by one. Ultrasonic sensors, a CCD camera and a laser are used to locate the teats.

Implementation in a Loose Housing Barn

An automatic milking system is more than a milking robot. It also includes the design of the milking place, and the routing of the cows towards this place.

A milking robot can be used in different ways. It can milk the cows two or three times a day at fixed intervals (Bottema, 1992), or the cows can visit the system voluntarily and be milked within predetermined intervals (Devir *et al.*, 1993a). In the first situation the cows have to be assembled in a collecting yard. The farmer has to bring the animals to the collecting area and they may have to remain there for some time.

To gain maximum benefit from an automatic milking system it has to be used 24 hours a day and the milking frequency must be variable and adjusted to each cow's parity, in accordance with her milk yield, milk composition, body weight and feed intake. A dairy control and management system on the farm must control this process (Devir *et al.*, 1993b).

In a 'self-service milking' situation the animals visit the milking robot more or less spontaneously several times a day. One way of achieving this is to combine milking with feeding. However, for optimal utilization of an automatic milking system the cow traffic will probably have to be regulated by setting up a compulsory routing through the cow house.

Effects on the Cow

Automatic milking will result in increased milking frequencies. This will result in a longer total time that the teat cups are attached to the cow, and therefore in a longer machine milking time (Table 13.1). Such an additional load on the cow has to be minimized by adequate milking technology. Consequently, automatic milking will increase the need for technology orientated more to the (individual) animal. This implies that the needs of the animal have to be formulated more accurately. In addition, effects of changes in milking frequency on milk yield, milk composition, milk quality, teat condition, animal health and

Table 13.1. Effect of frequency of milking and period in lactation on milk yield, fat and protein yield, dry matter intake, body fat score and machine milking time (Ipema and Benders, 1992).

Parameter	Frequency of milking	Period in lactation (weeks)					
		1-12	13-25	26-35	36-42	Dry	Total
Milk yield	2	37.9	30.2[1]	21.9[1]	17.1[1]		28.2
(kg d^{-1})	3	41.6	34.6[2]	26.0[2]	20.6		32.2
	4	41.7	33.8[2]	26.2[2]	21.9[2]		32.3
Fat + protein yield	2	2962	2476	1885[1]	1571[1]		2323
(g d^{-1})	3	3151	2677	2174	1869		2558
	4	3157	2618	2231[2]	1948[2]		2568
Dry matter	2	22.8	23.5	19.5	15.8	8.1	21.0
intake (kg d^{-1})	3	23.5	24.2	19.6	16.7	8.4	21.7
	4	23.8	24.3	20.0	16.7	8.7	21.9
Body fat score	2	3.45[1]	3.45[1]	3.65[1]	3.75[1]	3.90	3.60
	3	3.33	3.37	3.42[2]	3.55[2]	3.82	3.47
	4	3.28[2]	3.30[2]	3.39[2]	3.43[2]	3.76	3.40
Machine milking	2	15.9[1]	15.3[1]	12.6[1]	11.1[1]		14.1
time (min d^{-1})	3	22.3[2]	21.4[2]	17.5[2]	15.8[2]		19.8
	4	24.6[2]	22.8[2]	19.8[2]	19.3[3]		22.0

[1,2,3] statistically significant difference within same column ($P < 0.05$)

animal behaviour are of importance. These effects are also largely dependent on the functioning of the milking robot and the arrangement of the milking station and further routing. Therefore experimental data from various types of robots may differ significantly. Milking without direct human supervision requires also safeguards to prevent unfavourable effects on animal health and milk quality.

Milking technique

The more the milking technology has been adapted to the cow, the longer she can stand being milked. When cows are milked more frequently, matters such as optimum pulsation control and vacuum level increase in importance. It is known from experience that an adequate layout and adjustment of the milking machine allows cows to be milked three times a day. Research still has to indicate whether present milking technology is less suitable for higher milking frequencies with the consequently longer total milking times. Automatic milking offers new perspectives for techniques, like automatic stimulation and quarter take off.

Milking frequency

The number of cows to be milked in an automatic system in particular depends on whether the equipment is used efficiently and on the frequency of milking. Usually, some concentrates are fed in the milking station to stimulate cows to enter the system voluntarily. Some experiments have shown that cows are prepared to be milked voluntarily three or four times a day (Rabold, 1986; Ipema *et al.*, 1987). In another experiment (Ipema and Benders, 1992) cows in various lactation periods tried to visit the system on average 5.8 to 6.7 times a day. At De Waiboerhoeve, Lelystad, experience is that there is some hesitation among cows to come voluntarily to the milking robot.

The optimum milking frequency has not yet been determined, but it can be assumed that it is from two to four times a day. However, it may differ from cow to cow, depending on the milk yield. The number of automatic milking stalls needed (and thus the investment), the preferred milking frequency and the degree of freedom for the animal in the system will have to be balanced against each other. Also the choice between keeping cows in confinement and partly grazing will certainly affect the frequency of milking to be realized with a milking robot.

Production level

Both experimental and practical experiences have shown that increased milk yields are a consequence of a raised milking frequency. The effects of more frequent milking on milk yield have been analysed in a number of research projects in the last decade. Bruins (1983) summarized three times ($3\times$) daily milking results in traditional systems and Ipema *et al.* (1987) and Ipema and Benders (1992) studied the effects of milking frequency by simulating robotic milking. The increase in milk yield from $2\times$ to $3\times$ daily milking appeared to vary between 10 and 15%. It was also found that this increase was linked to a reduction in fat content. In one study the fat to protein ratio became narrower as a result. Furthermore, the effects of various milking frequencies appeared also to be related to the stages of lactation (Table 13.1).

The increase in milk yield is also dependent on the breed. Campos *et al.* (1994) found an increase of 15% with Holstein-Friesian cows and 6–7% with Jersey cows.

By milking half of the udder twice and the other half four times it was concluded that heifers and cows respond differently (Hillerton and Winter, 1992). It was suggested that stimulation of a greater peak yield could be the best strategy for the heifer.

Results with $4\times$ daily milking differ. In the experiment explained in Table 13.1, $4\times$ milking did not increase production significantly over $3\times$. However, experience in 16 herds milking $4\times$ daily in Arizona, USA, indicates that an additional 6–8% increase in production compared with 3× is possible (Armstrong, D.V., Arizona, 1992, personal communication). In these herds cows were comfortable staying in the milking parlour because of the high temperatures outside in this region.

Barnes *et al.* (1989) found a longer machine milking time, a lower milking speed and an increased yield with 3× milking instead of 2× in a traditional parlour.

Inefficient functioning of the milking robot, unfavourable design of the station and untrained animals influence the production results negatively. In one experiment (Kremer and Ordolff, 1992) the average frequency of milking was 2.8× per cow per day. Only 46% of all visits for milking were voluntarily. Uneasiness of the cows caused a lower production than the control group housed in a stanchion barn.

Milk quality

The milking interval can also affect milk quality. According to Rabold (1986) the somatic cell count (SCC) declines when the milking frequency increases. Research carried out at De Vijf Roeden, Duiven, showed the SCC for both test group (3× daily milking or more) and reference group (2× daily milking) to be at the same low level of an average of 150 000 per ml milk (Ipema *et al.*, 1987).

A risk of increasing the milking frequency is that of increased free fatty acid content (Verheij, 1992; Ipema and Schuiling, 1992). There can be wide variation among individual animals: for some cows increased lipolysis is already found when they are milked 3× daily (Jellema, 1986).

The total bacterial count, a major quality indicator, is hardly affected by the frequency of milking but can be affected by the frequency of cleaning. However, with robotic milking fresh milk is continuously passing through the circuit to the milk tank. Simulation and practical experiments have shown that the bacteriostatic action of fresh milk has a favourable effect on the bacterial count (Verheij, 1992). The required cleaning frequency depends on the ambient temperature. Under favourable conditions, twice daily cleaning is likely to suffice. The cleaning circuit in an automatic milking system can be much smaller than in a usual system, so that the water consumption per cleaning can be reduced.

In addition to the quality aspects in relation to a raised milking frequency, the absence of supervision has other consequences for the milk quality. Cleaning of the udder should ensure that the teats are really clean. Technical solutions for the cleaning of teats are promising (Schuiling, 1992), but this does not solve the problem that milk can be of poor quality because the udders are dirty. So far, automatic detection of dirty udders has not been possible. Therefore, automatic milking will require the hygienic conditions in the house to be optimal to ensure that each cow is milked with her udder being fairly clean. Under such conditions, a dry cleaning procedure of the udder is to be preferred.

Health and animal behaviour

Some research has indicated that stepping up the milking frequency to $3 \times$ daily does not affect udder health (Bruins, 1983). However, research in the USA (Elliot, 1961; Pearson *et al.*, 1979) indicates that $3 \times$ daily milking helps to prevent mastitis. Furthermore, wear and tear on the udder was reported to be less (Tilton, 1978; Ashfield, 1981). But raising the milking frequency to three times a day with traditional milking causes the risk of cross-infection via the milking machine to increase by 50% (Elliot, 1961; Pearson *et al.*, 1979). Cross-infection is less likely to happen with robotic milking because of the longer short milk tubes (for some types of robots no milk claw is needed).

As to teat condition, research results have to some extent been ambiguous, but it has been shown that a milking frequency of four times a day or more has undesirable effects (Ipema and Benders, 1992). On the other hand, a calf suckles the cow many times a day without any problems. Perhaps it is best to empty the udder only partly in the case of robotic milking. Rasmussen (1993) showed that such an approach improved the teat condition. However, incomplete milking and maintaining the same frequency of milking reduces total milk yield (Wheelock *et al.* 1965). In general, the expected higher milking frequency and yield will require a better health management.

With a higher frequency of milking the body fat reserves of the cows decrease more during the lactation (Table 13.1). The feeding regime of the animals requires more attention. When high yielders are milked at unequal intervals, such as 9 and 15 hours, they appear to lie down less during the last few hours before milking because of the greater pressure in the udder, especially during the long interval at night. When the milking frequency is increased, it may be expected that this inconvenience is eased (Ipema *et al.*, 1991).

So far, only limited information is available about the behaviour of the animals using different types of automatic milking systems. Nevertheless, the behaviour is a determining factor for the success of a system of automatic milking.

Udder conformation

The udder conformation will certainly become more important than it already is. Some animals will have to be culled because of unfavourable characteristics in udder shape. Developments such as embryo transfer and perhaps embryo cloning can help increase the uniformity of the herd. More detailed information is needed to find out which exterior traits will be required for cows in automatic milking systems. Considering the high heritability of udder characteristics, breeding programmes have the potential to correct these characteristics over time.

Economics

Up till now, much less emphasis in agricultural research has been put on the economic and social aspects of the milking robot on the dairy farm. However, it is clear that an automatic milking system will require a very substantial capital investment and will alter the operation of the farm. The decision as to whether an investment will be made in automatic milking arises in most cases when the old milking parlour has to be replaced or renovated. Also changes in annual yields and costs may occur when an automatic milking system is chosen. In most cases such a system will require adaptations in not only the housing but in the whole farm organization. Calculations by Kuipers and Van Scheppingen (1992) estimate the economic consequences of these adaptations and the returns of automatic milking compared to traditional milking systems.

The main factors studied were:

1. increase in milk yield per cow and the change in milk composition;
2. alteration in grazing and feeding strategy; and
3. the choice of the traditional milking parlour.

An increase in production per cow because of more frequent milking contributes positively to net returns from automatic milking. Transition from an unlimited grazing system to feeding cows all year round indoors to allow better access to the milking robot affects returns in a strongly negative way. The investment level the farmer chooses for a traditional parlour also affects the feasibility of investing in a milking robot.

The profitability of an automatic milking system can be expressed as the 'maximum acquisition value' (Kuipers and Van Scheppingen, 1992). This is the amount of capital which may be invested in the automatic system to achieve the same net farm result as with a traditional milking parlour. In the following equation the maximum acquisition value (MAV) is calculated by accumulating the returns from the increase in milk yield on a yearly basis (R_{my}), the costs of changing the grazing system on a yearly basis (C_{gs}) and the savings in annual costs by not investing in a traditional parlour (AC_{tp}) and subsequently dividing this total by the estimated percentage of annual costs (depreciation, maintenance and interest) of the automatic milking system (PAC_{ams}):

$$MAV = [\,(R_{my} - C_{gs} + AC_{tp})/(PAC_{ams}/100)\,].$$

The maximum acquisition value was calculated for different farm situations.

With the choice of a higher investment level for the conventional milking parlour, the maximum acquisition value for the automatic milking system will also increase. As a consequence, individual wishes as to the layout of the milking parlour play an important role in the profitability of the automatic milking system.

The annual costs influence the perspectives considerably. However, maintenance costs of milking robots are not yet known, but will affect the final level of the annual costs. Large herds will require lower investments to be made per cow because of the effect of the larger numbers; furthermore the capacity of the milking robots may be used more efficiently with more cows.

A very important point is the labour economy that can be achieved with a fully autonomously operating system. At the present it is likely that no significant saving of labour can be achieved. Then, calculated maximum acquisition values for automatic milking will be less than the prices mentioned for milking robots at the time. Eventually, it could be possible to reduce the work requirements as the milking robot can take over a great deal of the physical effort. In this stage the herdsmate's task will shift more towards supervision. When the system has established itself, a real labour economy seems to be feasible. However, saving of labour can only be capitalized in larger farms with more employees.

For the American situation, Armstrong *et al.* (1992) concluded that it is unlikely that the capitally intensive robotic milking system will be a viable alternative to a large-scale, efficient production line milking system in the short term.

Prospects

The prospects of automatic milking systems depend on technical, economical and social factors. Aspects that may influence the implementation of milking robots are now described.

Implementation

A working committee in The Netherlands studied the prospects of automatic milking systems (Kuipers and Van Scheppingen, 1992). The following factors were considered of importance for the introduction of the system in practice:

1. *Size of dairy herds.* Smaller farms (below 40 cows) are less likely to be interested in automatic milking systems, due to investment level and possible inefficiencies in utilizing the capacity of the robots.
2. *Housing system.* Milking robots currently in operation require self entrance of the cow. Cows also need to be allowed to enter the milking stall several times a day. This implies that loose-housing is required.

3. *Production level*. High production levels are associated with more frequent milking, and thus with robotic milking. More frequent milking may also contribute to the idea of animal welfare provision.
4. *Labour costs*. When an automatic milking system results in labour savings, high labour costs will stimulate the introduction of the system.
5. *Grassland management system*. Automatic milking requires the cows to be near the robot. If cows are kept indoors all year round, society may react in a negative way to the animal welfare and landscape implications.
6. *Contact with animals*. Less contact with the animals can be a negative factor in control of the herd. However, more experience is needed to determine to what extent a management programme and sensors can take over this task from the farmer. Also some farmers may consider less direct contact with the animals not to be in agreement with their professional ethics.
7. *Ease of working*. It is expected that automatic milking will reduce physical labour. The herdman's function will transfer more to supervisory work. This demands new skills from the farmer.
8. *Attachment to farm*. In the long run it is expected that the farmer (and his family) is less tied to the farm, because automatic milking requires only incidental attendance by the farmer. This may place the profession of dairy farmer more in line with general developments in society.
9. *Capital position and income*. The financial position of the dairy farm(s) is important when deciding upon such a considerable investment as is required for automatic milking. Therefore, the trends in farm incomes will also greatly influence the rate of introduction of automatic milking systems in practice.

Seabrook (1992) found that stockpeople who believed they were livestock orientated appeared more demeaned by the prospects of automatic milking than those who believed they were machinery orientated. Probably the interest of the dairyman in automatic milking will vary from country to country, depending on the circumstances and attitudes.

Perspectives

The introduction of the automatic milking on dairy farms will probably occur in two stages (Kuipers and Van Scheppingen, 1992). In the first stage, use of a milking robot means a further automation of the milking process. The cows are milked without manual assistance of the farmer or milking personnel. The presence of the farmer is, however, regularly needed because of the selection of abnormal cows, especially cows with mastitis. Also some 'difficult' cows need additional supervision in entering the robot and during the milking process. In this stage of introduction there is no labour saving expected.

In the second stage, a complete automatic milking system will be developed. A management programme with various sensors takes over most of the control tasks of the farmer in the traditional milking parlour. After the automatic attachment of teat cups, detection and separation of abnormal and dirty cows or less hygienic milk is another large challenge for the automatic milking system. However, the degree to which this goal will be reached is presently difficult to assess.

In the long run, these developments may result in labour saving which can contribute significantly to the economic returns of automatic milking under certain conditions. Social aspects may be as important as economic factors. When the dairy farmer (and his family) is less tied to the milking process and therefore to the farm, this development will be seen as a very important achievement for the profession of the dairy farmer.

Saving of labour will be easiest to realize in large herds with several employees. On family farms with usually 1–1½ family members employed, less attachment to the dairy operation will be appreciated most. Therefore, introduction of automatic milking systems seems to be most attractive for family farms with a full labour input and also large dairy farms with more than two employees.

References

Armstrong, D.V., Daugherty, L.S., Galton, D.M. and Knoblauch, W.A. (1992) Analysis of capital investment in robotic milking systems for US dairy farms. *Proceedings of the International Symposium on Prospects for Automatic Milking*. European Association for Animal Production, publication no. 65, Wageningen, The Netherlands, pp. 432–439.

Artmann, R. (1992) Status, results and further developments of automatic milking system. *Proceedings of the International Symposium on Prospects for Automatic Milking*. European Association for Animal Production, publication no. 65, Wageningen, The Netherlands, pp. 23–33.

Ashfield, G. (1981) More are trying 3*x* milking. *Dairy Herd Management*, May 1981, pp. 22–24.

Barnes, M.A., Kazmer, G.W. and Pearson, R.E. (1989). Influence of sire selection and milking frequency on milking characteristics in Holstein cows. *Journal of Dairy Science* 72, 2153–2160.

Bottema, J. (1992) Automatic milking: reality. *Proceedings of the International Symposium on Prospects for Automatic Milking*. European Association for Animal Production, publication no. 65, Wageningen, The Netherlands, pp. 63–72.

Bramley, A.J., Dodd, F.H., Mein, G.A. and Bramley, J.A. (1992) *Machine Milking and Lactation*. Insight Books, Berkshire, England and Vermont, USA.

Bruins, W.J. (1983) *Drie keer daags melken op de Waiboerhoeve*. Research Station for Cattle, Sheep and Horse Husbandry, Lelystad, The Netherlands, Report 89.

Campos, M.S., Wilcox, C.J., Head, H.H., Webb, D.W. and Hagen, J. (1994) Effects on production of milking three times daily on first lactation Holsteins and Jerseys in Florida. *Journal of Dairy Science* 77, 770–773.

Devir, S., Ipema, A.H. and Huismans, P.J.M. (1993a) Automatic milking and concentrates supplementation system based on the cow's voluntary visits. *Livestock Environment IV*, American Society of Agricultural Engineers, St. Joseph, Michigan, pp. 195–204.

Devir, S., Renkema, J.A., Huirne, R.B.N. and Ipema, A.H. (1993b) A new dairy control and management system in the automatic milking farm: concept and components. *Journal of Dairy Science* 76, 3607–3616.

Dück, M. (1992) Evolution of Düvelsdorf milking robot. *Proceedings of the International Symposium on Prospects for Automatic Milking*. European Association for Animal Production, publication no. 65, Wageningen, The Netherlands, pp. 49–155.

Elliot, G.M. (1961) The effect on milk yield of three times a day milking and of increasing the level of residual milk. *Journal of Dairy Research* 28, 209–219.

Hillerton, J.E. and Winter, A. (1992) The effects of frequent milking on udder physiology and health. *Proceedings of the International Symposium on Prospects for Automatic Milking*. European Association for Animal Production, publication no. 65, Wageningen, The Netherlands, pp. 201–212.

Hogewerf, P.H., Huijsmans, P.J., Ipema, A.H., Janssen T. and Rossing, W. (1992) Observation of automatic teat cup attachment in an automatic milking system. *Proceedings of the International Symposium on Prospects for Automatic Milking*. European Association for Animal Production, publication no. 65, Wageningen, The Netherlands, pp. 80–190.

Ipema, A.H. and Benders, E. (1992) Production, duration of machine milking and teat quality of dairy cows milked 2, 3 or 4 times daily with variable intervals. *Proceedings of the International Symposium on Prospects for Automatic Milking*. European Association for Animal Production, publication no. 65, Wageningen, The Netherlands, pp. 244–252.

Ipema, A.H. and Schuiling, E. (1992) Free fatty acids; influence of milking frequency. *Proceedings of the International Symposium on Prospects for Automatic Milking*. European Association for Animal Production, publication no. 65, Wageningen, The Netherlands, pp. 491–496.

Ipema, A.H., Benders, E. and Rossing, W. (1987) Effect of more frequent milking on production and health of dairy cattle. *Proceedings of the Third Symposium Automation in Dairying*. Institute for Agricultural and Environmental Engineering, Wageningen, pp. 283–293.

Ipema, A.H., Ketelaar-de Lauwere, C.C. and Metz-Stefanowska, J. (1991) The influence of six times per day milking on milk production, technology and cow behaviour. Institute for Agricultural and Environmental Engineering, Wageningen, The Netherlands, Report 91–20.

Jellema, A. (1986) Some factors affecting the susceptibility of raw milk to lipolysis. *Milchwissenschaft* 41, 553–558.

Kremer, J.H. and Ordolff, D. (1992) Experiences with continuous robot milking with regard to milk yield, milk composition and behaviour of cows. *Proceedings of the International Symposium on Prospects for Automatic Milking*. European Association for Animal Production, publication no. 65, Wageningen, The Netherlands, pp. 253–260.

Kuipers, A. and Scheppingen, A.T.J. van (1992) Dairy Farming and Automatic Milking. Research Station for Cattle, Sheep and Horse Husbandry, Lelystad, The Netherlands, Report 141.

Linden, R. van der and Lubberink, J. (1992) Robot milking system (RMS): design and performance. *Proceedings of the International Symposium on Prospects for Automatic Milking*. European Association for Animal Production, publication no. 65, Wageningen, The Netherlands, pp. 55–63.

Marchal, Ph., Rault, G., Collewet, Ch. and Wallian, L. (1992) Mains project–automatic Milking. *Proceedings of the International Symposium on Prospects for Automatic Milking*. European Association for Animal Production, publication no. 65, Wageningen, The Netherlands, pp. 33–40.

Pearson, R.E., Fulton, L.A., Thompson, P.D. and Smith, J.W. (1979) Three times a day milking during the first half of lactation. *Journal of Dairy Science* 62, 1941–1950.

Rabold, K. (1986) Vollautomatisches Melken. Erste Ergebnisse aus Tierhaltungs-und physiologischen Untersuchungen. *Landtechnik* 41–5, 224–226.

Rasmussen, M.D. (1993) Influence of switch level of automatic cluster removers on milking performance and udder health. *Journal of Dairy Research* 60, 287–297.

Schön, H., Artmann, R. and Worstorff, H. (1992) The automatic milking as a key issue in future oriented dairy farming. *Proceedings of the International Symposium on Prospects for Automatic Milking*. European Association for Animal Production, publication no. 65, Wageningen, The Netherlands, pp. 7–22.

Schuiling, E. (1992) Teat cleaning and stimulation. *Proceedings of the International Symposium on Prospects for Automatic Milking*. European Association for Animal Production, publication no. 65, Wageningen, The Netherlands, pp. 164–168.

Seabrook, M.F. (1992) The perception by stockpersons of the effect on their esteem, self-concept and satisfaction of the incorporation of automatic milking into their herds. *Proceedings of the International Symposium on Prospects for Automatic Milking*. European Association for Animal Production, publication no. 65, Wageningen, The Netherlands, pp. 409–413.

Street, M.J., Hall, R.C., Spencer, D.S., Wilkin, A.L., Mottram, T.T. and Allen, C.J. (1992) Design features of the Silsoe automatic milking system. *Proceedings of the International Symposium on Prospects for Automatic Milking*. European Association for Animal Production, publication no. 65, Wageningen, The Netherlands, pp. 40–149.

Tilton, L. (1978) Three times milking making waves out west. *Dairy Herd Management*, Feb. 1978, pp. 14–48.

Verheij, J.G.P. (1992) Cleaning frequency of automatic milking equipment. *Proceedings of the International Symposium on Prospects for Automatic Milking*. European Association for Animal Production, publication no. 65, Wageningen, The Netherlands, pp. 175–178.

Wheelock, J.V., Rook, J.A.F. and Dodd, F.H. (1965) The effect of incomplete milking or of an extended milking interval on the yield and composition of cows milk. *Journal of Dairy Research* 32, 237–248.

Modification of Milk Protein Composition by Gene Transfer

14

J.L. Vilotte and P.J. l'Huillier[1]

*Laboratoire de Génétique Biochimique et de Cytogénétique,
INRA-CRJ, 78352 Jouy-en-Josas Cedex, France*
[1]*Permanent address: AgResearch, Ruakura Agricultural
Centre, Hamilton, New Zealand*

Summary

Manipulation of milk composition for the production of high value pharmaceuticals, or improved nutritional or processing benefits has recently been achieved in transgenic mice, rats, rabbits, sheep, pigs, goats and cattle. The approaches used for the production of transgenic animals and the results obtained in recent experiments are described. This work has highlighted the potentiality of such manipulations but also has identified areas that need improvement to increase the efficiency and reliability of this technology. A range of new approaches and refinements for obtaining transgenic mammals and control of transgene expression that have recently been developed are discussed.

Introduction

Milk plays a major role as an important source of nutritional protein in the diet of humans, and is the sole food of newborn mammals. Protein derived from the milk of cattle constitutes about 30% of the total human dietary protein intake in developed nations (Hambreus, 1992). Milk composition is characterized by the occurrence of large quantities of a few specific proteins; the three calcium-sensitive caseins (α_{S1}, α_{S2} and β) that associate with κ-casein to form micelles, β-lactoglobulin, and α-lactalbumin are the major proteins in ruminant species.

Milk protein composition varies qualitatively and quantitatively between and within species. These individual genetic variations have enabled traditional selection techniques to be used for the improvement of breeds specialized for dairy production. Conversely, with the same approach the potential modifications of milk protein composition are limited.

281

In 1987, Simons *et al.* micro-injected the ovine β-lactoglobulin gene into mouse eggs and produced transgenic mice that specifically expressed this gene at very high levels in their milk. This work clearly demonstrated the utility of transgenic technology for delivery and expression of foreign genes in the mammary gland of mammals, and also the feasibility of modifying the protein composition of milk.

Since this time, many workers have utilized similar techniques to express milk protein or foreign genes under the control of milk protein regulatory sequences in the mammary gland of mammals. The aim of the majority of this work has been either to use lactating animals for the production of high value pharmaceuticals, or to modify the composition of milk for improved nutritional or processing benefits. Much of the work has been carried out in mice due to its experimental nature and for reasons related to cost and time. However, some workers have, in the past, or are now beginning to use larger animals such as sheep, goats and even cattle for this work. In this chapter, we will look at the current status of development of techniques for generation of transgenic mammals and the results achieved to date.

Gene Transfer Technology

Pronuclear and cytoplasmic micro-injection

Pronuclear micro-injection into fertilized eggs is the most frequently used and efficient technique for introducing foreign DNA in mice and producing germ-line transgenic mammals (see Rexroad, 1992; Ebert and Schindler, 1993; Mullins and Mullins, 1993; Pursel and Rexroad, 1993 for recent reviews). Transgenic cows, sheep, goats, pigs and rabbits have been obtained using this approach, although pig, cow and to a lesser extent goat ova are optically opaque and require centrifugation for improved visualization of their pronuclei.

The process of integration of the transgene into the genome is still not well understood (Bishop and Smith, 1989; Hamada *et al.*, 1993). However, it is clear that one to several hundred copies of the transgene are usually integrated at a unique and apparently random site. At least in mice, although these injections are performed in the pronuclei of fertilized eggs, it appears that the majority of founder animals are mosaic, suggesting that the DNA often integrates after the first cell division (Whitelaw *et al.*, 1993). Furthermore, insertional mutations, including visible mutations and prenatal lethals, have been estimated to be between 5 and 10% (Meisler, 1992).

The percentage of livestock born that are identified as transgenic varies widely between experiments and appears overall slightly lower than that observed in mice (Table 14.1; see Houdebine, 1994 for review). However, these values can be as high as 30% in pigs (Vize *et al.*, 1988; Velander *et al.*, 1992) and rabbits (Ernst *et al.*, 1991), 25% in sheep (Rexroad *et al.*, 1988), and around 10% in cattle (Krimpenfort *et al.*, 1991) and goats (Ebert *et al.*, 1991). Nethertheless, this methodology implies the use of a large number of embryos due to their poor survival after micro-injection (Table 14.1).

The high cost of transgenesis in farm animals compared with mice therefore does not appear to reflect a difference in the efficiency of the micro-injection technology applied to these species. It is rather related to the number of fertilized ova obtainable per female, the availability and the number of recipients needed for the micro-injected eggs, to the length of the gestion time, the generation interval, and the maintenance costs of these animals (Table 14.1). In addition, for physiological reasons, oestrus synchronization seems more difficult to achieve in farm animals as compared with mice.

The number of ova obtainable per female, even after superovulation, is relatively small in cows, sheep and goats (Table 14.1). The availability of embryos in these species can be increased and the cost of handling reduced through *in vitro* maturation and fertilization of oocytes derived from ovaries collected in slaughterhouses (Crozet *et al.*, 1987; Ectors *et al.*, 1988; Gordon, 1989; De Smedt *et al.*, 1992). This technology has already been applied successfully in cattle to produce transgenic animals (Krimpenfort *et al.*, 1991; Hyttinen *et al.*, 1994).

The opacity of the cytoplasm might be circumvented if transgenic animals could be efficiently obtained by a cytoplasmic microinjection technique. Cytoplasmic injection of DNA alone was shown to be very inefficient (Palmiter and Brinster, 1986). However, cytoplasmic injection of mouse embryos with a polylysine/DNA mixture resulted in a rate of transgenic pups of 13%. This compared favourably with pronuclear injection of DNA alone which resulted in 22% of pups born being transgenic (W.H. Velander, unpublished results). Optimization of this new approach might improve efficiency further and provide an attractive alternative to pronuclear injection for the production of transgenic animals.

The cost of producing transgenic livestock could be lowered if developmentally-competent and/or transgenic embryos only were transferred to donor females. This would reduce the number of synchronized recipients required, and therefore, in the case of ruminants lower the cost considerably. One approach for selection of viable eggs is to mature embryos from the one–two cell stage to morulae or blastocysts either in rabbit oviducts or in *in vitro* culture systems (Krimpenfort *et al.*, 1991; Hyttinen *et al.*, 1994). This approach enables the taking of biopsies for sexing and transgene analysis by polymerase chain reaction (PCR) (Fajfar-Whetstone *et al.*, 1993; Bowen *et al.*, 1994; Hyttinen *et al.*, 1994). However, whilst theoretically this should ensure that only transgenic-positive embryos of known sex are transferred to recipients, the use of transgene-targeted primers (single-stranded oligonucleotides) for detection of transgenic positive embryos has been found to result in a high number of false-positive identifications (Bowen *et al.*, 1993, 1994; Horvat *et al.*,

J.L. Vilotte and P.J. l'Huillier

Table 14.1. Comparison of reproductive and lactational performance, and transgenic yield in different species.

Species	No. of embryos per super-ovulation	Rate of pregnancy (%)	Rate of born embryos (%)	Transgenic animals (% of born)	Age at sexual maturity (months)	Gestation length (days)	Milk yield per lactation (litres)
Mouse	15	50	10–20	15	1.5	20	0.02
Rabbit	25	65	10	10	4–5	30	5
Pig	15	40	5–8	10	8	114	400
Sheep	8	40	15	5	6–7	150	700
Goat	15	40	15	5	6–7	150	800–900
Cow	5	20	10	5	14–15	270	9000

1993; Hyttinen *et al.*, 1994). More interestingly, recent experiments suggest that the percentage of new born transgenic calves was similar when either all micro-injected or only the resulting FISH (fluorescent *in situ* hybridization) indentified transgenic embryos were transferred to recipients (F.R. Pieper, personal communication).

The injection of *dam*-methylated DNA followed by *Dpn*I digestion of the biopsy-derived DNA prior to PCR analysis does not appear to improve significantly the reliability of the approach (Hyttinen *et al.*, 1994). Furthermore, such manipulations of the embryos reduce their viability. Alternatively, it might be possible to assess the occurrence of an integrated transgene either by identifying and sequencing PCR-amplified junction fragments (Rohan *et al.*, 1990; Hamada *et al.*, 1993) or by co-injection of a reporter gene under the transcriptional control of integration-dependent, preimplantation active regulatory sequences (Thompson *et al.*, 1994). This latter approach might result in insertion of transgenes with different tissue specificity at a single locus. Proper expression of such genes has already been observed in transgenic mice (Einat *et al.*, 1987). Finally, amniotic fluid analysis performed during early gestation should avoid the carrying to term of pregnancies of animals transferred with non-transgenic embryos (Hyttinen *et al.*, 1994; F.R. Pieper, personal communication).

Embryonic stem cell manipulation

Embryonic stem cells (ES cells) are totipotent cells derived directly from the embryonic inner cell mass of preimplantation embryos (see Smith, 1991 for review). They can be propagated and manipulated *in vitro* while retaining their ability to fully participate in embryonic development when re-introduced into blastocysts (see Baribault and Kemler, 1989; Capecchi, 1989; Evans, 1989; Bradley, 1991 for reviews) or co-cultured with eight-cell embryos (Wood *et al.*, 1993). Although ES cells have mostly been used for deletion of target genes by homologous recombination (see Capecchi, 1989; Evans, 1989; Lemarchandel and Montagutelli, 1990; Rastan, 1990; Bradley, 1991 for reviews), introduction of subtle mutations could also be achieved (Valancius and Smithies, 1991; Askew *et al.*, 1993; Stacey *et al.*, 1994; Wu *et al.*, 1994). However, this very powerful technology appears still limited to mice as true ES-cells have not yet been established for other species (see Anderson, 1992, for review). In hamsters (Doetschman *et al.*, 1988) and pigs (Notarianni *et al.*, 1990) ES-like cells that can be maintained in culture and undergo differentiation have been described. Similarly in rats, ES-like cells have been shown to produce chimeras (Iannaccone *et al.*, 1994), but their contribution to the germ-line remains to be demonstrated. Short-term *in vitro* culture of cells derived from the inner cell mass of blastocysts from cattle and the subsequent production of calves by transfer of nuclei from these cells into enucleated oocytes has also recently been reported (Sims and First, 1994).

Production of animals by transferring the nuclei derived from the inner mass of cells to enucleated oocytes will be an essential step to avoid the chimeric generation and thus reduce the time required for breeding up animals (Wilmut *et al.*, 1990). This approach has already been successfully used to produce calves and lambs (Marx, 1988; Smith and Wilmut, 1989; Sims and First, 1994). The recent derivation of ES cells from murine primordial germ cells might also open up an alternative strategy for manipulating the genome of farm animals (Matsui *et al.*, 1992).

Other alternative strategies

Germ-line transformation of animals can also be achieved using engineered retroviruses. In mammals, the use of such vectors has mainly been restricted to mice (see Hogan *et al.*, 1986; Bremel *et al.*, 1989; Wilmut *et al.*, 1990 for reviews) and has not been widely applied to livestock. The possibility of obtaining transgenic animals using spermatozoa to carry the foreign DNA (Lavitrano *et al.*, 1989; Gagné *et al.*, 1991) remains doubtful (Brinster *et al.*, 1989).

Finally, it is possible that foreign DNA could be introduced into mammary epithelial cells using various techniques applicable to gene therapy. This includes direct injection of naked DNA (Wolff *et al.*, 1990) or liposome-encapsulated DNA (Zhu *et al.*, 1993; Nabel *et al.*, 1993), recombinant viruses (Ragot *et al.*, 1993) and DNA-coated microprojectiles (Williams *et al.*, 1991; Cheng *et al.*, 1993). Recently a replication-defective retrovirus was infused into the mammary gland of goats during a period of hormonal-induced mammogenesis. Fourteen days later in the first few days of lactation, expression of the reporter gene human growth hormone was detected in the milk at a peak level of 60 ng ml^{-1} and declined to a plateau of 12 ng ml^{-1} from day nine to twelve of lactation (Archer *et al.*, 1994). In a recent experiment we have delivered liposomal-encapsulated plasmid DNA encoding bovine α-lactalbumin directly to the mammary artery of mice but in this case failed to detect α-lactalbumin expression in the lactating gland (J.L. Vilotte, P.J. l'Huillier, S. Soulier and M.G. Stinnakre, unpublished results). The practical applicability of many of these approaches is dependent on further developments that improve the efficiency of uptake of DNA by cells and its longevity within the cell.

Targeting Transgene Expression to the Mammary Gland

Several reviews have focused on transgenic technology in livestock agriculture and its potential use to modify milk composition (Bremel *et al.*, 1989; Wilmut *et al.*, 1990; Rexroad, 1992; Ebert and Schindler, 1993; Houdebine, 1994).

Control of gene expression in transgenic animals

Genes encoding the casein and major whey proteins have been cloned and analysed structurally in several species (see Mercier and Vilotte, 1993 for review; Hansson *et al.*, 1994b). To identify regulatory elements involved in their tissue-specificity and to assess their potentiality for driving expression of foreign genes in the mammary gland, several of these genes have been injected into mice and in a limited number of cases in rats and pigs (Table 14.2).

In the first study carried out Simons *et al.* (1987) observed very high levels of expression of sheep β-lactoglobulin in transgenic mice. This result demonstrated that milk composition could be significantly altered (Table 14.2). Furthermore, β-lactoglobulin transgenes encompassing 4.3–0.4 kb of 5' and 1.9 kb of 3' flanking regions were developmentally regulated and expressed in a position-independent, copy-number related fashion (Harris *et al.*, 1990; Whitelaw *et al.*, 1992). High-level, position-independent expression has similarly been observed for the rat WAP (whey acidic protein) gene in mice (Dale *et al.*, 1992). However in this case, the transgene was expressed earlier during pregnancy than its endogenous counterpart (Dale *et al.*, 1992). Similar high expression has also been observed for the mouse WAP transgene in mice (Burdon *et al.*, 1991a) and pigs (Shamay *et al.*, 1991; Wall *et al.*, 1991). Interestingly, the expression of this WAP transgene led to impaired mammary development and a milchlos phenotype (Burdon *et al.*, 1991b; Shamay *et al.*, 1992).

Tissue-specific and developmentally-regulated expression of other milk protein genes has been observed for rat and goat β-casein, and bovine and goat α-lactalbumin genes in mice (Lee *et al.*, 1988; Vilotte *et al.*, 1989; Persuy *et al.*, 1992; Roberts *et al.*, 1992; Soulier *et al.*, 1992). However, in these experiments, expression levels were dependent on the integration site and independent from the number of copies of the transgene integrated. Similar observations were made for bovine α-lactalbumin expression in rats (Hochi *et al.*, 1992). These results reflect, in general terms the expression patterns of almost all mammary-targeted transgenes (Houdebine, 1994).

Although high levels of expression in the mammary gland of the transgenes have been observed in the experiments described above, this expression was not restricted to just this tissue. Minute amounts of RNA have been detected in some other tissues of animals from several lines, suggesting weak ectopic expression related to the integration site. High expression of a guinea-pig α-lactalbumin transgene was also observed in the sebaceous gland of lactating mice (Maschio *et al.*, 1991). However, these authors also described similar high expression of endogenous milk protein genes which has never been confirmed since (Vilotte and Soulier, 1992; Persuy *et al.*, 1992).

The location of *cis*-regulatory sequences involved in milk protein gene regulation *in vivo* has been determined, in general terms, by injection of constructs with truncated 5' or 3' flanking sequences or where intronic sequences had been removed. These sequences are located between nucleotides 477–220

Table 14.2. Expression of milk protein transgenes in other species.

Origin of 5' regulatory sequence	Species	No. of transgenic lines expressing[1]	Highest level of expression (mg ml^{-1})	Reference
Ovine β-lactoglobulin	Mouse	7/7	23	Simons *et al.*, 1987
Bovine α$_{S1}$-casein	Mouse	8/8	4	Rijnkels, pers. comm.
Bovine β-casein	Mouse	12/12	20	Rijnkels, pers. comm.
Bovine α$_{S2}$-casein	Mouse	0/3	ND	Rijnkels, pers. comm.
Bovine κ-casein	Mouse	0/7	ND	Rijnkels, pers. comm.
Bovine α-lac	Mouse	3/6	0.45	Vilotte *et al.*, 1989
Bovine α-lac	Rat	11/17	2.4	Hochi *et al.*, 1992
Goat α-lac	Mouse	5/10	3.7	Soulier *et al.*, 1992
Rat β-casein	Mouse	3/5	~1% of endogenous β-casein	Lee *et al.*, 1988
Goat β-casein	Mouse	9/9	24	Persuy *et al.*, 1992
Rat WAP	Mouse	16/18	~500% of endogenous WAP	Dale *et al.*, 1992
Mouse WAP	Pig	5/5	1.5	Shamay *et al.*, 1991
Rat WAP	Mouse	8/9	95% of endogenous WAP	Bayna and Rosen, 1990
Mouse WAP	Pig	3/3	1	Wall *et al.*, 1991

[1] No. of lines expressing the transgene compared with the no. of transgenic animals analysed for expression. ND, expression of transgene not detected; WAP, whey acidic protein.

and 406–146 in the 5' flanking region of bovine α-lactalbumin and ovine β-lactoglobulin respectively (Soulier *et al.*, 1992; Whitelaw *et al.*, 1992). For the rat WAP gene, elements have been identified at nucleotides 853–729 in the 5' flanking region and within the 3' untranslated region (Dale *et al.*, 1992). Experiments dealing with DNA/protein interactions have confirmed or even further defined the location of *cis* sequences involved in milk protein gene regulation (see Mercier and Vilotte, 1993 for a recent review). Similarly, intronic sequences have been shown to enhance sheep β-lactoglobulin expression (Whitelaw *et al.*, 1991). Unexpectedly, addition of 5'-flanking sequences to sheep β-lactoglobulin (5.5 kb of 5' sequences as compared to 4.3 kb as used by Simons *et al.* (1987)) resulted in a position-dependent expression of the gene in transgenic mice (Shani *et al.*, 1992).

Expression of milk protein transgenes has been enhanced by addition of regulatory sequences from other genes. For example, addition of a fragment derived from the mouse mammary tumor virus (MMTV) long-terminal repeat, containing four hormone response elements, 5' to a rat β-casein gene appeared to enhance its expression in mice (Greenberg *et al.*, 1991). Similarly, addition of the chicken lysozyme A-element to the mouse WAP gene improved its genetic control as it conferred position independent hormonal and developmental regulation of this gene in four out of five lines analysed (McKnight *et al.*, 1992).

Targeting expression of hybrid transgenes

The regulatory sequences of milk protein genes have been used to target expression of foreign proteins in mammary epithelial cells during lactation, either to study the effect of their expression on the mammary epithelial cell activity or to obtain secretion of protein in the milk. We will refer only to these latter experiments.

Targeting expression of foreign cDNAs
Production of a protein by the use of a cDNA rather than the complete transcription unit of a gene has often been necessary because the gene was unavailable or its large size meant manipulation was difficult. Such cDNA-based constructs can be subdivided into three broad categories: (i) near-complete replacement of the milk protein gene transcription unit with a foreign cDNA (intron-less constructs, Table 14.3 and Fig. 14.1B); (ii) insertion of the foreign cDNA into the 5' non-coding region of a near-intact transcription unit of a milk protein gene (Table 14.3 and Fig. 14.1A); (iii) replacement of some 'internal' exons and introns of the milk protein transcription unit with the foreign cDNA (Fig. 14.1C).

Intron-less constructs have been found to be poorly expressed in the majority of cases, despite the use of different 5' flanking regions (Table 14.3 and Fig. 14.1B). The levels of recombinant proteins detected in milk have rarely

Table 14.3. Expression profiles of foreign cDNA sequences under the control of milk protein regulatory sequences.

Origin of 5′ regulatory sequences	Foreign cDNA sequence	Species	Insertion site	No. of transgenic lines expressing*	Highest level of expression	Reference
Transgene (A)						
Ovine β-lactoglobulin	α$_1$-antitrypsin	Sheep	exon 1	2/3	18 μg ml^{-1}	McClenaghan *et al.*, 1991
Ovine β-lactoglobulin	Human serum albumin	Mouse	exon 1	6/40	2.5 mg ml^{-1}	Shani *et al.*, 1992
Mouse WAP	Human protein C	Pig	exon 1	4/6	1 mg ml^{-1}	Velander *et al.*, 1992
Transgene (B)						
Ovine β-lactoglobulin	Anti-haemophilic factor IX	Sheep	exon 1	–	25 ng ml^{-1}	Clark *et al.*, 1989
Mouse WAP	Human t-PA	Mouse	exon 1	4/6	50 μg ml^{-1}	Pittius *et al.*, 1988
Mouse WAP	Human t-PA	Goat	exon 1	1/2	6 μg ml^{-1}	Ebert *et al.*, 1991
Mouse WAP	Human CD4	Mouse	exon 1	4/6	200 ng ml^{-1}	Yu *et al.*, 1989
Mouse WAP	Human t-PA	Mouse	exon 1	3/3	400 ng ml^{-1}	Gordon *et al.*, 1987
Bovine α-lac	Ovine trophoblast interferon	Mouse	exon 1	1/3	1 μg ml^{-1}	Stinnakre *et al.*, 1991

Transgene (C)

Mouse WAP	Superoxide dismutase	Mouse	exons 1 and 3	2/4	0.7 mg ml^{-1}	Hansson et al., 1994a
Ovine β-lactoglobulin	Superoxide dismutase	Mouse	exons 1 and 4 (3′ WAP)	1/3	8 ng ml^{-1}	Hansson et al., 1994a
Ovine β-lactoglobulin	Human γ-interferon	Mouse	exons 1 and 6	2/8	20 ng ml^{-1}	Dobrovolsky et al., 1993
Ovine β-lactoglobulin	α$_1$-antitrypsin	Mouse	exons 1 and 5	7/9	600 μg ml^{-1}	Clark et al., 1992
Ovine β-lactoglobulin	Human factor IX	Mouse	exons 1 and 5	7/11	900 ng ml^{-1}	Clark et al., 1992
Bovine α$_{S1}$-casein	Human lactoferrin	Mouse	exons 2 and 18	6/6	36 μg ml^{-1}	Platenburg et al., 1994
Bovine β-casein	Human lysozyme	Mouse	exon 2 and intron 8	1/8	RNA only by RT-PCR	Maga et al., 1994
Bovine α$_{S1}$-casein	Human lysozyme	Mouse	exon 2 and intron 12	2/5	RNA 116% of endogenous WAP	Maga et al., 1994
Goat β-casein	Human t-PA	Goat	exons 2 and 7	1/1	3 mg ml^{-1}	Ebert et al., 1994
Goat β-casein	Human CFTR	Mouse	exons 2 and 7	2/4	not measured	DiTullio et al., 1992

*, No. of lines expressing the transgene compared with the no. of transgenic animals analysed for expression; ND, not detected; -, not measured. For description of transgenes see Fig. 14.1.

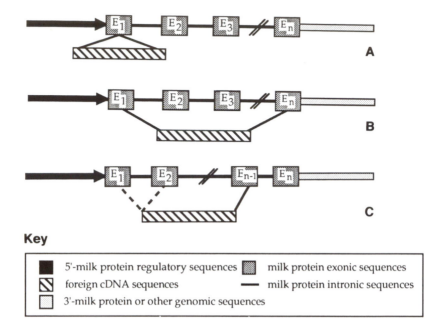

Key

■ 5'-milk protein regulatory sequences	▨ milk protein exonic sequences
▨ foreign cDNA sequences	— milk protein intronic sequences
□ 3'-milk protein or other genomic sequences	

Fig. 14.1. Stylized construct designs of transgenes. (A) Transgenes in which the foreign cDNA was inserted into the complete milk protein transcription unit at one restriction endonuclease cleavage site, usually in 5' untranslated sequence of exon 1. (B) Transgenes in which the foreign cDNA replaced the entire milk protein gene transcription unit except for a small portion of the non-coding exon 1. In some cases the 3' sequences were not derived from the respective milk protein gene but rather from, for example, SV40. (C) Transgenes in which the foreign cDNA replaced some of the internal exons and introns of the transcription unit of the milk protein gene.

exceeded just a few μg ml^{-1}. In addition, the percentage of expressing lines has been consistently low. These experiments strongly suggested that intronic sequences are necessary to enhance expression of transgenes.

Insertion of cDNA sequences into truncated or partially-deleted milk protein transcription units has resulted in highly variable levels of expression. Utilizing the sheep β-lactoglobulin gene, a set of constructs were designed with a cDNA inserted within the 5'-untranslated region of exon 1 of the gene while all, or at least the region encompassing the last two exons of its transcription unit were intact (Table 14.3 and Fig. 14.1A, C). These constructs were poorly expressed in the mammary gland, both in terms of levels of expression and percentage of expressing lines (Clark *et al.*, 1989; McClenaghan *et al.*, 1991; Shani *et al.*, 1992; Dobrovolsky *et al.*, 1993). Similar experiments performed using the mouse WAP gene resulted again in a relatively low percentage of expressing lines but a few of these expressed the recombinant protein in the milk at high levels (1.0 and 0.7 mg ml^{-1}; Velander *et al.*, 1992; Hansson *et al.*, 1994a respectively).

A similar set of constructs using foreign cDNAs but based on the α_{S1} and β-casein genes have also been evaluated. These constructs are characterized by the occurrence of 5' and 3' exons encoding untranslated regions from the respective casein genes and cDNAs inserted within a 'middle' exon. The cDNA is therefore surrounded by part of the transcription unit of the casein gene (Table 14.3 and Fig. 14.1c). The β-casein gene sequences (bovine or goat) were used to target expression of human t-PA (tissue plasminogen activator) in goats (Denman *et al.*, 1991; Ebert *et al.*, 1994), and human cystic fibrosis transmembrane conductance regulator (CFTR; DiTullio *et al.*, 1992) and lysozyme (Maga *et al.*, 1994) in mice. Human t-PA was found to be highly expressed in the milk of a transgenic goat whereas only a very low level of human lysozyme mRNA was detected in the lactating mammary gland of mice from one out of eight transgenic lines analysed. Similarly, the expression of CFTR was 'modest' but the protein was found to be associated with the milk fat globule membranes in two out of the four transgenic lines. Bovine α_{S1}-casein gene sequences have also been used to target expression of human lysozyme (Maga *et al.*, 1994) and lactoferrin (Platenburg *et al.*, 1994) in mice. For lactoferrin, only two lines were analysed and both expressed a low level of the protein in their milk. However, only two out of five lines expressed human lysozyme but with high amounts of mRNA. The same lactoferrin construct has been injected into cows and one transgenic bull has been obtained but levels of expression are not yet known (Krimpenfort *et al.*, 1991).

Finally, the MMTV promoter has been found to direct high level expression of the bovine α_{S1}-casein cDNA linked to SV40 splice and polyadenylation sequences in two out of the three lines of transgenic mice (Yom *et al.*, 1993). It should be mentioned that in several of the experiments described above, occurrence in the mammary gland of transgene-specific transcripts of various and unexpected sizes was reported. These RNAs probably resulted from aberrant splicings of the pre-mRNAs.

Targeting expression of foreign genes or minigenes
Most of these constructs are designed such that the entire transcription unit of the foreign gene or a minigene derived from it is under the control of a milk protein gene promoter (Table 14.4). The 3'-flanking region is often that of the reporter gene. It appears that these constructs are generally more efficiently expressed both in terms of percentage of expressing lines and the highest level of expression (Table 14.4) observed than cDNA-based constructs Table 14.3 and Fig. 14.1), although such comparisons are very difficult to assess.

Expression of several minigenes encoding human serum albumin under the control of sheep β-lactoglobulin 5'-flanking sequences have been analysed in transgenic mice (Shani *et al.*, 1992). Again, intronic sequences appear to enhance expression of the transgene although no general rule could be deduced from this experiment.

Expression of several recombinant proteins in the milk of transgenic animals at levels close to, or sometimes higher than that of the endogenous gene

Table 14.4. Expression of foreign genes under the control of milk protein regulatory sequences.

Origin of 5' regulatory sequence	Reporter gene	Species	No. of transgenic lines expressing[1]	Highest level of expression (mg ml^{-1})	Reference
Ovine β-lactoglobulin	Human serum albumin	Mouse	-	10	Barash et al., 1994
Ovine β-lactoglobulin	α$_1$-antitrypsin	Mouse	7/13	7	Archibald et al., 1990
Ovine β-lactoglobulin	α$_1$-antitrypsin	Sheep	3/3	35	Wright et al., 1991
Bovine α$_{S1}$-casein	hGH	Mouse	2/7	6.4	Ninomiya et al., 1994
Bovine α$_{S1}$-casein	Human urokinase	Mouse	-	1-2	Meade et al., 1990
Rabbit β-casein	Human interleukin-2	Rabbit	4/4	0.004	Bühler et al., 1990
Goat β-casein	Goat κ-casein	Mouse	8/8	20	M.A. Persuy, unpublished results
Bovine β-casein	hGH	Mouse	11/13	10.9	Ninomiya et al., 1994
Bovine α-lac	hGH	Mouse	5/6	4.4	Ninomiya et al., 1994
Bovine κ-casein	hGH	Mouse	0/10	ND	Ninomiya et al., 1994
Mouse WAP	hGH	Mouse	4/17	0.41	Reddy et al., 1991
Rabbit WAP	hGH	Mouse	6/13	22	Devinoy et al., 1994
Mouse WAP	hGH	Mouse	4/11	0.06	Ninomiya et al., 1991

[1] No. of lines expressing the transgene compared with the no. of transgenic animals analysed for expression. ND, expression of transgene not detected; hGH, human growth hormone. -, not available.

encoded by the milk protein promoter has demonstrated that such constructs can be efficiently expressed. Consequently, milk composition can be significantly modified using such transgenes. However, the host animal, the integration site and possibly interactions between the reporter and the milk protein gene sequences seem to influence the transgene expression as each construct acts differently. Similarly, no promoter appears significantly more reliable in terms of frequency of expressing lines and levels of expression with the probable exception of the bovine κ-casein which appears to be very inefficient (Ninomiya *et al.*, 1994; RijnKels, personal communication). However, the choice of the regulatory sequences appears important for the tissue-specificity and developmental regulation of the transgene.

Tissue specificity and developmental regulation of hybrid transgenes
Rat and mouse WAP regulatory sequences have been found to induce mammary precocious expression of transgenes in early gestation. These regulatory sequences have often been associated with ectopic expression of the transgene (Pittius *et al.*, 1988; Reddy *et al.*, 1991; Günzburg *et al.*, 1991; Bischoff *et al.*, 1992; Devinoy *et al.*, 1994; Ninomiya *et al.*, 1994; Paleyanda *et al.*, 1994). This 'weak' regulation has been observed with both foreign genes and the rat or mouse WAP genes, suggesting that it is related to the WAP sequences rather than the foreign gene sequences. It might therefore limit the usefulness of these regulatory sequences as ectopic expression of some recombinant proteins or leakage of the proteins into the serum from mammary epithelial cells in early gestation might be detrimental to the health of the animal.

Although expression of sheep β-lactoglobulin was found to be mammary specific and developmentally-regulated in transgenic mice (Simons *et al.*, 1987; Harris *et al.*, 1990; Whitelaw *et al.*, 1992), unexpected expression patterns of derived hybrid transgenes has been observed (Archibald *et al.*, 1990; Barash *et al.*, 1994). Sheep β-lactoglobulin 5′-flanking region fused to reporter genes confers expression in the lactating mammary gland, but also leads to novel, reporter-gene-related, tissue specificities.

The use of other milk protein gene regulatory sequences has also resulted in the detection of the recombinant protein in the serum of some of transgenic animals and/or in ectopic expression of the transgene. However, it appears that this is mostly related to effects of the site of integration on the transgene regulation. Expression of foreign genes exclusively in the mammary gland appears therefore to be difficult to achieve and might require either the identification of new milk protein gene regulatory elements or the addition of sequences to insulate the transgene from its chromatin environment (see Houdebine, 1994 for review and section below).

Besides the potential side-effects resulting from ectopic expression of the recombinant protein, such proteins might also affect the lactation process. For example, abrupt and unexpected involution with overexpression of bovine β-casein driven by bovine α-lactalbumin regulatory sequences (Bleck *et al.*,

1995), and precipitation in the milk of a t-PA variant when overexpressed in the mammary gland (Ebert *et al.*, 1994) have been reported.

Analysis of the recombinant proteins

Recombinant proteins produced so far in the milk of transgenic animals are mainly pharmaceuticals. The very high potential of this approach was highlighted many years ago (Lathe *et al.*, 1986) and since this time several companies have invested in this technology. However, for such proteins to be commericalized, several steps have yet to be achieved, including the production of sufficient protein in the milk: (i) the stability and biological activity of the recombinant protein, which are often related to co- and post-translational modifications, should mimic those of the protein from its original source; (ii) the protein synthesis should be consistent within and between lactations and from generation to generation; (iii) since many of these proteins require a high degree of purification from milk, additional research is required to characterize the conditions for isolation of the proteins.

Mammary epithelial cells possess the necessary cellular machinery to perform co- and post-translational modifications of recombinant proteins, such as O- and N-glycosylation, γ-carboxylation, signal peptide and pro-peptide cleavages. However, milk-derived recombinant proteins often differ from their 'wild-type' counterparts in terms of apparent molecular weight. Quantitative variations in their glycosylation process have been observed for some proteins (Denman *et al.*, 1991; Velander *et al.*, 1992; Drohan *et al.*, 1994; Hansson *et al.*, 1994a). The predominant form of recombinant t-PA found in the milk of transgenic goats was proteolytically cleaved into the 'two chain' form (Denman *et al.*, 1991), whereas several forms of recombinant human protein C, resulting from inefficient proteolytic processing and γ-carboxylation, were detected in the milk of transgenic mice (Drohan *et al.*, 1994). This heterogeneity might result in the occurrence of different protein fractions with different specific activities or biological properties.

Inheritance of transgenes and expression levels have been analysed in detail for transgenic mice and sheep expressing human α_1-antitrypsin (Carver *et al.*, 1993). Stable transmission of the transgene was observed in most of the lines studied. However, whilst expression levels were also inherited in transgenic sheep and consistent from one lactation to the next, in mice expression was very variable with a high incidence of the genetic background in several lines. Whether this variance is restricted to mice, as suggested by the authors, remains to be clearly established.

Purification of a few soluble proteins from milk, to an apparent high homogeneity has been achieved (Denman *et al.*, 1991; Wilkins and Velander, 1992). However, the observed heterogeneity in biological activity and biochemical properties of the recombinant proteins described earlier might further complicate these purification steps as separation between subfractions may be required (Drohan *et al.*, 1994). Furthermore, despite the extreme care that companies take with the choice of animals for transgenic experimentation and on breeding conditions to minimize possible contamination, the biological purity of these purified proteins will require close examination.

Modification of endogenous genes

Manipulation of expression of endogenous genes requires the use of ES cells (as mentioned earlier), or antisense and/or ribozyme techniques. No results have yet been published demonstrating the effectiveness of antisense and ribozymes against mammary-specific target sequences. In the case of ES cells, the application of this technology is limited to mice due to the unavailability of established ES cells for other species. Null alleles for the β-casein and the α-lactalbumin genes have been recently created using ES cells and gene targeting. Deficiency in β-casein resulted in reduced micelle sizes, total milk protein concentration and growth of pups feeding on this type of milk (Kumar *et al.*, 1994). Deficiency in α-lactalbumin resulted in secretion of a lactose-free, highly viscous milk, rich in fat and protein, that pups were unable to remove from the mammary gland (Stinnakre *et al.*, 1994). Structure of the mammary epithelial cells also appeared to be affected by this genotype.

   Homologous recombination in ES cells was also used to replace the mouse α-lactalbumin gene, along with short 5' and 3' sequences by the human counterpart, using a two-step process (Stacey *et al.*, 1994). In the middle step, a null allele was created and mice were derived from these cells. Phenotypic observation of these animals gave results that were virtually identical to those described above. More interestingly, animals heterozygous for both the mouse and the human genes expressed in their lactating mammary gland 15 times more α-lactalbumin mRNA for the human than mouse gene (A.J. Stacey, personal communication). This result suggests that the higher level of α-lactalbumin expression observed in humans compared with mice is due to different regulatory sequences located within or in close proximity to the transcription unit of the gene. Whether these sequences are involved in transcriptional regulation or related to the maturation and/or stability of the mRNA remains to be determined.

Potential Improvements, and Applications of Gene Transfer Technology to Milk Proteins

Improvement of tissue specificity and developmental regulation of transgene expression

Transgene expression appears to be influenced by the chromatin environment. An open chromatin configuration in the vicinity of the transgene in lactating mammary tissue can be obtained by coinjection of an efficiently expressed mammary gene. Thus, improvement of both the frequency and level of expression of two intron-less transgenes was observed when coinjected with sheep β-lactoglobulin gene (Clark *et al.*, 1992). Improvement could also be achieved if regulatory elements insulating the transgene were added at one or both ends of a gene construct. Such elements (MAR or SAR, matrix or scaffold attachment region) which correspond to DNA regions where chromatin loops are attached to the matrix have been located within the flanking regions of several genes from *Drosophila* or vertebrates. Their influence on transgene expression has also been investigated (Bonifer *et al.*, 1990). Coinjection of the chicken lysozyme MAR with a mouse WAP transgene resulted in position-independent hormonal and developmental regulation of the mouse gene (McKnight *et al.*, 1992).

Regulation of some genes or group of genes belonging to the same locus can be controlled by locus control regions or LCRs (Greaves *et al.*, 1989; Talbot *et al.*, 1989; Chamberlain *et al.*, 1991). The casein genes have been found to belong to a single locus of ~ 200 kb in cattle (Ferretti *et al.*, 1990; Threadgill and Womack, 1990), suggesting that their regulation might be controlled by such an element. However, LCRs have not yet been reported for any milk protein genes. Identification of a milk protein gene LCR would improve the efficiency of transgene expression in the mammary gland in terms of percentage of expressing lines, levels of expression and probably developmental regulation.

Yeast Artificial Chromosome (YAC) vectors have been successfully integrated into the mouse genome either by direct micro-injection (Gaensler *et al.*, 1993; Schedl *et al.*, 1993) or through ES cell manipulation (Strauss *et al.*, 1993). These mice usually expressed the transgene at levels comparable with that of their endogenous counterparts, with correct tissue specificity and developmental regulation (see Forget, 1993 for review). Although manipulation of YAC inserts is difficult, it might be possible by homologous recombination to specifically replace the transcription unit of a milk protein gene by that of the reporter gene of interest in such large inserts ≥ 100 kb). Therefore it is likely that such transgenes may express more efficiently and correctly in transgenic animals. Furthermore, although gene replacements have been performed in mice using ES cells (Stacey *et al.*, 1994), YAC vectors will enable such manipulations in any mammals.

Finally, integration of the transgene could be targeted to a locus transcriptionally active in lactating mammary epithelial cells, such as the casein locus, using ES cells. Compared with the gene replacement approach, this alternative might not modify endogenous milk protein production while placing a transgene in a potentially favourable chromatin environment.

Expression of pharmaceutical proteins

Reasonable levels of secretion of some recombinant proteins in the milk of transgenic animals have been reported in recent years (Fig. 14.1 and Tables 14.3

and 14.4). However, further characterization of these proteins has revealed that they might suffer from inefficient processing, heterogenous post-translational modifications. Such often apparently minor differences, might be detrimental to their use in therapeutical treatments. Theoretically, co- and post-translational modifications of these proteins might be improved using transgenic animals genetically modified to express in their mammary gland enzymes involved in these activities. Alternatively, these observations might be related to a saturation of the mammary epithelial cell enzymatic activity and/or reflect species or individual variations. Therefore, for some applications, low or moderate levels of transgene expression could result in a better final product. Alternatively, use of the animals carrying null alleles (see Ng-Kwai-Hang and Grosclaude, 1992 for review) may allow a more efficient processing of exogenous proteins. Similarly, lack of some major endogenous proteins might facilitate the purification from the milk of the recombinant proteins.

So far most of these proteins are of blood origin and are to be used for therapeutic purposes. Transgenesis could also be used to produce in the milk either monoclonal antibodies for diagnostic tests or to confer passive immunity against some intestinal pathogens, or as oral vaccines.

Other potential milk protein manipulations

Besides expression of pharmaceuticals, the composition of endogenous milk proteins could be modified to confer advantageous nutritional or processing properties. Again, several reviews have focused on this topic and readers are referred to these for more comprehensive descriptions (Mercier, 1986; Jimenez-Flores and Richardson, 1988; Yom and Bremel, 1993). Some modifications that may be benefical include: (i) overexpression of a milk protein gene, such as the κ-casein to reduce micelle size; (ii) expression of a mutated gene to produce a milk protein with modified properties such as a β-casein without phosphate groups to yield a higher moisture and thus softer cheese; (iii) reduction or even inhibition of gene expression, such as α-lactalbumin to reduce milk lactose content.

Overexpression of milk protein genes has been widely achieved. However, the majority of these experiments were performed for other purposes and were carried out in mice, thus the properties of the milk were not evaluated. Mice efficiently expressing goat κ-casein under the control of goat β-casein regulatory sequences have recently been obtained and the size of the resulting casein micelles is currently being investigated (M.A. Persuy, personnal communication).

Similarly, even if secretion of a mutated protein is not technically demanding, the occurrence in the milk of the endogenous protein might interfere with the expected outcome. Thus, only mutations with dominant or at least co-dominant effects should be investigated without manipulation of the level of expression or deletion in ES cells of the endogenous gene.

Alteration of expression of endogenous genes can be obtained using ES cells. However, alternative strategies exist to reduce protein synthesis levels such as the use of antisense or ribozyme constructs. Ribozymes (Efrat *et al.*, 1994; Larsson *et al.*, 1994) and antisense (Katsuki *et al.*, 1988; Pepin *et al.*, 1992; Moxham *et al.*, 1993; Richard *et al.*, 1993) transgenes have recently been shown to alter production of specific proteins encoded by the target RNA. Ribozymes targeted against bovine and mouse α-lactalbumin have been designed and successfully tested *in vitro* (L'Huillier *et al.*, 1992). Their potential effect on expression of α-lactalbumin in transgenic mice is currently being tested (L'Huillier, unpublished results). Unlike gene targeting in ES cells, the apparent low efficacy of many antisense and ribozyme constructs suggests that the levels of target RNA and protein are likely to be reduced only to some extent.

Finally, it should be emphasized that milk with different nutritional or behaviour properties already exist naturally in dairy animals (see Ng-Kwai-Hang and Grosclaude, 1992 for review). Transgenesis should be applied in conjunction with these 'natural models' as they provide important resources both for the study of gene function and as the basis for animal populations for subsequent manipulation.

Conclusions

Modification of milk protein composition by gene transfer has been successfully achieved. However, because of the high cost of this technology when applied to farm animals, it has so far been limited almost exclusively to the production of proteins of very high commercial value. Improvements in the efficency of production of transgenic ruminants along with a better understanding of gene regulation should enable the technology to be applied to other problems more related to livestock agriculture.

Acknowledgements

We are very grateful to Drs Houdebine, Lubon, Mercier, RijnKels, Pieper, Stacey and Velander for providing us with unpublished results.

References

Anderson, G.B. (1992) Isolation and use of embryonic stem cells from livestock species. *Animal Biotechnology* 3, 165–175.
Archer, J.S., Kennan, W.S., Gould, M.N. and Bremel, R.D. (1994) Human growth hormone (hGH) secretion in milk of goats after direct transfer of the hGH gene into the mammary gland by using replication-defective retrovirus vectors. *Proceedings of the National Academy of Sciences of the USA* 91, 6840–6844.

Archibald, A.L., McClenaghan, M., Hornsey, V., Simons, J.P. and Clark, A.J. (1990) High level expression of biologically active human α_1-antitrypsin in the milk of transgenic mice. *Proceedings of the National Academy of Sciences of the USA* 87, 5178–5182.

Askew, G.R., Doetschman, T. and Lingrel, J.B. (1993) Site-directed point mutations in embryonic stem cells: a gene-targeting tag-and-exchange strategy. *Molecular and Cellular Biology* 13, 4115–4124.

Barash, I., Faerman, A., Ratovitsky, T., Puzis, R., Nathan, M., Hurwitz, D.R. and Shani, M. (1994) Ectopic expression of β-lactoglobulin/human serum albumin fusion genes in transgenic mice: hormonal regulation and *in situ* localization. *Transgenic Research* 3, 141–151.

Baribault, H. and Kemler, R. (1989) Embryonic stem cell culture and gene targeting in transgenic mice. *Molecular Biology and Medicine* 6, 481–492.

Bayna, E.M. and Rosen, J.M. (1990) Tissue-specific, high level expression of the rat whey acidic protein gene in transgenic mice. *Nucleic Acids Research* 18, 2977–2985.

Bischoff, R., Degryse, E., Perraud, F., Dalemams, W., Ali-Hadji, D., Thepot, D., Devinoy, E., Houdebine, L.M. and Pavirani, A. (1992) A 17.6 kbp region located upstream of the rabbit WAP gene directs high level expression of a functional human protein variant in transgenic mouse milk. *FEBS Letters* 305, 265–268.

Bishop, J.O. and Smith, P. (1989) Mechanism of chromosomal integration of micro-injected DNA. *Molecular Biology and Medicine* 6, 283–298.

Bleck, G.T., Jiménez-Flores, R. and Bremel, R. (1995) Abnormal properties of milk from transgenic mice expressing bovine β-casein under control of the bovine α-lactalbumin 5' flanking region. *International Dairy Journal* 5, 619–632.

Bonifer, C., Vidal, M., Grosveld, F. and Sippel, A.E. (1990) Tissue specific and position independent expression of the complete gene domain for chicken lysozyme in transgenic mice. *The EMBO Journal* 9, 2843–2848.

Bowen, R.A., Reed, M., Schnieke, A., Seidel, G.E., Brink, Z. and Stacey, A. (1993) Production of transgenic cattle from PCR-screened embryos. *Theriogenology* 39, 194.

Bowen, R.A., Reed, M., Schnieke, A., Seidel, G.E., Stacey, A., Thomas, W.K. and Kajikawa, O. (1994) Transgenic cattle resulting from biopsied embryos: expression of c-ski in a transgenic calf. *Biology of Reproduction* 50, 664–668.

Bradley, A. (1991) Modifying the mammalian genome by gene targeting. *Current Opinion in Biotechnology* 2, 823–829.

Bremel, R.D., Yom, H.C. and Bleck, G.T. (1989) Alteration of milk composition using molecular genetics. *Journal of Dairy Science* 72, 2826–2833.

Brinster, R.L., Sandgren, E.P., Behringer, R.R. and Palmiter, R.D. (1989) No simple solution for making transgenic mice. *Cell* 59, 239–241.

Bühler, T.A., Bruyère, T., Went, D.F., Stranzinger, G. and Bürki, K. (1990) Rabbit β-casein promoter directs secretion of human interleukin-2 into the milk of transgenic rabbits. *Bio/Technology* 8, 140–143.

Burdon, T., Sankaran, L., Wall, R.J., Spencer, M. and Hennighausen, L. (1991a) Expression of a whey acidic protein transgene during mammary development: evidence for different mechanisms of regulation during pregnancy and lactation. *Journal of Biological Chemistry* 266, 6909–6914.

Burdon, T., Wall, R.J., Shamay, A., Smith, G.H. and Hennighausen, L. (1991b) Over-expression of an endogenous milk protein gene in transgenic mice is associated with impaired mammary alveolar development and a milchlos phenotype. *Mechanisms of Development* 36, 67–74.

Capecchi, M.R. (1989) The new mouse genetics: altering the genome by gene targeting. *Trends in Genetics* 5, 70–76.

Carver, A.S., Dalrymple, M.A., Wright, G., Cottom, D.S., Reeves, D.B., Gibson, Y.H., Keenan, J.L., Barrass, J.D., Scott, A.R., Colman, A. and Garner, I. (1993) Transgenic livestock as bioreactors: stable expression of human alpha-1-antitrypsin by a flock of sheep. *Bio/Technology* 11, 1263–1270.

Chamberlain, J.W., Vasavada, H.A., Ganguly, S. and Weissman, S.M. (1991) Identification of *cis* sequences controlling efficient position-independent tissue-specific expression of human major histocompatibility complex class I genes in transgenic mice. *Molecular and Cellular Biology* 11, 3564–3572.

Cheng, L., Ziegelhoffer, P.R. and Yang, N.S. (1993) *In vivo* promoter activity and transgene expression in mammalian somatic tissues evaluated by using particle bombardment. *Proceedings of the National Academy of Sciences of the USA* 90, 4455–4459.

Clark, A.J., Bessos, H., Bishop, J.O., Brown, P., Harris, S., Lathe, R., McClenaghan, M., Prowse, C., Simons, J.P., Whitelaw, C.B.A. and Wilmut, I. (1989) Expression of human anti-hemophilic factor IX in the milk of transgenic sheep. *Bio/Technology* 7, 487–492.

Clark, A.J., Cowper, A., Wallace, R., Wright, G. and Simons, J.P. (1992) Rescuing transgene expression by co-integration. *Bio/Technology* 10, 1451–1455.

Crozet, N., Huneau, D., Desmedt, V., Théron, M.C., Szöllösi, D., Torrès, S. and Sévellec, C. (1987) *In vitro* fertilization with normal development in the sheep. *Gamete Research* 16, 159–170.

Dale, T.C., Krnacik, M.J., Schmidhauser, C., Yang, C.L.Q., Bissell, M.J. and Rosen, J.M. (1992) High-level expression of the rat whey acidic protein gene is mediated by elements in the promoter and 3′ untranslated region. *Molecular and Cellular Biology* 12, 905–914.

Denman, J., Hayes, M., O'Day, C., Edmunds, T., Barlett, C., Hirani, S., Ebert, K.M., Gordon, K. and McPherson, J.M. (1991) Transgenic expression of a variant of human tissue-type plasminogen activator in goat milk: purification and characterization of the recombinant enzyme. *Bio/Technology* 9, 839–843.

De Smedt, V., Crozet, N., Ahmed-Ali, M., Martino, A. and Cognié, Y. (1992) *In vitro* maturation and fertilization of goat oocytes. *Theriogenology* 37, 1049–1060.

Devinoy, E., Thepot, D., Stinnakre, M.G., Fontaine, M.L., Grabowski, H., Puissant, C., Pavirani, A. and Houdebine, L.M. (1994) High level production of human growth hormone in the milk of transgenic mice: the upstream region of the rabbit whey acidic protein (WAP) gene targets transgene expression to the mammary gland. *Transgenic Research* 3, 79–89.

DiTullio, P., Cheng, S.H., Marshall, J., Gregory, R.J., Ebert, K.M., Meade, H.M. and Smith, A.E. (1992) Production of cystic fibrosis transmembrane conductance regulator in the milk of transgenic mice. *Bio/Technology* 10, 74–77.

Dobrovolsky, V.N., Lagutin, O.V., Vinogradova, T.V., Frolova, I.S., Kuznetsov, V.P. and Larionov, O.A. (1993) Human γ-interferon expression in the mammary gland of transgenic mice. *FEBS Letters* 319, 181–184.

Doetschman, T., Williams, P. and Maeda, N. (1988) Establishment of hamster blastocyst-derived embryonic stem (ES) cells. *Developmental Biology* 127, 224–227.

Drohan, W.N., Zhang, D.W., Paleyanda, R.K., Chang, R., Wroble, M., Velander, W. and Lubon, H. (1994) Inefficient processing of human protein C in the mouse mammary gland. *Transgenic Research* 3, 355–364.

Ebert, K.M. and Schindler, J.E.S. (1993) Transgenic farm animals: progress report. *Theriogenology* 39, 121–135.

Ebert, K.M., Selgrath, J.P., DiTullio, P., Denman, J., Smith, T.E., Memon, M.A., Schindler, J.E., Monastersky, G.M., Vitale, J.A. and Gordon, K. (1991) Transgenic production of a variant of human tissue-type plasminogen activator in goat milk: generation of transgenic goats and analysis of expression. *Bio/Technology* 9, 835–839.

Ebert, K.M., DiTullio, P., Barry, C.A., Schindler, J.E., Ayres, S.L., Smith, T.E., Pellerin, L.J., Meade, H.M., Denman, J. and Roberts, B. (1994) Induction of human tissue plasminogen activator in the mammary gland of transgenic goats. *Bio/Technology* 12, 699–702.

Ectors, F.J., van der Zwalmen, P., Touati, K., Beckers, J.F. and Ectors, F. (1988) Obtention de blastocystes après maturation et fécondation *in vitro* d'ovocytes de bovin: premiers résultats. *Annales de Médecine Vétérinaire* 132, 517–519.

Einat, P., Bergman, Y., Yaffe, D. and Shani, M. (1987) Expression in transgenic mice of two genes of different tissue specificity integrated into a single chromosomal site. *Genes and Development* 1, 1075–1084.

Efrat, S., Leiser, M., Wu, Y.J., Fusco-DeMane, Emran, O.A., Surana, M., Jetton, T.L., Magnuson, M.A., Weir, G. and Fleischer, N. (1994) Ribozyme-mediated attenuation of pancreatic b-cell glucokinase expression in transgenic mice results in impaired glucose-induced insulin secretion. *Proceedings of the National Academy of Sciences of the USA* 91, 2051–2055.

Ernst, L.K., Zakcharchenko, V.I., Suraeva, N.M., Ponomareva, T.I., Miroshnichenko, O.I., Prokofev, M.I. and Tikchonenko, T.I. (1991) Transgenic rabbits with antisense RNA gene targeted at adenovirus H5. *Theriogenology* 35, 1257–1289.

Evans, M.J. (1989) Potential for genetic manipulation of mammals. *Molecular Biology and Medicine* 6, 557–565.

Fajfar-Whetstone, C.J., Rayburn, A.L., Schook, L.B. and Wheeler, M.B. (1993) Sex determination of porcine preimplantation embryos via Y-chromosome specific DNA sequences. *Animal Biotechnology* 4, 183–193.

Ferretti, L., Leone, P. and Sgaramella, V. (1990) Long range restriction analysis of the bovine casein genes. *Nucleic Acids Research* 18, 6829–6833.

Forget, B.G. (1993) YAC transgenes: bigger is probably better. *Proceedings of the National Academy of Sciences of the USA* 90, 7909–7911.

Gaensler, K.M.L., Kitamura, M. and Wai Kan, Y. (1993) Germ-line transmission and developmental regulation of a 150-kb yeast artificial chromosome containing the human b-globin locus in transgenic mice. *Proceedings of the National Academy of Sciences of the USA* 90, 11381–11385.

Gagné, M.B., Pothier, F. and Sirard, M.A. (1991) Electroporation of bovine spermatozoa to carry foreign DNA in oocytes. *Molecular Reproduction and Development* 29, 6–15.

Gordon, I. (1989) Large-scale production of cattle embryos by *in vitro* culture methods. *AgBiotech News and Information* 1, 345–348.

Gordon, K., Lee, E., Vitale, J.A., Smith, A.E., Westphal, H. and Hennighausen, L. (1987) Production of human plasminogen activator in transgenic mouse milk. *Bio/Technology* 5, 1183–1187.

Greaves, D.R., Wilson, F.D., Lang, G. and Kioussis, D. (1989) Human CD2 3′-flanking sequences confer high-level, T cell-specific, position independent gene expression in transgenic mice. *Cell* 56, 979–986.

Greenberg, N.M., Reding, T.V., Duffy, T. and Rosen, J.M. (1991) A heterologous hormone response element enhances expression of rat β-casein promoter driven chloramphenicol acetyltransferase fusion genes in the mammary gland of transgenic mice. *Molecular Endocrinology* 5, 1504–1512.

Günzburg, W.H., Salmons, B., Zimmermann, B., Maller, M., Erfle, V. and Brem, G. (1991) A mammary specific promoter directs expression of growth hormone not only to the mammary gland but also to Bergman glia cells in transgenic mice. *Molecular Endocrinology* 5, 123–133.

Hamada, T., Sasaki, H., Seki, R. and Sakaki, Y. (1993) Mechanism of chromosomal integration of transgenes in microinjected mouse eggs: sequence analysis of genome-transgene and transgene-transgene junctions at two loci. *Gene* 128, 197–202.

Hambreus, L. (1992) Nutritional aspects of milk proteins. In: Fox, P.F. (ed.) *Advanced Dairy Chemistry-1: Proteins.* Elsevier Science Publishers Ltd, Essex, England, pp. 457–490.

Hansson, L., Edlund, M., Edlund, A., Johansson, T., Marklund, S.L., Fromm, S., Strömqvist, M. and Törnell, J. (1994a) Expression and characterization of biologically active human extracellular superoxide dismutase in milk of transgenic mice. *Journal of Biological Chemistry* 269, 5358–5363.

Hansson, L., Edlund, A., Johansson, T., Hernell, O., Strömqvist, M., Lindquist, S., Lönnerdal, B. and Bergström, S. (1994b) Structure of the human β-casein encoding gene. *Gene* 139, 193–199.

Harris, S., McClenaghan, M., Simons, J.P., Ali, S. and Clark, A.J. (1990) Gene expression in the mammary gland. *Journal of Reproduction and Fertility* 88, 707–715.

Hochi, S.I., Ninomiya, T., Waga-Homma, M., Sagara, J. and Yuki, A. (1992) Secretion of bovine α-lactalbumin into the milk of transgenic rats. *Molecular Reproduction and Development* 33, 160–164.

Hogan, B., Costantini, F. and Lacy, E. (1986) *Manipulating The Mouse Embryo: A Laboratory Manual.* Cold Spring Harbor Laboratory, New York, USA.

Horvat, S., Medrano, J.F., Behboodi, E., Anderson, G.B. and Murray, J.D. (1993) Sexing and detection of gene construct in microinjected bovine blastocysts using the polymerase chain reaction. *Transgenic Research* 2, 134–140.

Houdebine, L.M. (1994) Production of pharmaceutical proteins from transgenic animals. *Journal of Biotechnology* 34, 269–287.

Hyttinen, J.M., Peura, T., Tolvanen, M., Aalto, J., Alhonen, L., Sinervirta, R., Halmekytö, M., Myöhänen, S. and Jänne, J. (1994) Generation of transgenic dairy cattle from transgene-analysed and sexed embryos produced *in vitro*. *Bio/Technology* 12, 606–608.

Iannaccone, P.M., Taborn, G.U., Garton, R.L., Caplice, M.D. and Brenin, D.R. (1994) Pluripotent embryonic stem cells from the rat are capable of producing chimeras. *Developmental Biology* 163, 288–292.

Jimenez-Flores, R. and Richardson, T. (1988) Genetic engineering of the caseins to modify the behavior of milk during processing: a review. *Journal of Dairy Science* 71, 2640–2654.

Katsuki, M., Sato, M., Kimura, M., Yokoyama, M., Kobayashi, K. and Nomura, T. (1988) Conversion of normal behavior to shiverer by myelin basic protein antisense cDNA in transgenic mice. *Science* 241, 593–595.

Krimpenfort, P., Rademakers, A., Eyestone, W., van der Schans, A., van den Broek, S., Kooiman, P., Kootwijk, E., Platenburg, G., Pieper, F., Strijker, R. and de Boer, H. (1991) Generation of transgenic dairy cattle using '*in vitro*' embryo production. *Bio/Technology* 9, 845–847.

Kumar, S., Clarke, A.R., Hooper, M.L., Horne, D.S., Law, A.J.R., Leaver, J., Springbett, A., Stevenson, E. and Simons, J.P. (1994) Milk composition and lactation of β-casein-deficient mice. *Proceedings of the National Academy of Sciences of the USA* 91, 6138–6142.

Larsson, S., Hotchkiss, G., Andäng, M., Nyholm, T., Inzunza, J., Jansson, I. and Ährlund-Richter, L. (1994) Reduced β2-microglobulin mRNA levels in transgenic mice expressing a designed hammerhead ribozyme. *Nucleic Acids Research* 22, 2242–2248.

Lathe, R., Clark, A.J., Archibald, A.L., Bishop, J.O., Simons, P. and Wilmut, I. (1986) Novel products from livestock. In: Smith, C., King, J.W.B. and McKay, J.C. (ed.) *Exploiting New Technologies in Animal Breeding. Genetic Developments*. Oxford University Press, UK, pp. 91–102.

Lavitrano, M., Camaioni, A., Fazio, V.M., Dolci, S., Farace, M.G. and Spadafora, C. (1989) Sperm cells as vectors for introducing foreign DNA into eggs: genetic transformation of mice. *Cell* 57, 717–723.

Lee, K.F., DeMayo, F.J., Atiee, S.H. and Rosen, J.M. (1988) Tissue-specific expression of the rat β-casein in transgenic mice. *Nucleic Acids Research* 16, 1027–1041.

Lemarchandel, V. and Montagutelli, X. (1990) La recombinaison homologue: de nouvelles perspectives pour la transgénèse chez les mammifères. *Médecine/Sciences* 6, 18–29.

L'Huillier, P.J., Davis, S.R. and Bellamy, A.R. (1992) Cytoplasmic delivery of ribozymes leads to efficient reduction in α-lactalbumin mRNA levels in C127I mouse cells. *The EMBO Journal* 11, 4411–4418.

McClenaghan, M., Archibald, A.L., Harris, S., Simons, J.P., Whitelaw, C.B.A., Wilmut, I. and Clark, A.J. (1991) Production of human α_1-antitrypsin in the milk of transgenic sheep and mice: targeting expression of cDNA sequences to the mammary gland. *Animal Biotechnology* 2, 161–176.

McKnight, R.A., Shamay, A., Sankaran, L., Wall, R.J. and Hennighausen, L. (1992) Matrix-attachment regions can impart position-independent regulation of a tissue-specific gene in transgenic mice. *Proceedings of the National Academy of Sciences of the USA* 89, 6943–6947.

Maga, E.A., Anderson, G.B., Huang, M.C. and Murray, J.D. (1994) Expression of human lysozyme mRNA in the mammary gland of transgenic mice. *Transgenic Research* 3, 36–42.

Marx, J.L. (1988) Cloning in sheep and cattle embryos. *Science* 239, 463–464.

Maschio, A., Brickell, P.M., Kioussis, D., Mellor, A.L., Katz, D. and Craig, R.K. (1991) Transgenic mice carrying the guinea-pig α-lactalbumin gene transcribe milk protein genes in their sebaceous glands during lactation. *Biochemical Journal* 275, 459–467.

Matsui, Y., Zsebo, K. and Hogan, B.L.M. (1992) Derivation of pluripotential embryonic stem cells from murine primordial germ cells in culture. *Cell* 70, 841–847.

Meade, H., Gates, L., Lacy, E. and Lonberg, N. (1990) Bovine alpha $_{S1}$-casein gene sequences direct high level expression of active human urokinase in mouse milk. *Bio/Technology* 8, 443–446.

Mercier, J.C. (1986) Genetic engineering applied to milk producing animals: some expectations. In: Smith, C., King, J.W.B. and McKay, J.C. (ed.) *Exploiting New Technologies in Animal Breeding. Genetic Developments.* Oxford University Press, UK, pp. 122–131.

Mercier, J.C. and Vilotte, J.L. (1993) Structure and function of milk protein genes. *Journal of Dairy Science* 76, 3079–3098.

Meisler, M.H. (1992) Insertional mutation of 'classical' and novel genes in transgenic mice. *Trends in Genetics* 8, 341–348.

Moxham, C.M., Hod, Y. and Malbon, C.C. (1993) Induction of Gαi2-specific antisense RNA *in vivo* inhibits neonatal growth. *Science* 260, 991–995.

Mullins, J.J. and Mullins, L.J. (1993) Transgenesis in nonmurine species. *Hypertension* 22, 630–633.

Nabel, G.J., Nabel, E.G., Yang, Z.Y., Fox, B.A., Plautz, G.E., Gao, X., Huang, L., Shu, S., Gordon, D. and Chang, A.E. (1993) Direct gene transfer with DNA-liposome complexes in melanoma: expression, biological activity, and lack of toxicity in humans. *Proceedings of the National Academy of Sciences of the USA* 90, 11307–11311.

Ng-Kwai-Hang, N.G. and Grosclaude, F. (1992) Genetic polymorphism of milk proteins. In: Fox, P.F. (ed.) *Advanced Dairy Chemistry-1: Proteins.* Elsevier Science Publishers LTD, Essex, England, pp. 405–456.

Ninomiya, T., Hirabayashi, M., Sagara, J. and Yuki, A. (1994) Functions of milk protein gene 5′ flanking regions on human growth hormone gene. *Molecular Reproduction and Development* 37, 276–283.

Notarianni, E., Laurie, S., Moor, R.M. and Evans, M.J. (1990) Maintenance and differentiation in culture of pluripotential embryonic cell lines from pig blastocysts. *Journal of Reproduction and Fertility* 41, 51–56.

Paleyanda, R.K., Zhang, D.W., Hennighausen, L., McKnight, R. and Lubon, H. (1994) Regulation of human protein C gene expression by the mouse WAP promoter. *Transgenic Research,* 3, 335–343.

Palmiter, R.D. and Brinster, R.L. (1986) Germ-line transformation of mice. *Annual Review of Genetics* 20, 465–499.

Pepin, M.C., Pothier, F. and Barden, N. (1992) Impaired type II glucocorticoid receptor function in mice bearing antisense RNA transgene. *Nature* 355, 725–728.

Persuy, M.A., Stinnakre, M.G., Printz, C., Mahe, M.F. and Mercier, J.C. (1992) High level expression of the β-casein gene in transgenic mice. *European Journal of Biochemistry* 205, 887–893.

Pittius, C.W., Hennighausen, L., Lee, E., Westphal, H., Nicols, E., Vitale, J. and Gordon, K. (1988) A milk protein gene promoter directs the expression of human tissue plasminigen activator cDNA to the mammary gland in transgenic mice. *Proceedings of the National Academy of Sciences of the USA* 85, 5874–5878.

Platenburg, G.J., Kootwijk, E.P.A., Kooiman, P.M., Woloshuk, S.L., Nuijens, J.H., Krimpenfort, P.J.A., Pieper, F.R., de Boer, H.A. and Strijker, R. (1994) Expression of human lactoferrin in milk of transgenic mice. *Transgenic Research* 3, 99–108.

Pursel, V.G. and Rexroad, C.E. (1993) Recent progress in the transgenic modification of swine and sheep. *Molecular Reproduction and Development* 36, 251–254.

Ragot, T., Vincent, N., Chafey, P., Vigne, E., Gilgenkrantz, H., Couton, D., Cartaud, J., Briand, P., Kaplan, J.C., Perricaudet, M. and Kahn, A. (1993) Efficient adenovirus-mediated transfer of human minidystrophin gene to skeletal muscle of mdx mice. *Nature* 361, 647–650.

Rastan, S. (1990) Czech mouse. *Trends in Genetics* 6, 233–236.

Reddy, V.B., Vitale, J.A., Wei, C., Montaya-Zavala, M., Stice, S.L., Balise, J. and Robl, J.M. (1991) Expression of human growth hormone in the milk of transgenic mice. *Animal Biotechnology* 2, 15–29.

Rexroad, C.E. (1992) Transgenic technology in animal agriculture. *Animal Biotechnology* 3, 1–3.

Rexroad, C.E., Behringer, R.R., Bolt, D.J., Miller, K.F., Palmiter, R.D. and Brinster, R.L. (1988) Insertion and expression of a growth hormone fusion gene in sheep. *Journal of Animal Science* 66, 267.

Richard, D., Chapdelaine, S., Deshaies, Y., Pépin, M.C. and Barden, N. (1993) Energy balance and lipid metabolism in transgenic mice bearing an antisense GCR gene construct. *American Journal of Physiology* 265, R146–R150.

Roberts, B., DiTullio, P., Vitale, J., Hehir, K. and Gordon, K. (1992) Cloning of the goat β-casein-encoding gene and expression in transgenic mice. *Gene* 121, 255–262.

Rohan, R.M., King, D. and Frels, W.I. (1990) Direct sequencing of PCR-amplified junction fragments from tandemly repeated transgenes. *Nucleic Acids Research* 18, 6089–6095.

Schedl, A., Montoliu, L., Kelsey, G. and Schütz, G. (1993) A yeast artificial chromosome covering the tyroinase gene confers copy number-dependent expression in transgenic mice. *Nature* 362, 258–261.

Shamay, A., Solinas, S., Pursel, V.G., McKnight, R.A., Alexander, L., Beattie, C., Hennighausen, L. and Wall, R.J. (1991) Production of the mouse whey acidic protein in transgenic pigs during lactation. *Journal of Animal Science* 69, 4552–4562.

Shamay, A., Pursel, V.G., Wilkinson, E., Wall, R.J. and Hennighausen, L. (1992) Expression of the whey acidic protein in transgenic pigs impairs mammary development. *Transgenic Research* 1, 124–132.

Shani, M., Barash, I., Nathan, M., Ricca, G., Searfoss, G.H., Dekel, I., Faerman, A., Givol, D. and Hurwitz, D.R. (1992) Expression of human serum albumin in the milk of transgenic mice. *Transgenic Research* 1, 195–208.

Simons, J.P., McClenaghan, M. and Clark, A.J. (1987) Alteration of the quality of milk by expression of sheep β-lactoglobulin in transgenic mice. *Nature* 328, 530–532.

Sims, M. and First, N.L. (1994) Production of calves by transfer of nuclei from cultured inner cell mass cells. *Proceedings of the National Academy of Sciences of the USA* 90, 6143–6147.

Smith, A.G. (1991) Culture and differentiation of embryonic stem cells. *Journal of Tissue Culture Methods* 13, 89–94.

Smith, L.C. and Wilmut, I. (1989) Influence of nuclear and cytoplasmic activity on the development *in vivo* of sheep embryos after nuclear transfer. *Biology of Reproduction* 40, 1027–1035.

Soulier, S., Vilotte, J.L., Stinnakre, M.G. and Mercier, J.C. (1992) Expression analysis of ruminant α-lactalbumin in transgenic mice: developmental regulation and general location of important cis-regulatory elements. *FEBS Letters* 297, 13–18.

Stacey, A., Schnieke, A., McWhir, J., Cooper, J., Colman, A. and Melton, D.W. (1994) Use of double-replacement gene targeting to replace the murine α-lactalbumin gene with its human counterpart in embryonic stem cells and mice. *Molecular and Cellular Biology* 14, 1009–1016.

Stinnakre, M.G., Vilotte, J.L., Soulier, S., L'Haridon, R., Charlier, M., Gaye, P. and Mercier, J.C. (1991) The bovine α-lactalbumin promoter directs expression of ovine trophoblast interferon in the mammary gland of transgenic mice. *FEBS Letters* 284, 19–22.

Stinnakre, M.G., Vilotte, J.L., Soulier, S. and Mercier, J.C. (1994) Creation and phenotypic analysis of α-lactalbumin-deficient mice. *Proceedings of the National Academy of Sciences of the USA* 91, 6544–6548.

Strauss, W.M., Dausman, J., Beard, C., Johnson, C., Lawrence, J.B. and Jaenisch, R. (1993) Germ line transmission of a yeast artificial chromosome spanning the murine a1(I) collagen locus. *Science* 259, 1904–1907.

Talbot, D., Collis, P., Antoniou, M., Vidal, M., Grosveld, F. and Greaves, D.R. (1989) A dominant control region from the human β-globin locus conferring integration site-independent gene expression. *Nature* 338, 352–355.

Thompson, E.M., Christians, E., Stinnakre, M.G. and Renard, J.P. (1994) Scaffold attachment regions stimulate HSP70.1 expression in mouse preimplantation embryos but not in differentiated tissues. *Molecular and Cellular Biology* 14(7), 4694–4703.

Threadgill, D.W. and Womack, J.E. (1990) Genomic analysis of the major bovine milk protein genes. *Nucleic Acids Research* 18, 6935–6942.

Valancius, V. and Smithies, O. (1991) Testing an 'in–out' targeting procedure for making subtle genomic modifications in mouse embryonic stem cells. *Molecular and Cellular Biology* 11, 1402–1408.

Velander, W.H., Johnson, J.L., Page, R.L., Russell, C.G., Subramanian, A., Wilkins, T.D., Gwazdauskas, F.C., Pittius, C. and Drohan, W.N. (1992) High-level expression of a heterologous protein in the milk of transgenic swine using the cDNA encoding human protein C. *Proceedings of the National Academy of Sciences of the USA* 89, 12003–12007.

Vilotte, J.L. and Soulier, S. (1992) Isolation and characterization of the mouse α-lactalbumin-encoding gene: interspecies comparison, tissue- and stage-specific expression. *Gene* 119, 287–292.

Vilotte, J.L., Soulier, S., Stinnakre, M.G., Massoud, M. and Mercier, J.C. (1989) Efficient and tissue-specific expression of bovine α-lactalbumin in transgenic mice. *European Journal of Biochemistry* 186, 43–48.

Vize, P.D., Michalska, A.E., Ashman, R., Lloyd, B., Stone, B.A., Quinn, P., Wells, J.R.E. and Seamark, R.F. (1988) Introduction of a porcine growth hormone fusion gene into transgenic pigs promotes growth. *Journal of Cell Science* 90, 295–300.

Wall, R.J., Pursel, V.G., Shamay, A., McKnight, R.A., Pittius, C.W. and Hennighausen (1991) High-level synthesis of a heterologous milk protein in the mammary glands of transgenic swine. *Proceedings of the National Academy of Sciences of the USA* 88, 1696–1700.

Whitelaw, C.B.A., Archibald, A.L., Harris, S., McClenaghan, M., Simons, J.P. and Clark, A.J. (1991) Targeting expression to the mammary gland: intronic sequences can enhance the efficiency of gene expression in transgenic mice. *Transgenic Research* 1, 3–13.

Whitelaw, C.B.A., Harris, S., McClenaghan, M., Simons, J.P. and Clark, A.J. (1992) Position-independent expression of the ovine β-lactoglobulin gene in transgenic mice. *Biochemical Journal* 286, 31–39.

Whitelaw, C.B.A., Springbett, A.J., Webster, J. and Clark, J. (1993) The majority of G0 transgenic mice are derived from mosaic embryos. *Transgenic Research* 2, 29–32.

Wilkins, T.D. and Velander, W. (1992) Isolation of recombinant proteins from milk. *Journal of Cellular Biochemistry* 49, 333–338.

Williams, R.S., Johnston, S.A., Riedy, M., DeVit, M.J., McElligott, S.G. and Sanford, J.C. (1991) Introduction of foreign genes into tissues of living mice by DNA-coated microprojectiles. *Proceedings of the National Academy of Sciences of the USA* 88, 2726–2730.

Wilmut, I., Archibald, A.L., Harris, S., McClenaghan, M., Simons, J.P., Whitelaw, C.B.A. and Clark, A.J. (1990) Methods of gene transfer land their potential use to modify milk composition. *Theriogenology* 33, 113–123.

Wolff, J.A., Malone, R.W., Williams, P., Chong, W., Acsadi, G., Jani, A. and Felgner, P.L. (1990) Direct gene transfer into mouse muscle *in vivo*. *Science* 247, 1465–1468.

Wood, S.A., Allen, N.D., Rossant, J., Auerbach, A. and Nagy, A. (1993) Non-injection methods for the production of embryonic stem cell-embryos chimeras. *Nature* 365, 87–89.

Wright, G., Carver, A., Cottom, D., Reeves, D., Scott, A., Simons, J.P., Wilmut, I., Garner, I. and Colman, A. (1991) High level expression of active human alpha-1-antitrypsin in the milk of transgenic sheep. *Bio/Technology* 9, 830–834.

Wu, H., Liu, X. and Jaenisch, R. (1994) Double replacement: strategy for efficient introduction of subtle mutations into the murine Colla-1 gene by homologous recombination in embryonic stem cells. *Proceedings of the National Academy of Sciences of the USA* 91, 2819–2823.

Yom, H.C. and Bremel, R.D. (1993) Genetic engineering of milk composition: modification of milk components in lactating transgenic animals. *American Journal of Clinical Nutrition* 58, 299S–306S.

Yom, H.C., Bremel, R.D. and First, N.L. (1993) Mouse mammary tumor virus promoter directs high-level expression of bovine αs1 casein in the milk of heterozygous and homozygous mice. *Animal Biotechnology* 4, 89–107.

Xu, S.H., Deen, K.C., Lee, E., Hennighausen, L., Sweet, R.W., Rosenberg, M. and Westphal, H. (1989) Functional human CD4 protein produced in milk of transgenic mice. *Molecular Biology and Medicine* 6, 255–261.

Zhu, N., Liggitt, D., Liu, Y. and Debs, R. (1993) Systemic gene expression after intravenous DNA delivery into adult mice. *Science* 261, 209–211.

Autocrine Regulation of Milk Secretion 15

C.J. Wilde, C.H. Knight and M. Peaker

Hannah Research Institute, Ayr KA6 5HL, UK

Introduction

Advances in biotechnology are creating new opportunities to increase the efficiency of milk production, and improve animal welfare. Conventional breeding or nutritional intervention to improve animal performance can now be weighed against the use of gene transfer for rapid production of genetically-superior animals, and endocrine or immunological manipulation to achieve short term improvement in productivity. The scientific rationale behind innovative approaches to the control of lactation has come not from traditional dairy science although, as we will demonstrate, it may have its origins there. As in other areas, experimental physiology in dairy animals is increasingly underpinned by knowledge of the molecular and cellular events in tissue development and function, which enable manipulation of not so much animal performance, but rather the function of a target tissue in animals that are otherwise normal.

Strategies for manipulating lactation performance

Gene targetting, exploiting tissue-specific regulatory elements, has already proved successful in generating animals with altered mammary function. This technique can be used to obtain expression and secretion of high-value therapeutic proteins in milk (Wright *et al.*, 1991) or may be applied to endogenous milk protein genes, in order to change milk composition or rate of secretion (Kumar *et al.*, 1994; Stinnakre *et al.*, 1994; Stacey *et al.*, 1995). However, this technology has not been universally successful: in one case, over-expression of an endogenous milk protein resulted in lactational failure (Burdon *et al.*, 1991).

Shorter-term intervention can be achieved by endocrine manipulation, an approach whose tissue specificity obviously depends on that of the hormone involved. For endocrine manipulation of lactation, the most obvious candidate is growth hormone, whose galactopoietic actions are well established (Bauman *et al.*, 1985). However, endocrine intervention need not necessarily be limited by the properties of the native hormone, since its effects can be achieved indirectly, by an immunological approach. The immune system's response to xenogenic antibodies can be turned to advantage in the production of anti-idiotypic antibodies which mimic the structure and function of hormones. Anti-idiotypes have been produced from a number of hormones including insulin (Sege and Petersen, 1978), β-adrenergic agents (Schreiber *et al.*, 1980), thyrotropin (Farid *et al.*, 1982) and also growth hormone (Gardner *et al.*, 1990). In the case of growth hormone, the anti-idiotypes mimicked the hormone both structurally and functionally. This approach has two notable advantages over hormone therapy. First, it lends itself to a vaccination approach, and the half-life of such responses is in the order of weeks compared with 20–30 minutes for injected growth hormone. Second, anti-idiotypes mimicking growth hormone reproduced some but not all the actions of the native hormone. This raises the possibility that immunization of lactating animals with monoclonal antibodies, which should mimic individual epitopes of the hormone, may allow a particular endocrine response to be elicited selectively. Alternatively, endocrine and immunological approaches may be combined by administration of hormone pre-complexed with an appropriate monoclonal antibody (Holder *et al.*, 1985; Beattie and Holder, 1994). In this paper we describe for the first time the application of immunological techniques to control lactation in dairy cows.

Specificity of manipulation may, of course, be conferred by tissue-specificity of the targetted function. Clearly, then, the most attractive approach to control of lactational performance would be one in which the actual process of milk secretion is targetted, particularly if the strategy took advantage of a local, i.e. autocrine or paracrine rather than endocrine, mechanism regulating this process specifically. In the remainder of this paper we describe a local, in this case autocrine, mechanism which regulates the rate of milk secretion acutely during lactation, and a number of experimental approaches, both direct and indirect, which we have devised to exploit this mechanism and control lactational performance in dairy animals.

Feedback Inhibition of Milk Secretion

Characterization of autocrine inhibition

Strategic endocrine regulation of milk secretion is modulated by a short-term tactical control mechanism operating within each mammary gland. This intra-mammary mechanism responds to the frequency and completeness of milk removal to set the actual rate of milk secretion in each gland. The nature of the local regulation was established in a series of physiological experiments over more than 30 years, which have been reviewed elsewhere (Peaker, 1994). Briefly, these showed that local regulation of milk secretion by milking frequency or efficiency was wholly dependent on milk removal from the gland (Linzell and Peaker, 1971), but was not due to the physical effect of milk removal on gland distension (Henderson and Peaker, 1984). This pointed to the involvement of a chemical mechanism, and initiated a search for a milk constituent able to regulate milk secretion. A mammary tissue bioassay was used, initially to screen goat's milk constituents (Wilde *et al.*, 1995a), and subsequently cow's milk (Addey *et al.*, 1991), human milk (Wilde *et al.*1995b), even milk from a macropodid marsupial (Hendry *et al.*, 1992). Since milk removal stimulates milk secretion and the effect of frequent milking is realized within hours (Linzell and Peaker, 1971), the tissue culture bioassay was designed to identify factors acting acutely to inhibit synthesis of milk constituents rapidly and reversibly. In each milk studied thus far, tissue culture bioassay has identified the inhibitor as a small milk protein of M_r 6000–8000, the actual M_r varying between species. The same caprine and bovine proteins also inhibited milk protein synthesis and secretion in mouse mammary acini (Rennison *et al.*, 1993; A. Sudlow, R.D. Burgoyne and C.J. Wilde, unpublished work). There is good reason to believe that results obtained in tissue and cell culture bioassays reflect the activity of these milk proteins *in vivo*. The caprine inhibitor, a glycoprotein of M_r 7600, which inhibited casein and lactose synthesis in explant culture, also inhibited milk secretion in a temporary, dose-dependent manner when introduced back into a gland of lactating goats through the teat canal (Wilde *et al.*, 1995a). Amino acid analysis of this protein showed no sequence homology with other milk proteins, nor any recorded sequence. Therefore, on the basis of this novelty and its biological activity, the protein has been named FIL, Feedback Inhibitor of Lactation.

Mechanism of autocrine inhibition

The ability of goat FIL to inhibit milk secretion in a concentration-dependent manner both *in vitro* and *in vivo* suggests that it is a physiological regulator of the rate of milk secretion. We have yet to confirm directly the effect of bovine FIL *in vivo*, but such a role is supported by the effect of indirect immunological manipulation of the protein's activity in dairy cows (see below). No other locally-active factor has been identified as competent to regulate secretion of milk constituents coordinately, i.e. without affecting milk composition in established lactation. Transforming growth factor-β (TGF-β) has been proposed as a regulator of milk secretion in late pregnancy (Robinson *et al.*, 1993), and has been identified in milk of dairy cows and lactating mice (Maier *et al.*, 1991; Letterio *et al.*, 1994), suggesting that it could also perform the same function during lactation. However, TGF-β is more likely to act as a chronic inhibitor of functional differentiation (Robinson *et al.*, 1993; Yamamoto *et al.*, 1994), since it does not acutely regulate casein secretion from lactating mammary acini (Sudlow *et al.*, 1995). *De novo* lipid synthesis in rat mammary tissue is regulated *in vivo* and *in vitro* by medium chain fatty acids, which are a normal component of rat milk, and it has been proposed that these may act to feedback-inhibit milk lipid synthesis (Heesom *et al.*, 1992). However, whether medium-chain fatty acid levels change in response to physiological stimuli is not known, and how such a mechanism would coordinately regulate other milk constituents is unclear.

FIL, on the other hand, acts on the mammary epithelial cell in a manner which may explain both the coordinate regulation of milk secretion by milk removal and the multipotent actions of the inhibitor *in vivo*. FIL treatment of suspension cultures of mouse mammary acini produced dramatic but fully-reversible dispersion of the Golgi apparatus and vesiculation of the endoplasmic reticulum (Rennison *et al.*, 1993). Disruption of secretory protein trafficking at this stage appears, in turn, to inhibit milk protein synthesis, a retrograde effect mimicked by the Golgi-disrupting fungal drug brefeldin A in the same cultures (Rennison *et al.*, 1993), and by membrane-disrupting treatments in other secretory cells (Kuznetsov *et al.*, 1992). *De novo* synthesis of milk lipid may also depend on endoplasmic reticulum integrity (Keenan *et al.*, 1992; Keon *et al.*, 1993), so it seems possible that disruption of endoplasmic reticulum–Golgi integrity could account for FIL's concerted regulation of all milk constituents, i.e. its effect on milk yield but not milk composition, when administered by intra-ductal injection in lactating goats (Wilde *et al.*, 1995a). However, FIL's coordinate regulation of lipid secretion together with other milk constituents *in vivo* contrasts with experiments in tissue and cell culture where goat FIL had no effect on [^{14}C]acetate incorporation into total lipid (Wilde *et al.*, 1987a; Rennison *et al.*, 1993). This dichotomy is as yet unresolved, but may be explained by the absence in culture of some factor or mechanism mediating FIL's effect on lipid synthesis. For instance, triacylglycerol synthesis *in vitro* takes place in the absence of transcellular fatty acid transport, whereas the two processes appear to be interdependent in the lactating tissue (Robinson and Williamson, 1978; Heesom *et al.*, 1992).

Kinetics of autocrine inhibition

Based on FIL's dramatic but readily-reversible effects on epithelial cell ultra-structure and secretory activity, we have concluded that the FIL proteins identified in goat's and cow's milk mediate regulation of the rate of milk secretion by frequency and completeness of milking. If so, the concentration-dependent inhibition of milk secretion exerted *in vitro* (Wilde *et al.*, 1987a) and *in vivo* (Wilde *et al.*, 1995a) by the caprine inhibitor suggests that a change in milking practice and its resultant effect on milk secretion, should be associated with changes in FIL's milk concentration. However, the response of milk secretion to hourly milking (Linzell and Peaker, 1971), or after dilution of stored milk by inert solution (Henderson and Peaker, 1987) or, indeed, to FIL administration (Wilde *et al.*, 1995a) is rapid, occurring within hours. This indicates that changes in autocrine inhibition may be equally rapid, and that the diurnal cycle of milk accumulation and removal is accompanied by a cyclical pattern of FIL accretion and depletion. Preliminary measurements of FIL concentration in lactating dairy cows support this prediction. Milk accumulation was accompanied by a significant increase in FIL concentration not observed over a similar period when the gland was drained continuously (P. Irving, K. Stelwagen, S.R. Davis, and C.J. Wilde, unpublished work). As yet, little is known about the mechanism by which changes in FIL's milk concentration are engineered. At this stage, it seems unlikely that this is the result of a changing rate of FIL synthesis alone, if for no other reason than this would require that gene expression, synthesis or secretion of this protein be regulated independently of other milk constituents. Differential regulation of milk protein synthesis has been observed in a number of situations, perhaps the most dramatic being in the tammar wallaby (Bird *et al.*, 1994). On the other hand, these and the differential changes in milk proteins observed in ruminants (e.g. Goodman and Schanbacher, 1991) are developmental responses occurring at the level of gene expression, and therefore are unlikely to accommodate the acute response to milk removal observed *in vivo*. Post-transcriptional selection of milk proteins for translation and secretion is possible: synthesis and secretion of an ovine β-lactoglobulin transgene in hemizygous mice is at the expense of endogenous milk proteins, whose milk concentrations are reduced proportionately (Wilde *et al.*, 1992). However, this may in part reflect the relative abundance of milk protein messenger RNAs (McClenaghan *et al.*, 1995). A more plausible explanation is that FIL is subject to processing after secretion into the alveolar lumen of the gland. Several inactive, but immunologically-related variants of the FIL protein have been detected in goat's and cow's milk (Wilde *et al.*, 1995a; C.V.P. Addey and C.J. Wilde, unpublished work). Putative precursor-product relationships between these protein species have yet to be explored, but their presence raises the possibility that FIL is secreted as a pro-inhibitor and activated in the alveolar lumen, or conversely, that the FIL-related proteins are processed, inactivated forms of FIL which is secreted as the bioactive protein. Computer modelling indicates that, in either case, first order rate of metabolism in the alveolar lumen would bring about an increase in active inhibitor during milk accumulation, even though FIL was being secreted at a constant rate relative to other milk constituents. Whatever the mechanism involved, FIL's concentration-dependent inhibition of milk secretion *in vivo* and *in vitro*, and the change in FIL concentration with milk accumulation, suggest that any treatment which affects the concentration of active inhibitor in milk will regulate the rate of milk secretion. We have approached manipulation of autocrine inhibition in two ways.

Manipulation of Autocrine Control

Several features of this regulatory mechanism make it an attractive target for manipulation. Above all, it is an autocrine mechanism: FIL is synthesized by secretory epithelial cells in the mammary gland on which it subsequently exerts its effect (Wilde *et al.*, 1995a), a mode of action facilitated by the unusual, virtually unique capacity of this exocrine gland to store its secretion extracellularly. Another attraction is that FIL is unlikely to act in an endocrine fashion in other secretory tissues during lactation. We have failed to detect the protein in lactating goat serum (C.V.P. Addey and C.J. Wilde, unpublished work). Experiments with goat mammary epithelial cells cultured on reconstituted basement membrane also suggest that FIL is secreted preferentially, *in vivo* perhaps exclusively, into milk. On this culture substratum, which is derived from the Engelbreth-Holm-Swarm tumour (EHS matrix) the cells form mammospheres similar to those formed by mouse mammary epithelial cells (Barcellos-Hoff *et al.*, 1989; Hurley *et al.*, 1994). DAPI (4,6-diamidino-2-phenylindole) nuclear staining shows cells arranged peripherally in these structures surrounding a central lumenal space, in a manner reminiscent of alveoli in the lactating tissue *in vivo*. At the same time, the epithelial cells differentiate, become polarized, and secrete milk proteins vectorially, with milk proteins being secreted across the apical membrane into the mammosphere lumen (Wilde *et al.*, 1995a). Immunoblotting of lumenal extract (apical secretion) and culture medium (basolateral secretion) demonstrated that FIL secretion followed the same pattern as casein and α-lactalbumin, that is, FIL was detected in the lumenal secretion and was barely detectable in basolateral secretion. These observations, together with the local, within-gland, response to manipulation of milking frequency are indicative of a mammary-specific action. Cloning of caprine FIL is in progress, but it is not yet known if the protein is expressed only in mammary tissue. In any case, since it is the secreted protein which is active, i.e. inhibition is autocrine rather than intracrine, it is likely that feedback inhibition is a receptor-mediated process. Specific [^{125}I]-FIL binding to apical membrane preparations from lactating goat mammary tissue have provided preliminary evidence for a FIL receptor (M. Lang, C.H. Knight and C.J. Wilde, unpublished work). Therefore, FIL-like activity in extra-mammary tissue would depend not only on local production of inhibitor, but also on the presence of a receptor and its associated intracellular signal transduction system. Bioassay of FIL's effect on two other secretory cell types, one secreting constitutively (hepatocytes), the other secreting intermittently by regulated exocytosis (adrenal chromaffin cells), have indicated that the receptor has at least some tissue specificity; FIL had no effect on secretory activity in either cell type.

Intraductal injection of inhibitor

The obvious, direct approach to manipulation of autocrine inhibition *in vivo* is to increase FIL's concentration in milk stored in the gland. This approach was tested in lactating goats by injecting FIL purified by anion exchange chromatography (Wilde *et al.*, 1995a) through the teat canal into the gland cistern and alveoli. FIL injection produced a temporary ipsilateral inhibition of milk secretion in the treated gland (Wilde *et al.*, 1995a). Simultaneous introduction of carrier solution into the contralateral gland did not change milk yield, indicating that the injection procedure had no adverse effect. Treatment of other animals with much larger amounts of other milk proteins also did not affect milk yield (Wilde *et al.*, 1988a), showing that the effect was attributable to FIL, and not a non-specific effect of increasing milk protein concentration. Both the degree and duration of inhibition were dose-dependent. Lower doses of 100 μg or 250 μg of FIL inhibited milk secretion for 24–48 h after injection, whereas with higher doses, inhibition persisted for up to 3 days. In each case, FIL regulated the rate of milk secretion without affecting gross milk composition, i.e. inhibition was exerted on all milk constituents equally. The persistent effect of higher doses of inhibitor would appear inconsistent with the autocrine mechanism's acute regulation of milk secretion discussed above. It might be anticipated that exogenous inhibitor introduced into the gland after the afternoon milking, as was the case in this experiment, would act during the subsequent period of milk accumulation but, once removed at the next morning's milking, would make no further contribution to the degree of autocrine inhibition and the resultant rate of milk secretion. On the other hand, it is well known that milking normally does not empty the mammary gland. The presence of residual milk is likely to influence significantly the kinetics of autocrine inhibition since, according to our predictions, it should contain a high concentration of inhibitor. This high concentration of FIL in residual milk cannot, however, persist indefinitely, since it would prevent cyclical changes in autocrine inhibition and milk secretion rate as milk accumulates and is removed. Processing of FIL in the alveolar lumen, a mechanism that would obviate the need for independent regulation of FIL synthesis and secretion (see above), could also account for the reversal of autocrine inhibition in residual milk. A mechanism involving FIL inactivation would rapidly neutralize the effect of a small pool of residual inhibitor. In the event of pro-inhibitor activation, neutralization of this residual inhibitor would be slower, since it would depend on the dilution of residual bioactive protein. Whatever the mechanism involved, introduction of a bolus of active FIL into the gland immediately after milking would impair this process and, in so doing, should act not so much to accelerate the re-attainment of a high inhibitor concentration, but to delay relief of the autocrine inhibition exerted by residual milk. Furthermore, since we postulate that FIL metabolism is a first order process, i.e. one whose rate depends only on the concentration of protein in milk, one would then predict that higher doses of exogenous FIL could render the inactivation process ineffective, to the extent that several cycles of milk accumulation and removal would be required to re-establish a normal pattern of FIL secretion, processing and removal.

Clearly, the same situation could arise if normal milking is so ineffective as to leave a volume of milk and mass of FIL which cannot be neutralized by post-secretion processing. This was demonstrated experimentally by catheter milking, a technique that removes only cisternal milk, leaving the alveoli of the gland full. Frequent (hourly) milking of lactating goats by catheter failed to stimulate the rate of milk secretion (Henderson and Peaker, 1987), presumably because the treated glands experienced no significant change in the degree of autocrine inhibition. In normal circumstances, completeness of milking, and its influence on milk secretion rate depends on both milk let-down and on mammary gland anatomy, in particular the distribution of milk between alveolar and cisternal storage spaces. The importance of gland anatomy in determining susceptibility to autocrine inhibition and the productivity of dairy animals is discussed later. However, it is worthy of note that those species which lack the capacious cistern of the ruminant, and consequently store milk primarily in the alveolar lumena, exhibit a rate of milk secretion that is especially dependent on the completeness of gland emptying at suckling. This is illustrated by elegant studies of breastfeeding mothers, which demonstrated a significant relationship between the degree of breast emptying at each feed and the subsequent rate of milk secretion (Daly *et al.*, 1993). This is probably due to operation of an autocrine mechanism similar to that discovered in ruminants: techniques used to identify the caprine and bovine FILs have identified an immunologically-related protein with similar activity in tissue culture bioassay. (C.J. Wilde and A. Prentice, unpublished work).

Immunological manipulation of autocrine inhibition

An alternative, indirect approach to manipulation of autocrine inhibition *in vivo* is to actively immunize lactating animals against endogenous FIL. This approach was adopted in dairy cows, where limited availability of purified protein, together with the quantity required to increase the concentration of endogenous FIL significantly in high-producing animals, made intraductal injection impractical. The theory behind the immunological approach is that active immunization against FIL and generation of a systemic anti-FIL antibody titre may be accompanied by secretion of antibody in milk, where it could act to neutralize the effect of endogenous FIL. This should result in a stimulation of milk secretion, or relative insensitivity to any manipulation that tended to increase the degree of autocrine feedback, e.g. once-daily milking. So far, this technique has been applied to only a small number of dairy cows, but the results have been encouraging. Auto-immunization of cows on three occasions with hapten-conjugated FIL and a suitable adjuvant, the first vaccination being around peak milk yield at week 8 of lactation, produced a circulating antibody titre against the inhibitor in two out of three animals tested. The immune response was apparent after the second immunization and increased after the third, at which time antibody against FIL was also detected in milk. The antibody titre persisted for 8 weeks after the third immunization, and during this period milk yield, which would have been expected to decline progressively, was maintained at a level higher than that predicted from the animal's pre-immunization yield and its lactation curve in the preceding lactation. In the two responding cows, cumulative milk yield was calculated to be more than 40% higher than predicted values (Fig. 15.1). The third animal did not develop detectable antibody and consequently showed no change in milk yield. This was also the pattern in two of three sham-immunized control cows. The third control cow appeared to develop an autoimmune response against the inhibitor, and milk yield was higher than predicted values. However, comparison of milk yield of experimental animals with other healthy cows in the herd again demonstrated the positive response of the two immunized animals. These results indicated that the presence in milk of antibodies specific for FIL had conferred some protection against autocrine inhibition, with the result that milk yield was temporarily increased.

The effect of auto-immunizing dairy cows was similar, but more pronounced and longer-lived than that observed in lactating goats (C.J. Wilde, C.V.P. Addey and M. Peaker, unpublished), which may reflect the more frequent immunization schedule (4-weekly intervals in cows instead of 8 weeks in goats). Goats were immunized after peak lactation and, as expected, milk yield in sham-immunized animals declined progressively. The change in milk yield of immunized goats was initially similar to controls, but after the second and third immunizations, the decline in milk yield slowed considerably. Immunized goats also had a greater tolerance of once daily milking than sham-immunized animals: the decrease in milk yield was significantly lower than in controls. Both effects on milk yield are consistent with partial immuno-neutralization of the increase in autocrine feedback accompanying infrequent milking.

Practical application of the autocrine technology

In practice, immunological manipulation of autocrine inhibition would depend on the animals' ability to store a greater volume of milk even at peak lactation. Scope for this approach may therefore be limited to, for example, situations where husbandry is less intensive. New Zealand animals are reported to be able to accommodate 24-hours of milk accumulation without adverse effect (Carruthers *et al.*, 1991), although a recent study has cast doubt on this (Stelwagen *et al.*, 1993). On the other hand, there is now evidence that the

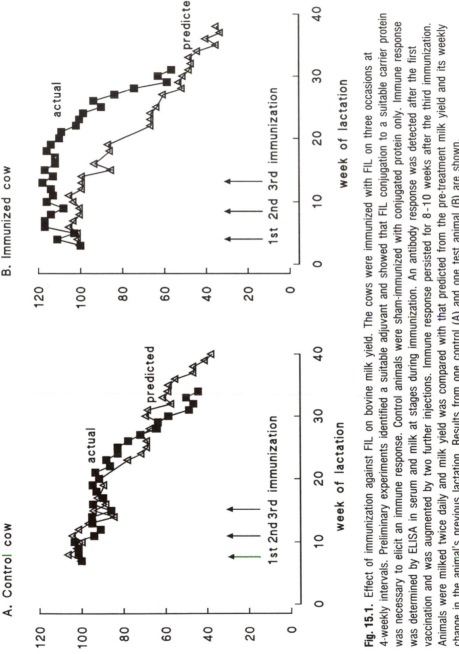

Fig. 15.1. Effect of immunization against FIL on bovine milk yield. The cows were immunized with FIL on three occasions at 4-weekly intervals. Preliminary experiments identified a suitable adjuvant and showed that FIL conjugation to a suitable carrier protein was necessary to elicit an immune response. Control animals were sham-immunized with conjugated protein only. Immune response was determined by ELISA in serum and milk at stages during immunization. An antibody response was detected after the first vaccination and was augmented by two further injections. Immune response persisted for 8–10 weeks after the third immunization. Animals were milked twice daily and milk yield was compared with that predicted from the pre-treatment milk yield and its weekly change in the animal's previous lactation. Results from one control (A) and one test animal (B) are shown.

cistern may grow during the course of lactation and in successive lactations (Dewhurst and Knight, 1993), and can adapt with repeated short periods of once daily milking to increase cisternal storage capacity (J.R. Brown and C.H. Knight, unpublished work). The next section will consider the influence of mammary storage characteristics on individual gland susceptibility to autocrine feedback, and the interaction of these factors to determine the gland's rate of milk secretion and responsiveness to changes in milking practice.

A more general application of the autocrine inhibitor technology may be in preventing excessive milk accumulation during drying off. In this situation, FIL injection through the teat canal would simply be an adjunct to dry cow antibiotic treatment. Indeed, by alleviating the discomfort of milk engorgement and promoting rapid drying-off, this treatment could act to reduce the dairy cow's relatively high susceptibility to mastitic infection during the early part of the dry period.

Mammary Gland Anatomy

Reference has already been made to the mammary gland's unusual attribute of extracellular storage of its product. The degree to which this occurs varies between species: minutes in rodents versus days in tree shrews, for instance. Not surprisingly, storage anatomy also varies; at one extreme storage is restricted to the spaces within alveoli and the small ducts that carry the milk directly to the nipple whilst, at the other extreme, the small ducts lead into larger galactophores which end in a number of large, elastic sinuses. Dairy species epitomize the latter. The sinuses are referred to as the cistern or, more correctly, gland cistern, since the teat contains a second storage cistern which can be quite large, particularly in goats. In the following discussion we shall refer only to dairy species, particularly the cow and goat.

Storage characteristics

A simplistic model of milk storage would view the gland as a two compartment structure; secreted milk is stored either within the lumen of secretory alveolar tissue (alveolar milk) or within the large ducts and the gland and teat cisterns (cisternal milk). The important distinction is that FIL is active in the former but not in the latter, simply because it is then remote from its site of action. Distribution and movement of milk between the two compartments can be measured by fractional milking. This entails first removing just the cisternal milk, leaving the alveolar milk within the gland for subsequent extraction. Alveolar milk is only obtainable with active participation from the cow in the form of the milk ejection reflex, therefore, the two approaches to removing only cisternal milk are to either actively inhibit or else passively avoid the ejection reflex.

The established procedure for pharmacological blockade of milk ejection is adrenaline injection (Gorewit and Aromando, 1985), operating through a poorly characterized combination of central inhibition of oxytocin secretion and peripheral restriction of mammary blood flow. The disadvantage of adrenaline, apart from its well-known effects on the cardiovascular system, is that the action is relatively long lasting, so removal of alveolar milk may be compromised for some time (S.R. Davis, personal communication). We have developed a second approach, using a biologically inert analogue of oxytocin (CAP: Ferring, Malmo, Sweden) to competitively block the action of endogenous oxytocin at the receptor level (Knight et al., 1994). Following intravenous injection of CAP, the goat or cow is milked as normal but no milk ejection occurs so only cisternal milk is obtained. CAP has a short biological half-life so a second milking approximately 10 minutes later yields alveolar milk.

Cisternal milk can also be obtained by careful drainage of milk through a catheter inserted intraductally. Catheter drainage relies on passive avoidance of the milk ejection reflex; arguably it has a welfare advantage since no pharmacological treatment is required, and, like CAP treatment but unlike adrenaline, it does not compromise subsequent collection of alveolar milk. Confirmation that milk ejection does not occur can be provided by drainage from one gland with simultaneous recording of intramammary pressure in a second gland (Knight and Dewhurst, 1994a). A direct comparison of adrenaline, CAP and drainage methods in goats gave very similar cistern milk volumes with CAP and drainage but significantly larger values with adrenaline, suggesting that the latter failed to prevent milk ejection (Knight et al., 1994).

Milk distribution has been measured at various intervals after milking in goats (Peaker and Blatchford, 1988) and in cows (Zaks, 1962; Knight and Dewhurst, 1994a). Three important characteristics have been identified in the cow. The alveolar compartment fills progressively and approximately linearly from immediately after milking until at least 12 h post milking. Cisternal filling, by contrast, is in two distinct stages. Soon after milking, a relatively small amount of milk (approximately 500 ml) descends into the cistern but no further cisternal filling occurs until after 4 h post-milking, and from 6 h onwards the cistern fills rapidly. Clearly, by this time pressure within the alveolar compartment has reached the point where milk is forced into the cistern, causing it to expand. The goat is rather different, in that cisternal and alveolar compartments both increase linearly for the first 6 h after milking. Thereafter, alveolar milk remains constant whilst cisternal milk continues its linear increase. In both species, the cistern starts to fill well before the alveolar compartment reaches its maximum volume, hence cisternal filling is not simply by overflow from alveolar tissue.

Cisternal:alveolar distribution characteristics alter during the course of lactation. It is well known that milk yield declines gradually after peak lactation, but it is now apparent that cisternal milk volume either does not decrease at all (in heifers) or decreases at a much slower rate than total milk yield (cows), hence the proportion of milk stored in the cistern (the cisternal fraction) increases. Although the cistern is often thought of as 'the' storage area, more milk is actually stored within secretory tissue; the 8 h post-milking alveolar fraction comprised 84% of total milk in peak lactation heifers, and even at 20 h post-milking in late lactation, 48% of the milk was still retained within secretory tissue. Parity also influences the distribution; cisternal fraction is typically higher in mature cows than heifers (e.g. at peak lactation, 21% versus 12%: Dewhurst and Knight, 1994).

Relationship to FIL action

The mathematical models alluded to earlier to explain post-milking FIL inactivation inversely correlate the degree of neutralization with the absolute amount of FIL protein still present in the alveolar lumen a short time after milking. Comparison of peak and late lactation cows revealed that the initial efflux of milk from alveolar lumen to cistern was significantly greater in the latter, despite their lower secretion rate. The reason for this is that cisternal compliance increases as lactation progresses (Dewhurst and Knight, 1993), the consequence is that FIL inactivation is likely to be greater in late lactation cows than in peak lactation cows, because the absolute amount of FIL (and milk) present in the alveolar lumen will be less. There is another factor to consider. Residual milk is that milk which remains in the gland after a normal milking which can be removed by the action of exogenous oxytocin, i.e. stimulation of the milk ejection reflex. Typically it constitutes some 10% of the total milk volume, but is highly variable between cows. Surprisingly, the amount of milk appearing in the cistern at 1 h post-milking did not correlate with residual milk volume (Knight and Dewhurst, 1994a). In other words, inefficient milking (poor milk ejection) is not compensated for by more rapid efflux of milk, so FIL inactivation is compromised. High cisternal compliance will compound this problem, so from a practical point of view efficient milking is particularly important at peak lactation.

The second critical event in the mechanics of FIL action is the time, post-milking, at which an effective inhibitory concentration of FIL is re-established. Typically, dairy cows are milked $2 \times$ daily at intervals varying from 8:16 h to 12:12 h, and where $3 \times$ daily milking is used, equal 8 h intervals are the norm. Milk distribution during the interval between 8 h and 16 h is of practical interest, therefore. Cisternal:alveolar milk distribution measured at 8 h was closely correlated (across cows) with the same distribution measured at 12 h and 20 h but not with values measured at short post-milking intervals, so the 8 h distribution is relevant to both $3 \times$ daily and once-daily milking scenarios.

Relationship with different milking frequencies

If FIL is inactive in cisternal milk, then three simple predictions can be made. 'Large-cisterned' cows (i.e. those with a high cisternal:alveolar storage ratio) should be more efficient producers of milk (kg of milk per kg of secretory tissue) than those with small cisterns, should be more tolerant of infrequent milking but should be less responsive to frequent milking. It transpires that 8 h cisternal fraction does indeed vary quite considerably between individual cows (e.g. from 53% down to 2%), so the predictions can be tested. We shall consider them in reverse order.

Cisternal fraction and milk yield responses to one week of half-udder 3 × daily milking were determined in 14 cows; response was assessed relative to yield of the normally milked half udder (Dewhurst and Knight, 1994). There was a significant negative relationship between cisternal fraction and thrice-daily milking response ($r = 0.70$, $P < 0.01$), confirming that greater cisternal storage equates with poorer reponse to more frequent milking, exactly as predicted. This has also been demonstrated in goats (Knight *et al.*, 1989). In a separate experiment, ten cows were milked once daily for one week. The immediate decrease in yield (i.e. during the first 48 h of once-daily milking) was again highly correlated with cisternal milk proportion ($r = 0.81$, $P < 0.01$), the lowest decrease being associated with the highest cistern proportion (Knight and Dewhurst, 1994b). In neither experiment was there any relationship between either pretreatment milk yield or residual milk volume and the yield response, so the effects were truly related to storage characteristics.

Experiments in goats have confirmed the third prediction: cisternal fraction is positively and significantly correlated with amount of milk produced per amount of secretory tissue (Peaker and Blatchford, 1988). Secretory tissue mass was estimated by determination of gross udder volume, a reliable technique in goats (Linzell, 1966). We have validated a polyurethane-foam casting technique for determination of gross udder volume in cows (Dewhurst *et al.*, 1993). Gross size is measured with great accuracy; secretory tissue mass, unfortunately, is not. The method overestimates secretory tissue in the large, pendulous udders of older cows, and underestimates it in the heifer. As a consequence, we have yet to determine whether storage and secretion efficiency are related in cows.

Manipulation of storage characteristics

The cyclical pattern of proliferation, secretion and involution of mammary parenchyma is well known. We suspect that the connective tissues making up the cistern also show dynamic changes during successive lactation cycles but, in truth, our understanding of the subject is very limited. We do know that cisternal compliance increases during the course of lactation and with parity, and the question we are now addressing experimentally is, can this process be manipulated? An essential feature of the cistern is its elasticity. A logical first step, therefore, is to simply stretch the cistern by milking infrequently in early lactation and examine the longer term consequences.

In the first experiment, five heifers were milked $2 \times$ daily as normal except that on two consecutive days each week half of the udder (the test half) was milked only once. This continued for the first nine weeks of lactation. In the tenth and eleventh weeks once-daily milking was performed as before, except on both halves of the udder. In all five heifers, the udder half which had not previously been exposed to once-daily milking decreased in yield by more than the test half; the difference was significant ($P < 0.01$). Although this cannot be positively ascribed to altered storage characteristics, the data are compatible with cisternal stretching in the test half. In a second experiment, goats were milked $3 \times$ daily on one gland (frequent) and once daily on the other gland (infrequent) throughout the first 6 weeks of lactation, and thereafter both glands were milked $2 \times$ daily. Half of the goats were in their first lactation and, as expected, these goats gave more milk from the frequent gland during the treatment period. Post-treatment, during bilateral $2 \times$ daily milking, yield of the frequent gland decreased whilst that of the infrequent gland increased, although the frequent gland continued to produce the most milk. All this was much as expected. The yield difference during treatment is undoubtedly explained by operation of the autocrine control mechanism (FIL), and since one consequence of frequent milking is enhanced mammary development, the carry over effect is also easily explained by the same local system. The other goats were older, and gave surprising results. During treatment, these goats yielded large amounts of milk from both frequent and infrequent glands; overall, there was no significant difference between the two. Post-treatment, yield of the frequent gland actually fell below the infrequent gland and remained lower throughout the rest of the experiment. Goats in general and older goats in particular have exceptionally large cisterns (exemplified by commercial practice in the Canary Islands where goats are routinely milked only once per day; J. Capote, personal communication). Our working hypothesis to explain the unexpected data is that the relatively compliant cistern of the older goats very rapidly adapted to the once-daily milking, counteracting FIL action to a very considerable extent. Subsequently, on $2 \times$ daily milking, these glands were able to maintain a better yield than the frequent glands because of their enhanced cisternal storage capability. The data is preliminary, and ongoing analysis of storage characteristics in these goats should prove or refute the hypothesis.

Local Control of Mammary Development

Epithelial differentiation

Immunological manipulation in dairy cows increased milk yield only temporarily: convergence of actual and predicted yields as the immune response declined indicated that the galactopoietic effect of FIL neutralization was readily reversible once immunization was discontinued. Similarly, changes in milk secretion rate elicited by frequent or infrequent milking were maintained only for as long as the treatment was applied; yield returned to pre-treatment levels when normal milking was re-introduced (Henderson et al., 1983; Wilde and Knight, 1990). The transient nature of the galactopoietic response suggests that neither immunological nor milking-induced changes in autocrine inhibition affect mammary development, or at least not in a way that confers persistency of effect after longer-term treatment. On the other hand, we have found that some developmental adaptation in the tissue is necessary to maintain the milk yield response; indeed both cell differentiation and cell number may be affected, depending on the duration of treatment. In lactating goats and cows, sustained changes in frequency or completeness of milking were consistently accompanied by modulation of secretory cell differentiation (reviewed by Wilde et al., 1990).

Stimulation of differentiation was shown principally by changes in key mammary enzyme activities, but was also demonstrated by increased concentrations of messenger RNAs encoding those enzymes and the principal milk proteins (Wilde et al., 1990; J.M. Bryson, C.V.P. Addey and C.J. Wilde, unpublished work). It is likely that this adaptation is a consequence of autocrine modulation of the epithelial cells' responsiveness to circulating hormones. Milking 3 × instead of 2 × daily milking increased total prolactin binding in goat mammary tissue after 4 weeks (McKinnon et al., 1988) and conversely, incomplete milking decreased prolactin binding and secretory cell differentiation (Wilde et al., 1989). This in turn may be a consequence of FIL's primary effect on membrane trafficking in the secretory epithelial cells. Disruption of membrane trafficking is likely to inhibit not only the secretory pathway but also the cycling of receptors between the cell surface and intracellular pools, with consequent accumulation of receptors at intracellular sites. Both occupied and unoccupied prolactin receptors are present in the Golgi (Djiane et al., 1981), indicating that they are recycled via this route. Therefore, FIL's unusual ability to block endoplasmic reticulum to Golgi transport may account for its ability to down-regulate prolactin receptors. Evidence supporting such a mechanism has come from experiments in vitro and in vivo. FIL decreased cell-surface prolactin binding in primary cultures of mouse mammary cells (Bennett et al., 1990; C.N. Bennett, C.H. Knight and C.J. Wilde, unpublished), and also decreased total prolactin binding in rabbit mammary gland when injected through the teat canal (C.N. Bennett, C.J. Wilde and C.H. Knight, unpublished). In each case the effect was associated with a decrease in cell differentiation (Wilde *et al.*, 1988b, 1991).

Mammary growth

As discussed above, changes in mammary differentiation, while important for maintaining the galactopoietic response to changes in milking practice or FIL manipulation, do not persist once the stimulus is removed. Persistency of local manipulation may be achievable only if the number rather than the differentiated state of the secretory epithelial cells is affected. This appears practicable. Unilateral $3 \times$ instead of $2 \times$ daily milking of lactating goats increased cell number compared with the contralateral control gland, but only when maintained for virtually an entire lactation (Wilde *et al.*, 1987b). Therefore, mammary cell number is also under local intramammary control during lactation. However, neither the identity of the local regulator nor its mechanism of action has yet been identified. The unilateral increase in mammary cell number was associated with an increase in DNA synthesis, suggesting that the active factor is a locally-produced growth factor. A number of candidate growth factors could potentially fulfil this role: insulin-like growth factors, epidermal growth factor and transforming growth factor-β are synthesized in the tissue, and are reported to have mammogenic effects *in vivo* and/or *in vitro* (reviewed by Collier *et al.*, 1995). On the other hand, cell proliferation may be stimulated by a release from inhibition associated with frequent milk removal. It may therefore be more pertinent to consider a potential role for transforming growth factor-β, which inhibits cell division (Daniel and Robinson, 1992) and is secreted in bovine and human milk (Maier *et al.*, 1991; Letterio *et al.*, 1994). Another potential inhibitor of mammary cell proliferation is mammary-derived growth inhibitor (MDGI), which has been identified in bovine mammary tissue (Böhmer *et al.*, 1987) and inhibits cell proliferation in mammary primary cultures and epithelial cell lines (Grosse and Langen, 1990).

The local response to milking frequency in lactating goats was obtained after peak lactation when mammary cell number was declining progressively (Wilde and Knight, 1989), and it may be that the greater cell number in the $3 \times$ milked gland was in part due to a lower rate of cell death. Mammary cell death at involution occurs by apoptosis (Walker *et al.*, 1989; Sheffield and Kotolski, 1992), and we have recently found that this process is responsible for mammary cell loss in late-lactating goats (Quarrie *et al.*, 1994). Moreover, apoptosis can be induced ipsilaterally when milking is discontinued in one gland, indicating that it, too, is under local, intramammary regulation (Quarrie *et al.*, 1994). Therefore, it may be equally appropriate to search for milk-borne factors whose removal by frequent milking would reduce the rate of cell death during lactation. We have no reason to implicate FIL in the control of mammary cell death, nor indeed in the local regulation of cell proliferation, but screening of goat's milk constituents has identified another bioactive milk fraction able to influence cell longevity in caprine mammary cell culture (Wendrinska *et al.*, 1993). The active component of this fraction, and its involvement in regulation of cell longevity *in vivo* is the subject of further investigation. These observations raise the possibility that, as with the autocrine regulation of milk secretion, mammary cell number may also prove controllable through local mechanisms operating with each gland. If so, this offers the long-term prospect of other novel approaches to the manipulation of ruminant lactation, based on increased persistency rather than higher milk production in a fixed-term lactation.

Acknowledgements

This work was funded by the Scottish Office Agriculture and Fisheries Department, and by project grants from the British Technology Group and the Agriculture and Food Research Council. We are indebted to Caroline Addey, Crispin Bennett, David Blatchford, Jane Bryson, Shirley Connor, Lynn Finch, Marian Kerr, Kay Hendry, Lynda Quarrie and other colleagues who contributed to this work. We are also grateful to those who have collaborated on aspects of the research programme, particularly Bob Burgoyne (University of Liverpool), Steve Davis and Kerst Stelwagen (Agresearch, New Zealand), and Walt Hurley (University of Illinois).

References

Addey, C.V.P., Peaker, M. and Wilde, C.J. (1991) Protein inhibitor which controls secretion of milk. *UK Patent Application* GB 2 238 052 A1.

Barcellos-Hoff, M.H., Aggeler, J., Ram, T.G. and Bissell, M.J. (1989) Functional differentiation and alveolar morphogenesis of primary mammary cultures on reconstituted basement membrane. *Development* 105, 223–235.

Bauman, D.E., Eppard, P.J., DeGeeter, M.J. and Lanza, G.M. (1985) Responses of high-producing dairy cows to long-term treatment with pituitary somatotropin and recombinant somatotropin. *Journal of Dairy Science* 68, 1352–1362.

Beattie, J. and Holder, A.T. (1994) Location of an epitope defined by an enhancing monoclonal antibody to growth hormone: some structural details and biological implications. *Molecular Endocrinology* 8, 1103–1110.

Bennett, C.N., Knight, C.H. and Wilde, C.J. (1990) Regulation of mammary prolactin binding by secreted milk proteins. *Journal of Endocrinology* 127, S141.

Bird, P.H., Hendry, K.A.K., Shaw, D., Wilde, C.J. and Nicholas, K.R. (1994) Progressive changes in milk protein gene expression in lactating tammars (*Macropus eugenii*). *Journal of Molecular Endocrinology* 13, 117–125.

Böhmer, F.D., Kraft, R., Otto, A., Wernstedt, C., Hellman, U., Kurtz, A., Muller, T., Rohde, K., Etzold, G., Lehmann, W., Langen, P., Heldin, C.H. and Grosse, R. (1987) Identification of a polypeptide growth inhibitor from bovine mammary gland. Sequence homology to fatty-acid and retinoid-binding proteins. *Journal of Biological Chemistry* 262, 15137–15143.

Burdon, T., Wall, R.J., Shamay, A., Smith, G.H. and Hennighausen, L. (1991) Over-expression of an endogenous milk protein gene in transgenic mice is associated with impaired mammary alveolar development and a *milchlos* phenotype. *Mechanisms of Development* 36, 67–74.

Carruthers, V.R., Davis, S.R. and Norton, D.H. (1991) The effects of oxytocin and bovine somatotropin on production of cows milked once a day. *Proceedings of the New Zealand Society of Animal Production* 51, 197–201.

Collier, R.J., Byatt, J.C., McGrath, M.F. and Eppard, P.J. (1995) Role of bovine placental lactogen in intercellular signalling during mammary growth and lactation. In: Wilde, C.J., Peaker, M. and Knight, C.H. (eds) *Intercellular Signalling in the Mammary Gland*. Plenum Publishing Company, London, pp. 13–24.

Daly, S.E.J., Owens, R.A. and Hartmann, P.E. (1993) The short term synthesis and infant-regulated removal of milk in lactating women. *Experimental Physiology* 78, 209–220.

Daniel, C.W. and Robinson, S.D. (1992) Regulation of mammary growth and function by TGF-β. *Molecular Reproduction and Development* 32, 145–151.

Dewhurst, R.J. and Knight, C.H. (1993) An investigation of the changes in sites of milk storage in the bovine udder over two lactation cycles. *Animal Production* 57, 379–384.

Dewhurst, R.J. and Knight, C.H. (1994) Relationship between milk storage characteristics and the short-term response of dairy cows to thrice-daily milking. *Animal Production* 58, 181–187.

Dewhurst, R.J., Mitton, A.M. and Knight, C.H. (1993) Calibration of a polyurethane foam casting technique for estimating the weight of bovine udders. *Animal Production* 56, 444.

Djiane, J., Houdebine, L.-M. and Kelly, P.A. (1981) Down-regulation of prolactin receptors in rabbit mammary gland: differential subcellular localization. *Proceedings of the Society for Experimental Biology and Medicine* 168, 378–381.

Farid, N.R., Pepper, B., Urbina-Briones, R. and Islam, N.R. (1982) Biologic activity of antithyrotropin anti-idiotypic antibody. *Journal of Cellular Biochemistry* 19, 305–312.

Gardner, M.J., Morrison, C.A., Stevenson, L.Q. and Flint, D.J. (1990) Production of anti-idiotypic antisera to rat GH antibodies capable of binding to GH receptors and increasing body weight gain in hypophysectomised rats. *Journal of Endocrinology* 125, 53–59.

Goodman, R.E. and Schanbacher, F.L. (1991) Bovine lactoferrin mRNA: sequence, analysis and expression in the mammary gland. *Biochemical and Biophysical Research Communications* 180, 75–84.

Gorewit, R.C. and Aromando, M.C. (1985) Mechanisms involved in the adrenalin-induced blockade of milk ejection in dairy cattle. *Proceedings of the Society for Experimental Biology and Medicine* 180, 340–347.

Grosse, R. and Langen, P. (1990) Mammary-derived growth inhibitor. In: Sporn, M.B. and Roberts, A.B. (eds.) *Peptide growth factors and their receptors*, Part II. Springer Verlag, Berlin, pp. 249–275.

Heesom, K.J., Souza, P.F.A., Ilic, V. and Williamson, D.H. (1992) Chain-length dependency of interactions of medium-chain fatty acids with glucose metabolism in acini isolated from lactating rat mammary glands. A putative feedback to control milk lipid synthesis. *Biochemical Journal* 281, 273–278.

Henderson, A.J. and Peaker, M. (1984) Feedback control of milk secretion in the goat by a chemical in milk. *Journal of Physiology, London* 351, 39–45.

Henderson, A.J. and Peaker, M. (1987) Effects of removing milk from the the mammary ducts and alveoli, or of diluting stored milk, on the rate of milk secretion in the goat. *Quarterly Journal of Experimental Physiology* 72, 13–19.

Henderson, A.J., Blatchford, D.R. and Peaker, M. (1983) The effect of milking thrice instead of twice daily on milk secretion in the goat. *Quarterly Journal of Experimental Physiology* 68, 645–652.

Hendry, K.A.K., Wilde, C.J., Nicholas, K.R. and Bird, P.H. (1992) Evidence for an inhibitor of milk secretion in the tammar wallaby. *Proceedings of the Australian Society for Biochemistry and Molecular Biology* 24, POS-2-3.

Holder, A.T., Aston, R., Preece, M.A. and Ivanyi, J. (1985) Monoclonal antibody mediated enhancement of growth hormone activity *in vivo*. *Journal of Endocrinology* 107, R9–R12.

Hurley, W.L., Blatchford, D.R., Hendry, K.A.K. and Wilde, C.J. (1994) Extracellular matrix and mouse mammary cell function: comparison of substrata in culture. *In Vitro Cell Developmental Biology* 30A, 529–538.

Keenan, T.W., Dylewski, D.P., Ghosal, D. and Keon, B.H. (1992) Milk lipid globule precursor release from endoplasmic reticulum reconstituted in a cell-free system. *European Journal of Biochemistry* 57, 21–29.

Keon, B.H., Ghosal, D. and Keenan, T.W. (1993) Association of cytosolic lipid with fatty acid synthase from lactating mammary gland. *International Journal of Biochemistry* 25, 533–543.

Knight, C.H. and Dewhurst, R.D. (1994a) Milk accumulation and distribution in the bovine udder during the interval between milkings. *Journal of Dairy Research* 61, 167–177.

Knight, C.H. and Dewhurst, R.D. (1994b) Once daily milking of dairy cows; relationship between yield loss and cisternal milk storage. *Journal of Dairy Research* 61, 441–449.

Knight, C.H., Brosnan, T., Wilde, C.J. and Peaker, M. (1989) Evidence for a relationship between gross mammary anatomy and the increase in milk yield obtained during thrice daily milking in goats. *Journal of Reproduction and Fertility Abstract Series* 3, 32.

Knight, C.H., Stelwagen, K., Farr, V.C. and Davis, S.R. (1994). Use of an oxytocin analogue to determine cisternal and alveolar milk pool sizes in goats. *Journal of Dairy Science* 77 (Suppl. 1), 84.

Kumar, S., Clarke, A.R., Hooper, M.L., Horne, D.S., Law, A.J.R., Leaver, J., Sprinbett, A., Stevenson., A. and Simons, J.P. (1994) Milk composition and lactation of β-casein-deficient mice. *Proceedings of the National Academy of Sciences, USA* 91, 6138–6142.

Kuznetsov, G., Brostrom, M.A. and Brostrom, C.O. (1992) Demonstration of a calcium requirement for secretory protein processing and export. Differential effects on calcium and dithiothreitol. *Journal of Biological Chemistry* 267, 3932–3939.

Letterio, J.J., Geiser, A.G., Kulkarni, A.B., Roche, N.S., Sporn, M.B. and Roberts, A.B. (1994) Maternal rescue of transforming growth factor β_1 null mice. *Science* 264, 1936–1938.

Linzell, J.L. (1966) Measurement of udder volume in live goats as an index of mammary growth and function. *Journal of Dairy Science* 49, 307–311.

Linzell, J.L. and Peaker, M. (1971) The effects of oxytocin and milk removal on milk secretion in the goat. *Journal of Physiology, London* 216, 717–734.

McClenaghan, M., Springbett, A., Wallace, R., Wilde, C.J. and Clark, A.J. (1995) Secretory proteins compete for production in the mammary gland of transgenic mice. *Biochemical Journal* (in press).

McKinnon, J., Knight, C.H., Flint, D.J. and Wilde, C.J. (1988) Effect of milking frequency and efficiency on goat mammary prolactin receptor number. *Journal of Endocrinology* 119 (Suppl.), 167.

Maier, R., Schmid, P., Cox, D., Bilbe, G. and McMaster, G.K. (1991) Localization of transforming growth factor-β_1, β_2 and β_3 gene expression in bovine mammary gland. *Molecular and Cellular Endocrinology* 82, 191–198.

Peaker, M. (1994) Autocrine control of milk secretion: development of the concept. In: Wilde, C.J., Peaker, M. and Knight, C.H. (eds) *Intercellular Signalling in the Mammary Gland*. Plenum Publishing Company, London, pp. 193–202.

Peaker, M. and Blatchford, D.R. (1988) Distribution of milk in the goat mammary gland and its relation to the rate and control of milk secretion. *Journal of Dairy Research* 55, 41–48.

Quarrie, L.H., Addey, C.V.P. and Wilde, C.J. (1994) Local regulation of mammary apoptosis in the lactating goat. *Biochemical Society Transactions* 22, 178S.

Rennison, M.E., Kerr, M.A., Addey, C.V.P., Handel, S.E., Turner, M.D., Wilde, C.J. and Burgoyne, R.D. (1993) Inhibition of constitutive protein secretion from lactating mammary epithelial cells by FIL (feedback inhibitor of lactation), a secreted milk protein. *Journal of Cell Science* 106, 641–648.

Robinson, A.M. and Williamson, D.H. (1978) Control of glucose metabolism in isolated acini of the lactating mammary gland of the rat. Effect of oleate on glucose utilization and lipogenesis. *Biochemical Journal* 170, 609–613.

Robinson, S.D., Roberts, A.B. and Daniel, C.W. (1993) TGFβ suppresses casein synthesis in mouse mammary explants and may play a role in controlling milk levels during pregnancy. *Journal of Cell Biology* 120, 245–251.

Schreiber, A.B., Courgud, P.D., Andre, C.L., Vray, B. and Strosberg, A.D. (1980) Anti-alprenolol anti-idiotypic antibodies bind to β-adrenergic receptors and modulate catecholamine-sensitive adenylate cyclase. *Proceedings of the National Academy of Sciences, USA* 77, 7385–7389.

Sege, K. and Petersen, P.A. (1978) Use of anti-idiotypic antibodies as cell surface receptor probes. *Proceedings of the National Academy of Sciences, USA* 77, 2443–2447.

Sheffield, L.G. and Kotolski, L.C. (1992) Prolactin inhibits programmed cell death during mammary gland involution. *FASEB Journal* 6, A1184.

Stacey, A., Schnieke, A., Kerr, M., Scott, A., McKee, C., Cottingham, I., Cooper, J., Binas, B., Wilde, C. and Colman, A. (1995) Lactation is disrupted by α-lactalbumin deficiency and can be restored by human α-lactalbumin gene replacement in mice. *Proceedings of the National Academy of Sciences, USA* 92, 2835–2839.

Stelwagen, K., Knight, C.H., Farr, V.C., Davis, S.R., McFadden, T.B. and Prosser, C.G. (1993) Continuous versus once daily milk drainage during a 24-h period: cisternal capacity may be limiting during once daily milking (ODM). *Journal of Dairy Science* 76 (Suppl.), P120.

Stinnakre, M.G., Vilotte, J.L., Soulier, S. and Mercier, J.C. (1994) Creation and phenotypic analysis of α-lactalbumin-deficient mice. *Proceedings of the National Academy of Sciences, USA* 91, 6544–6548.

Sudlow, A.W., Wilde, C.J. and Burgoyne, R.D. (1995) Transforming growth factor β_1 inhibits casein secretion from differentiating mammary gland explants but does not regulate secretion from lactating mammary acini. *Biochemical Journal* 304, 333–336.

Walker, N.I., Bennett, R.E. and Kerr, J.F.R. (1989) Cell death by apoptosis during involuting of the lactating breast in mice and rats. *American Journal of Anatomy* 185, 19–32.

Wendrinska, A., Addey, C.V.P., Orange, P.R., Boddy, L.M., Hendry, K.A.K. and Wilde, C.J. (1993) Effect of a milk fat globule membrane fraction on cultured mouse mammary cells. *Biochemical Society Transactions* 21, 220S.

Wilde, C.J. and Knight, C.H. (1989) Metabolic adaptations in mammary gland during the declining phase of lactation. *Journal of Dairy Science* 72, 1679–1692.

Wilde, C.J. and Knight, C.H. (1990) Milk yield and mammary function in goats during and after once-daily milking. *Journal of Dairy Research* 57, 441–447.

Wilde, C.J., Calvert, D.T., Daly, A. and Peaker, M. (1987a) The effect of goat milk fractions on synthesis of milk constituents by rabbit mammary explants and on milk yield *in vivo*. *Biochemical Journal* 242, 285–288.

Wilde, C.J., Henderson, A.J., Knight, C.H., Blatchford, D.R., Faulkner, A. and Vernon, R.G. (1987b) Effect of thrice daily milking on mammary enzyme activity, cell population and milk yield in the goat. *Journal of Animal Science* 64, 533–539.

Wilde, C.J., Addey, C.V.P., Casey, M.J., Blatchford, D.R. and Peaker, M. (1988a) Feedback inhibition of milk secretion: the effect of a fraction of goat milk on milk yield and composition. *Quarterly Journal of Experimental Physiology* 73, 391–397.

Wilde, C.J., Calvert, D.T. and Peaker, M. (1988b) Effect of a fraction of goat milk serum proteins on milk accumulation and enzyme activities in rabbit mammary gland. *Biochemical Society Transactions* 15, 916–917.

Wilde, C.J., Blatchford, D.R., Knight, C.H. and Peaker, M. (1989) Metabolic adaptations in goat mammary tissue during long term incomplete milking. *Journal of Dairy Research* 56, 7–15.

Wilde, C.J., Knight, C.H., Addey, C.V.P., Blatchford, D.R., Travers, M., Bennett, C.N. and Peaker, M. (1990) Autocrine regulation of mammary cell differentiation. *Protoplasma*, 159, 112–117.

Wilde, C.J., Blatchford, D.R. and Peaker, M. (1991) Regulation of mammary cell differentiation by extracellular milk proteins. *Experimental Physiology* 76, 379–387.

Wilde, C.J., Clark, A.J., Kerr, M.A., Knight, C.H., McClenaghan, M. and Simons, J.P. (1992) Mammary development and milk secretion in transgenic mice expressing the sheep β-lactoglobulin gene. *Biochemical Journal* 284, 717–720.

Wilde, C.J., Addey, C.V.P., Boddy, L.M. and Peaker, M. (1995a) Autocrine regulation of milk secretion by a protein in milk. *Biochemical Journal* 305, 51–58.

Wilde, C.J., Prentice, A. and Peaker, M. (1995b) Breastfeeding: matching milk supply with demand in human lactation. *Proceedings of the Nutrition Society* 54, 401–406.

Wright, G., Carver, A., Cottam., Reeves, D., Scott, A., Simons, J.P., Wilmut, I., Garner, I. and Colman, A. (1991) High level expression of active human α_1-antitrypsin in milk of transgenic sheep. *Bio/Technology* 9, 830–834.

Yamamoto, T., Komura, H., Morishige, K.-I., Tadokoro,C., Sakata, M., Kurachi, H. and Miyake, A. (1994) Involvement of autocrine mechanism of transforming growth factor-β in the functional differentiation of pregnant mouse mammary gland. *European Journal of Endocrinology* 130, 302–307.

Zaks, M.G. (1962) *The Motor Apparatus of the Mammary Gland*. Oliver and Boyd, Edinburgh.

Environment and Ethics

Organic Dairy Farming 16

R.F. Weller

Institute of Grassland and Environmental Research,
Trawsgoed, Aberystwyth, Dyfed SY23 4 LL, UK

Introduction

An organic dairy farming system can be defined as a system that aims to produce milk of a high nutritional quality by using farming methods that avoid the use of manufactured agrochemical inputs (compounded artificial fertilizers, pesticides, growth regulators and feed additives) and can be sustained by rotations and recycling of organic wastes. This both encourages biological cycling of nutrients and protects the environment and wildlife.

Interest in organic farming and the production of organic milk products has increased during the last few years in many countries, with research projects started both in Europe and North America (Lampkin, 1990). The main reasons for the increased interest are the adverse impact of many intensive farming systems on the environment, concerns about the standards of animal welfare and high culling rates in some production systems (Webster, 1993), overproduction of food in the European Union (EU) and consumer concerns about food and water safety. Nitrogen cycling on an average dairy farm is shown in Fig. 16.1. An imbalance between inputs of nutrients in the form of purchased fertilizers, concentrates and roughage and outputs of animal products can lead to the increased risks of environmental pollution (Aarts *et al.*, 1992). The surplus nitrogen from dairy farms with highly fertilized grass swards can be as high as 470 kg ha^{-1}, contributing to environmental pollution by ammonia volatilization, run-off from the fields, leaching and denitrification.

Losses of nitrogen through leaching and denitrification increase with increases in the quantity of nitrogen fertilizer applied (Wilkins, 1993). On an all-grass dairy farm the storage and disposal of animal wastes and silage effluent can also increase the risk of pollution. Future legislation to reduce pollution is currently being discussed within the EU and is likely to impose further

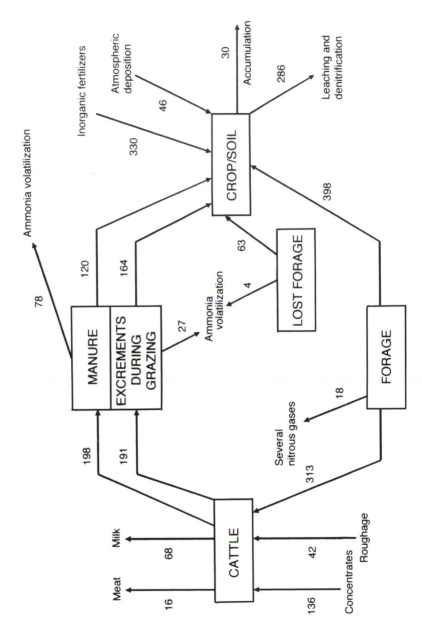

Fig. 16.1. Main N flows in an average dairy farming system. (kg N ha^{-1} yr^{-1}) (Aarts *et al.*, 1992).

restrictions on intensive production systems by limiting the emissions of ammonia and 'greenhouse' gases to the atmosphere and phosphates to water sources (Wilkins, 1993).

For many consumers some intensive farming systems include practices that are associated with unacceptably low standards of animal welfare, such as overcrowded housing, the use of hormones to stimulate production and the feeding of recycled animal wastes (e.g. bone meal) that may pose a risk to the health of both the animal to whom the waste is fed and to the consumer of animal products. There is also concern about the relationships between human health problems and chemical and nutrient contamination of water courses and chemical residues in food products (Newton, 1993). An organic farming system has the potential to improve the standards of food production by methods that are more acceptable to the consumer, reduce the risk of pollution and be a sustainable system without the requirement for large energy inputs from external sources.

Many suitable methods of crop and animal production for organic farms have been established and these techniques have been comprehensively reviewed by Lampkin (1990). However, in relation to organic dairy farming little research has been conducted, and the SAC and SARI Workshop (1989) concluded that further research is required such as the role of alternative forages, the effectiveness of silage additives, the role of forage herbs as a source of minerals and the clinical evaluation of the effect of individual alternative medicines (e.g. homeopathic, herbal) in the treatment of specific diseases. Further knowledge is also required to provide both physical and financial guidelines and practical management details for those undertaking the conversion of a dairy farm as these factors have a major effect on the management and profitability of a dairy farm (Houghton and Poole, 1990). In response to many of these requirements research projects were started in The Netherlands, Germany, Ireland and the UK in 1992 to assess the physical and financial effects of converting farms to organic dairy production, including monitoring changes in soil fertility status, herbage production, milk yield and composition, animal health and the use of alternative medicines. The results from the projects will provide valuable scientific evidence of the effects of changing to organic farming on soil structure, soil fertility, herbage production and animal performance. In the UK the work will be carried out both at the organic dairy farm of the Institute of Grassland and Environmental Research and also on eleven farms in England and Wales. The results from these projects will lead to the development of farm budget models to determine optimum strategies for converting farms to organic milk production and will also identify any major problems that may occur during the conversion period.

Prior to 1994 the total number of organic farms in the UK was only 800 (Nix and Hill, 1993), with only 65 organic dairy herds and organic milk and meat forming less than 1% of the total combined agricultural production of these products (Newton, 1993). These figures are low when compared with the statistics for Germany (1992 – 3450 organic farms) and other countries where farmers have received financial payments during the conversion period. However, the introduction in 1994 of an organic aid scheme for UK farmers, payable during the conversion period, has provided a modest incentive for more farmers to convert their dairy herds. Currently, for the majority of dairy farmers conversion of the land in stages over a period of 4–5 years is the most suitable option in terms of herbage production, stocking rates and financial implications, but this time period delays the production and marketing of milk products as organic compared with farms converting in a shorter period.

Standards for the Production of Organic Milk

The importance of organic farming in different countries around the world has been recognized by the formation of the International Federation of Organic Agriculture Movements (IFOAM) which now has a membership of over 50 countries (Lampkin, 1990). However, although IFOAM has defined national standards for the production of organic feeds the organization does not operate a scheme of regulatory and inspection procedures. Therefore individual producers are required to use the regulatory standards set by their own country.

For many years in the UK various sets of standards were set by the individual bodies responsible for increasing both the production and marketing of organic products (Lampkin, 1990). These differences often caused complications and confusion for both producers and consumers as either more than one definition was being used for the production and identification of organic products or products were incorrectly labelled as 'organic'. Major changes have recently occurred in both the standards and regulations required for the production and marketing of organic products with clear guidelines and definitions now established for both producers and consumers. The standards required for the production of organic milk are defined by the United Kingdom Register of Organic Food Standards (UKROFS, 1993). UKROFS was set up by MAFF in 1987 as a regulatory body for the approval of organizations that are responsible for inspecting and certifying organic farmers, growers and processors. In 1992 the European Union Organic Food Standards Regulation 2029/91 appointed UKROFS as the control authority in the UK responsible for the registration and supervision of organic certification bodies under EU regulations. Under the current EU regulations milk can only be sold as organic if the dairy herd has been registered and certified by an UKROFS approved organization.

The required conversion period is two years for the land followed by the conversion of the dairy cattle in the subsequent three-month period. Although up to 10% of the herd may be brought-in annually to either replace or increase the existing stock numbers (UKROFS, 1993), the objective on most dairy farms is to maintain a closed herd by ensuring sufficient numbers of replacements are reared on the farm each year.

At least 80% of the daily feed must be produced under UKROFS standards with forage comprising a minimum of 60% of the total diet as either fresh herbage or conserved silage, hay or straw. While the aim should be to use *only* feeds produced under UKROFS standards (100%), it is accepted that supplies of some organic concentrate feeds are not currently widely available in the UK. However, in other European countries where organic farming is more widely practiced, in particular Denmark and Germany, feed availability is not a problem and farmers are required to feed a greater proportion of organically-grown feeds (Lampkin, 1990). As an organic dairy herd within the EU cannot be fed animal byproducts, fishmeal or solvent-extracted feeds, balancing the ration for the organic dairy herd is more difficult and often more expensive.

Herbage and Arable Crop Production

A sustainable crop rotation is the principal method of increasing and maintaining soil structure and fertility and should include both crops with varying root systems and leguminous plants to supply nitrogen for use by subsequent crops (Lampkin, 1990; UKROFS, 1993). The rotation should aim to efficiently utilize the available organic wastes from the farm and avoid the build-up of pests and diseases associated with an imbalance of nutrients. As the main requirement is to provide enough forage both for grazing and conservation Houghton and Poole (1990) suggested a rotation based on perennial ryegrass (*Lolium perenne*), Italian ryegrass (*Lolium multiflorum*), white clover (*Trifolium repens*), red clover (*Trifolium pratense*) and cereals:

- 5 years – perennial ryegrass/white clover
- 2–3 years – Italian ryegrass/red clover
- 1 year – cereals

The quantity of cereals included in the rotation will depend on climatic constraints and whether the aim is to produce a major proportion of the required concentrate feeds on the farm or rely on purchased organic feeds. Growing cereals within a rotation would be beneficial as the fertility built up from previous grass/clover swards can be utilized and the straw requirements of the farm can be partially or fully satisfied. Another potential benefit is that surplus grain could be sold as the product is readily marketable and may achieve a premium price that is 30–50% greater than the price of conventional cereals (Nix and Hill, 1993).

The herbage production from swards may be reduced both during and after conversion depending on the previous farming system, the composition of the sward and level of nitrogen inputs. Holliday (1989) concluded that the potential milk output from grass/white clover swards was 85% of the output obtainable from heavily fertilized grass swards. However, as shown in Table 16.1, unless

Table 16.1. Yield of grass and mixed grass and white clover swards at different nitrogen fertilizer inputs.

Sward	Input of nitrogen (kg ha^{-1})	Dry matter yield (t h^{-1})
Grass/white clover	0	7.72
Grass	200	8.03
Grass	400	10.60

Source: Doyle (1984).

the farm has previously used large quantities of nitrogen the effect of conversion on herbage yield may be small providing that the white clover content of the sward is adequate. Newton (1993) reported that the average nitrogen fixation rates were as much as 180 kg ha^{-1} yr^{-1} (range 86–390) for white clover and 160 kg ha^{-1} yr^{-1} (range 103–249) for red clover.

The white clover content of permanent swards of mixed grasses and white clover are often below the preferred minimum level of 30% (Holliday, 1989; Houghton and Poole, 1990) during the period when fields are converted from a high nitrogen input system (300 kg N ha^{-1}) to a full organic system with no artificial fertilizer. This reduces both the quantity and quality of herbage produced during the growing season (Weller, 1993). This situation is less likely to occur when either the management prior to conversion has been based on a system of low nitrogen inputs (less than 160 kg ha^{-1} of N) or new swards of mixed species (including a high white clover content) can be established. Applying slurry to swards will increase the nitrogen input and improve yields without markedly affecting either the content or persistence of clover (Bax et al., 1993). The grazing dairy cow, which typically utilizes only 15–25% of the nitrogen ingested (Aarts et al., 1992), will excrete up to 250 g d^{-1} of N (Lampkin, 1990), of which 70% appears in the urine (Lampkin, 1990; Aarts et al., 1992). In permanent pastures where the introduction of a crop rotation is not possible, the change from a conventional nitrogen input system to no artificial fertilizer should encourage the natural build-up of white clover. While changing from grass swards receiving high quantities of nitrogen (350–400 kg ha^{-1}) to a grass/white clover system will reduce nitrogen losses and environmental pollution, it is important to appreciate that nitrogen losses will still occur and could be of a similar magnitude to grass swards receiving 170 kg N ha^{-1} (Wilkins, 1993).

The main problems associated with grass/white clover swards for dairy cattle grazing are the poor spring growth, low year-to-year reliability and increased risk of bloat when the clover content of the sward is high (Holliday, 1989; Mayne et al., 1991). White clover is also less suitable for growing on heavy soils that are prone to poaching than for more freely drained ones (Holliday, 1989).

It is essential that the pH value of the soil is over 6.0, and that adequate potassium, phosphorus and other minerals are available in the soil both during and after conversion (Houghton and Poole, 1990). If a deficiency occurs then approved fertilizers (e.g. rock phosphate, rock potash and seaweed) can be applied (UKROFS, 1993). It is essential that the available nutrients in the slurry and other farm wastes that are collected on the organic farm are used efficiently by matching the nutrient content and application rates of these wastes with the requirements of the individual crops (Lampkin, 1990).

Although perennial ryegrass is now the predominant species (more than 90%) of sown grass in the UK (Newton, 1993), the commercial varieties of perennial ryegrass currently available are normally selected on their perfor-mance in trials where relatively high inputs of nitrogen (300–350 kg ha^{-1}) are used. When Newton (1993) reviewed the results from trials comparing the herbage production from mixed grass/white clover swards he concluded that in systems based on low inputs of nitrogen, mixed grass species were preferable in both grass and grass/white clover swards. Other species, including cocksfoot (*Dactylis glomerata*), timothy (*Phleum pratense*), Yorkshire fog (*Holcus lanatus*) and red fescue (*Festuca rubra*), have the potential to produce similar or higher yields of herbage (grass only: 2.8–5.6; grass + clover: 5.2–9.3 t DM ha^{-1}) to swards based on perennial ryegrass (grass only: 2.4–5.7; grass + clover: 5.6–8.5 t DM ha^{-1}). Comparable liveweight gain (550–600 kg ha^{-1}) is also achievable on the species mentioned when no nitrogen fertilizer is applied. Newton (1993) also questioned the logic within an organic system of ploughing permanent pastures with low perennial ryegrass contents, as the role of grass in the sward needs to be determined for many factors, including time of onset of both early growth and maturity, winter hardiness, longevity, disease resis-tance, adaption to soil type, suitability for both grazing and conservation, ability to improve soil structure and compatibility with both legumes and herbs. Increasing species biodiversity within the sward by the inclusion of other grasses and in particular herbs in ryegrass/white clover leys may also have the potential to increase herbage palatability and improve nutritive balance of ingested herbage. Unless available herbage supplies are limited offering animals a choice of feeds may increase intake. However, further research is required to evaluate the effect of including other species in the sward on the nutritive quality and balance of the available sward and also on the dairy cow, in relation to the quantity and quality of milk produced. While little information is currently available on the effect of different herbs on either improving the soil structure or providing extra nutrients for the dairy herd, there is limited evidence to suggest that the inclusion of chicory (*Cichorium intybus*), salad burnet (*Poterium sanguisorba*), ribgrass (*Plantago lanceolata*), sheep's parsley (*Petroselinum crispum*) and yarrow (*Achillea millefolium*) in the sward may be beneficial (Newton, 1993), even if the herbs only survive in the sward for two to three years (Houghton and Poole, 1990).

Docks (*Rumex* spp.) and thistles (*Cirsium arvense*) are the main weed problems in grassland swards. Docks multiply by both seed germination (seed remaining viable for over 40 years) and tap root fragments and benefit from both open swards and the application of high quantities of slurry. Topping prior to seed dispersal is the normal control method, with severe infestations of docks reduced by ploughing and repeated cultivation during the summer. Thistles can also be controlled by topping in June or July at the early-flowering stage and the establishment of dense swards will prevent the rapid spread by creeping roots.

The forage requirements for the organic dairy farm should be produced within the farm, with forage providing at least 60% of the total daily feed requirements. The ratio of the area used annually for grazing and conservation may be similar to the 55:45 ratio suggested by Nix and Hill (1993) for conventional dairy farms, with a decline in the conservation area as herbage growth declines. Compared with a conventional dairy farm where the application of nitrogen fertilizer stimulates herbage growth in the spring, growth will be delayed on organic farms relying on white clover as the main source of nitrogen (Newton, 1993). Thus it is essential to ensure that adequate forage is conserved for the winter period. For example, silage requirements for a 190–210-day winter are between 2.5 and 3.5 tonnes DM per cow. A balance is required between cutting herbage to achieve maximum yield to ensure sufficient silage stocks are available for the winter period and conserving herbage of the highest quality which is required by cows in early lactation. Variations between seasons in herbage growth can be a major problem on all farms (Newton, 1993), with the problem potentially more acute for organic farmers both during and immediately after conversion. The high cost, impracticability and uncertain availability of purchased forage from another farm makes home production desirable. It is essential that a 'reserve' forage source in addition to the quantity of forage normally required is maintained on the farm at all times.

A major priority in grassland conservation on organic farms is to conserve adequate quantities of silage for the winter period at dry matter contents of >25%, to ensure large quantities of effluent are not produced. Wilkinson and Measures (1993) reported that many organic silages had poor fermentation quality with a high butyric acid content, low energy value due to a high stem-to-leaf ratio and low protein content. Frequently crops with a high stem content and a low legume content are ensiled. Although the lower dry matter and sugar content of clover can make the objective of good fermentation more difficult to achieve with mixed swards of grass and clover compared with pure grass swards (Thomas and Fisher, 1991), recent research demonstrates that high quality silage *can* be made on conventional farms from mixed herbage of grass and red or white clover (Bax *et al.*, 1993). The organic farmer is restricted to using bacterial inoculants, enzyme additives or molasses (UKROFS, 1993) to improve the fermentation process and is not allowed to use acid additives (e.g. formic acid) that are often used on conventional farms to improve the fermentation and intake potential of low dry matter herbage. High quality silages can be made on the organic farm with young legume-dominated swards by harvesting in good weather and wilting the herbage prior to ensiling. This increases the dry matter content (up to 35% is desirable), reduces protein degradation, improves the harvesting speed, improves preservation and reduces effluent production (Thomas and Fisher, 1991; Bax *et al.*, 1993; Wilkinson and Measures, 1993). With the exception of the types of additive that can be used, the principles for efficient silage making, as defined by Thomas and Fisher (1991), are the same for both organic and conventional farms.

Some forage crops may be suitable for growing on some organic farms, including forage maize, fodder beet and swedes, which require less than 170 kg ha^{-1} of nitrogen from soil organic matter and manure application (Aarts *et al.*, 1992). These crops provide high yielding feeds for the winter period that are both palatable and of high energy content (Lampkin, 1990) and are also less expensive to produce per MJ of metabolizable energy than grass silage (Nix and Hill, 1993). Introducing such crops may create problems in relation to weed control, the requirement for extra labour and specialist machinery, weather conditions at harvest and, with the exception of maize which can be stored in existing silos and either self-fed or mixed with the grass/clover silages, the need for a suitable storage area and mechanical method of feeding. Wilkinson and Measures (1993) suggested that another potential problem of these forage crops was that, within the constraints of the required organic standards, feeding these high energy feeds (11–14 MJ ME per kg DM) to an organic herd would exacerbate the problem of providing adequate protein in the diet. Growing kale for self-feeding would also help to increase the forage supplies but can create muddy conditions and an increased incidence of lameness.

Management of the Organic Dairy Herd

Milk production and milk composition

When comparisons are made between conventional and organic herds the lower milk yields of the organic herds (by 460–500 kg per cow) are generally attributable to lower inputs of concentrates. There are usually no significant differences in milk fat, protein or lactose concentrations, with the main differences being the lower stocking rate and milk production per hectare on organic farms (Houghton and Poole, 1990; Redman, 1992).

All breeds of dairy cattle appear to be equally suited to an organic system (Houghton and Poole, 1990) and throughout the world many different breeds are kept on organic dairy farms. Although in the UK Holstein-Friesian cattle are the main dairy breed, Macheboeuf *et al.* (1993) found that, compared with other breeds, the milk from Holstein cows had lower casein and calcium contents and poorer coagulation properties. The lower casein concentration increases the time required for achieving optimal clotting and curd firmness by about 12%. These results suggest that when the main objective is to produce organic milk for processing then breed type is a major factor to consider as the casein content (especially the β form of κ-casein) and ability of the milk to coagulate are important for efficient processing (Hardy and Skidmore, 1990). A comparison of milk from cows grazed on either grass or white clover swards suggests benefits of changing from grass to grass/clover swards in terms of significantly higher casein contents and improved curd firmness (Grandison *et al.*, 1983). Lund (1991) examined both conventional and organic milk from different breeds and concluded that despite the higher content of protein and vitamin C in the organic milk the differences between the two types of milk, in both technological properties and composition, were explained by differences in the diet of the two dairying systems.

A suitable diet for an organic dairy herd is high forage (>60%), nutritionally balanced and with high intake characteristics. This helps prevent the occurrence of health problems such as ketosis, acidosis, abomasal displacement, fatty liver, mastitis, infections of the reproductive tract and lameness. Many of these problems occur mainly in high yielding dairy cows fed imbalanced diets or diets containing high proportions of concentrates (Reid and Little, 1986; Webster, 1993). During the winter period the feeding on the organic dairy farms relies primarily on the forage component of the diet (e.g. grass/clover silage). The type and quantity of concentrates to be fed, for example cereal grains (oats, barley) or pulses, is best determined by the quality of the forage in relation to the concentrations of fermentable metabolizable energy and digestible undegraded protein (Wilkinson and Measures, 1993), level of intake and the stage of lactation of the cows. Both white and red clovers have lower cell wall contents than grasses and therefore have the potential to increase intake (Newton, 1993). For cows calving during the winter housing period, optimum use should be made of the available forage by feeding the highest quality forage to cows in early lactation when nutrient requirements are at their peak. If more than one forage is available for feeding during the winter period then offering the cows a choice of two forages allows cows to express individual forage and taste preference, leading to an increase in feed intake (+13%), milk production (+7%) and the yield of milk components (Weller and Phipps, 1986).

Feeding balanced high-forage rations will ensure that optimal milk production, milk composition, health status and reproductive performance will be achieved, but on some organic farms a potential problem is the inability to produce sufficient quantities of high energy forage (Houghton and Poole, 1990; Lampkin, 1990). Compared with silage made from grass swards, silage from mixed swards with 20–50% white clover content has been found to have higher protein, energy and mineral contents (Thomas and Fisher, 1991), higher silage intake characteristics (Castle *et al.*, 1983; Davies, 1991) with the potential to increase milk yield by up to 11% (Castle *et al.*, 1983; Davies, 1991) and improve milk composition (Thomas and Fisher, 1991). Red clover silage has also been found to increase intake with up to 13% more milk production (Bax *et al.*, 1993; Newton, 1993).

Feeding a high forage diet will also reduce the effect of the increased cost of purchased organic concentrate feed (12–16%) that reduces the profitability of many herds changing from conventional to organic concentrates. The results of a survey conducted by Redman (1992) comparing differences in milk yield between organic and conventional dairy herds showed that the lower milk yields recorded on the organic farms were not related to the change in farming system but were attributable to the lower inputs of concentrate feeds and higher quantity of forage consumed, with no significant differences in milk fat or milk protein.

Grazing management

The grazing management of the herd is critical if the available herbage is to be used efficiently during the growing season. Stocking rates will vary during the season depending on the growth and availability of herbage during the season, stage of lactation and milk yield of the herd and level of concentrate supplementation. Maximum milk production per cow is achieved at stocking rates lower than those required to achieve maximum milk yield and income per hectare (Mayne *et al.*, 1991).

While the priority in terms of farm profitability is to maximize the milk yield and income per hectare, it is important that stocking rates should also be determined in relation to the health and welfare of the dairy cow. It is essential to ensure that the body condition of the cows is maintained in the appropriate condition for the stage of lactation and no adverse effects on health occur due to a nutritional imbalance. These factors will help to ensure efficient reproductive performance of the herd is achieved. They are particularly important for the spring-calving herd during early lactation, as underfeeding during this period may result in both loss of body condition (Weller, 1993) and poor reproductive performance (Ducker, 1986).

Mayne *et al.* (1991) suggested that grazing pure grass swards down to either 8–10 cm for a rotational system with low sward density or 6–8 cm for a continuous system with greater sward density was a reasonable compromise between milk yield per cow and milk yield per hectare (sward heights being measured with a sward stick or ruler). Although there is currently little information on the optimum sward height for dairy cattle grazing grass/clover swards, Weller (1993), using a rising-plate meter, reported optimum milk yields per cow in rotational grazing at a residual grazing height of 6 cm (equivalent to a sward stick height of 8 cm), with optimum milk yield and income per hectare recorded at a residual sward height of 4.5 cm (sward stick 6.5 cm). Below these heights intake and milk yield will be reduced, while increased pasture residual heights will result in poor utilization of the available herbage, lower milk output per hectare and an increase in the proportion of grass in the sward. Nix and Hill (1993) estimated that increasing inputs of nitrogen fertilizer of 160, 220, 275 and 360 kg ha^{-1} on grass swards would sustain stocking rates of 1.65, 1.90, 2.20 and 2.50 dairy cows ha^{-1}. The results of the surveys by Houghton and Poole (1990) and Redman (1992) showed that the average stocking rate on organic dairy farms where white clover was the main source of nitrogen, was 1.6–1.8 cows ha^{-1} compared with an average of 2.2–2.4 cows ha^{-1} on the N-fertilized swards of conventional dairy farms.

When grazing swards of pure grass, cattle are not as selective as sheep. However, when herbage supplies are adequate some selectivity by cattle will occur even in swards of uniform height, allowing a greater proportion of leaf material to be consumed and increasing the digestibility of the diet (Phillips, 1993). Changing from pure grass swards to mixed swards of grass and white clover is unlikely to increase milk yield per cow (Mayne *et al.*, 1991) but will increase the potential for herbage selectivity and increased intake by the individual cow, even under a system of relatively tight grazing management.

For the high-yielding spring-calving cow the feeding value of low-quality grazed swards may be deficient in both energy and protein (Wilkinson and Measures, 1993), requiring the feeding of a concentrate supplement to ensure satisfactory levels of milk yield and protein content are achieved. Weller (1993) reported that with spring-calving dairy cows during the first year of conversion from conventional to organic farming, supplementing grazed swards of grass and white clover with 4.5 kg of concentrates per day increased the daily milk production per cow by 5.6 kg, income from milk sales per hectare by £212 and stocking rates by 1.2 cows ha^{-1}. However, the feeding of concentrates during the grazing season depends on the availability and quality of the herbage, since feeding concentrates for extra energy can substitute for herbage in the total DM intake at a time when available herbage is plentiful. By reducing fibre intake it may also reduce milk quality (Mayne *et al.*, 1991). Meijs (1986) reported a low substitution rate of herbage with a high-fibre concentrate (0.21 kg per kg of concentrate) compared with a starch-based concentrate (0.45 kg per kg of concentrate). Other factors to consider before a concentrate supplement is fed are the milk price during the grazing season and stage of lactation of the cows in the herd.

Mineral and vitamin intake should not be a problem on a well-established organic farm (Lampkin, 1990). Seaweed meal can be used to supplement the mineral intake of the herd if necessary. Although mineral imbalances have been reported during the conversion period, resulting in fertility problems (Lampkin, 1990), no problems during this period were recorded in the UK during the conversion of the dairy herd of the Institute of Grassland and Environmental Research to organic farming (R.F. Weller, 1994, unpublished data).

Animal welfare and health

In recent years, recommended guidelines have been published to improve the standards of welfare for dairy cattle. The standards of animal housing and welfare should be high on any dairy farm, whether conventional or organic, with good stockmanship based on disease prevention and the early detection of health problems. Webster (1993) stated that satisfactory production, health and welfare in the dairy cow could be met by the basic environmental requirements of thermal and physical comfort, hygiene, behavioural satisfaction and optimal productivity. In an organic dairy herd the emphasis is on creating an environment that leads to a lower incidence of disease, increased longevity and a reduced culling rate. The main reasons for culling cows in conventional herds are problems associated with mastitis, fertility and lameness, with more than 20% of the animals culled annually from many herds (Esslemont, 1993). These high culling levels, often associated with poor management and environmental problems (Lampkin, 1990) are unacceptable both in terms of farm profitability and animal welfare. In many countries cattle need to be housed during the winter period, without being constrained, in either a large straw-bedded area or individual straw-bedded cubicle stalls. Although housing density is higher and straw requirements lower with a system of cubicle stalls, there is an increased risk of lameness and the problem of slurry disposal.

In a three-year study in Germany, Klenke (1989) compared the health and performance of dairy cows in 27 organic herds with cows in 27 conventional herds and found that there were small but significant differences between organic and conventional herds in both average culling rate (20.8 vs 23.3%) and average age at culling (5.85 vs 5.43 years). Good standards of cleanliness, suitable housing during the winter period, early detection of health problems and the feeding of a balanced diet are vital. As shown in Fig. 16.2 there are important relationships between the environment in which the herd is kept, the health and condition of the individual cows and sources of infection. Poor standards of management and/or unhealthy cows can increase both stress levels and the risk of infection and disease. Cows need to build up a natural immunity to local disease risk and it is preferable therefore that all replacements are reared on the farm.

In terms of animal health the 40-year period between the introduction of penicillin for veterinary use in 1945 and the current availability of a large range of drugs available for treating and eradicating diseases represents a very rapid advancement in medical technology (Bramley and Dodd, 1984; Day, 1991). While conventional dairy herds routinely use conventional drugs to successfully control or eliminate specific disease problems without concern, in some herds increasing reliance on the routine use of prophylactic drugs, hormones and antibiotics to combat the problems of disease and poor fertility is unacceptable because of the high cost and often limited success in preventing disease problems associated with the more virulent strains of bacteria and

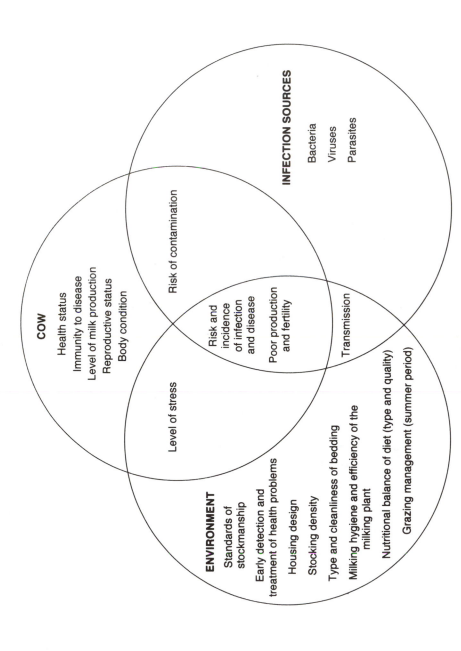

Fig. 16.2. The effects of the interaction between the environment and infection sources on the health and production of dairy cows.

viruses (Lampkin, 1990). For example, in many countries the bacterium *Staphylococcus aureus* is a major cause of mastitis, resulting in a chronic disease of long duration with only limited cure rate when treated with antibiotics (Bramley and Dodd, 1984). With many strains of *Staphylococcus aureus* now resistant to specific antibiotics, the cure rates are often as low as 25 and 50% for clinical and sub-clinical mastitis respectively (Bramley and Dodd, 1984). The increasing resistance of an individual bacterium strain to specific antibiotics is a serious problem that necessitates the continual development of new antibiotics to combat a wide range of diseases (Macleod, 1981; Day, 1991).

These factors have led to increased interest in other methods of treatment and the introduction of homeopathic remedies and herbal extracts as both preventative and curative remedies. However, for some veterinary surgeons in both European and North American countries the use of alternative remedies is controversial. The quality, safety and efficacy of the majority of alternative remedies, unlike conventional drugs such as antibiotics, have not been tested in clinical trials, are not registered by the Medicine Act and cannot therefore be classified as medicines (British Veterinary Association, 1991). Also, for farmers and herdsmen new to organic farming, alternative remedies require training to acquire the necessary skills and practical experience to ensure that the chosen remedy, potency level and the frequency and duration of treatment are appropriate for each specific health problem (Lampkin, 1990). Changing to alternative medicines is easier for those dairy farmers who have advice from a veterinary surgeon who is experienced in the use of these remedies.

In conventional herds, outbreaks of mastitis are normally treated with antibiotics, including the infusion of long-acting antibiotic during the dry period. Mastitis control in the organic dairy herd is the most difficult condition to prescribe for (Hansford, 1992) and early detection of the disease is the main aim. Treatments include the application of cold water to the udder, regular stripping-out of the affected quarters, the application of therapeutic creams and herbal remedies and the use of nosodes. The treatment of mastitis homeopathically is more complex than using antibiotics as the routine treatment depends both on the specific symptoms and also the response of individual cows to individual treatments (Lampkin, 1990). Among the remedies most commonly used are belladonna, apis mel, conium, bryonia, hepar sulph, phytolacca and calcarea fluorica.

As the experience and knowledge of some veterinary surgeons, farmers and herdsmen has increased, alternative remedies have been successfully used both in organic and conventional dairy herds. This has resulted in the publication of *some* literature on the prevention and treatment of diseases by alternative remedies (Macleod, 1981, 1983; Hansford, 1992). Both alternative remedies and conventional drugs are prescribed by members of the British Association of Homeopathic Veterinary Surgeons with many animals treated on a regular basis with alternative remedies, reserving conventional drugs for severe disease problems (British Veterinary Association, 1991). The clear priority for the organic dairy farmer is the welfare of the cow (UKROFS, 1993), the use of the most appropriate treatment for each specific ailment (i.e. either alternative remedies or conventional medicines) and obtaining the advice of a veterinary surgeon who is experienced in the concept and use of alternative medicines.

Future developments in organic dairy farming

Many aspects in relation to the physical and financial implications of converting to organic dairy farming are currently being evaluated, especially in Europe. The results from current research projects will help to provide practical guidelines that will be beneficial to dairy farmers both during and after the conversion period. High priority for future research include the effectiveness of alternative remedies in treating specific diseases and examination of a wider range of viable options for those undertaking conversion.

Converting to organic dairy farming is not a suitable option for those who cannot farm efficiently under conventional farming methods. Dairy farmers who consider converting are likely to be good farmers with a low input system, good quality land for forage production and in a reasonable financial state (Houghton and Poole, 1990). Depending on the previous farming system, for many farms converting to organic farming herbage production may be reduced and the choice will be either to increase the available land or reduce stock numbers.

The two major economic considerations in the comparison of products from conventional and organic units are the relative yields and price differences between the two systems (Nix and Hill, 1993). For an organic dairy farm to have acceptable profit margins comparable with conventional dairy farms, Houghton and Poole (1990) suggested a milk premium is required. In Denmark the number of organic farmers has increased rapidly in recent years, with some producers being paid a 40% premium for their organic milk, reflecting the willingness on the part of the consumer to pay up to 60% more for organic products (Dunn, 1991). A similar pattern has been found in Germany (Dunn, 1991) with premiums of 20% for many producers and consumers paying 30–50% more for organic dairy products. However, consumer concern about the use of artificial chemicals, animal welfare and environmental pollution needs to be balanced with the willingness and ability of the consumer to pay for a higher priced product. Two important questions to be answered are whether these increased prices for organic products will be maintained as the number of organic dairy producers increases and how much the market for organic products will expand in the next few years. In a UK survey by Redman (1992) the costings on 13 organic dairy farms were compared with conventional dairy farms recorded in the GENUS national herd recording scheme (Table 16.2). The payment of a premium markedly improved the margins per cow and per hectare, but the lower stocking rates recorded on the organic farms resulted

Table 16.2. Comparative margins for organic and national recorded herds in the UK (weighted average).

	National recorded herds[1]	Organic herds (no premium)	Organic herds (with premium)
Herd size	109	73	73
Yield ($l\,cow^{-1}\,yr^{-1}$)	5838	5436	5332
MMB milk price ($p\,l^{-1}$)	18.75	18.23	17.81
% milk sales with a premium	-	-	62
Premium paid ($p\,l^{-1}$)	-	-	5.21
Milk sales ($£\,cow^{-1}$)	1095	991	1123
Concentrate use			
$kg\,cow^{-1}$	1462	1141	1153
$kg\,l^{-1}$	0.25	0.21	0.22
Concentrate price ($£\,t^{-1}$)	132	132	154
Margin over concentrate ($£\,cow^{-1}\,yr^{-1}$)	902	840	945
Margin over purchased feed and fertilizer - MOPFF ($£\,cow^{-1}\,yr^{-1}$)	848	833	937
Stocking rate ($cows\,ha^{-1}$)	2.22	1.79	1.78
MOPFF ($£\,ha^{-1}\,yr^{-1}$)	1883	1490	1668
MOPFF ($£\,ha^{-1}\,yr^{-1}$) - assuming all organic milk receives premium			1854

[1] Genus Milkminder.
Source: Redman (1992).

in reduced returns per hectare. However, some of the differences in margins between the national recorded and the organic herds may be due to differences in herd size, as the financial performance of the individual organic herds improved with an increase in herd size.

The price of any premium paid is likely to vary according to the quality of the milk produced and the demand for organic milk. These factors will have a major affect on the accuracy of forecasting the potential profitability of an organic dairy herd.

Conclusions

The future income for a farmer converting to organic dairying will depend not only on the physical production and even level of supplies from the unit but also on the financial incentive of the proposed aid scheme during the conversion period, the level of premiums payable for organic milk products, successful organization of the marketing of the products and the purchasing of organic products by consumers. The introduction of further legislation in relation to environmental pollution and food safety and the policy for food production support in each country will have major impacts on the types of farming systems that can be operated in relation to both the practical and legal implications of managing different production systems, including organic systems. In the longer term the supply and price of renewable resources will also influence the type of system that is viable for the dairy farmer. Organic dairy farming has a role in the future of agriculture but expansion will depend not only on environmental factors but also on the financial viability of the system. Unless the producer receives a satisfactory premium for producing organic milk the profit margins on many farms will be unacceptable, except for those farmers who have diversified into other enterprises or part-time farmers for whom profit is not the motive for farming organically.

References

Aarts, H.F.M., Biewinga, A. and Van Keulen, H. (1992) Dairy farming systems based on efficient nutrient management. *Netherlands Journal of Agricultural Science* 40, 285–299.

Bax, J.A., Thomas, C. and Fisher, G.E.J. (1993) Practical experience of using legumes in silage. *Proceedings of the British Grassland Society Winter Meeting 1993*, Malvern, UK. BGS, Reading, pp. 63–73.

Bramley, A.J. and Dodd, F.H. (1984) Reviews of the progress of dairy science: mastitis control – progress and prospects. *Journal of Dairy Research* 51, 481–512.

British Veterinary Association (1991) BVA Council Report. *Veterinary Record* 128, 161–162.

Castle, M.E., Reid, D. and Watson, J.N. (1983) Silage and milk production: studies with diets containing white clover silage. *Grass and Forage Science* 38, 193–200.

Davies, O.D. (1991) The value of white clover in grass/clover silage fed to dairy cows. *Animal Production* 52, 589 (Abstract).

Day, C. (1991) An introduction to homeopathy for cattle. *New Farmer and Grower* Summer Issue, 21–23.

Doyle, C.J.(1984) Practical potential of legumes: An economic assessment. In: Thomson D.J. (ed.) *Forage Legumes*. British Grassland Society Occasional Symposium number 16. BGS, Reading, pp. 243–257.

Ducker, M.J. (1986) Herd reproductive performance. In: *Principles and Practice of Feeding Dairy Cows*. Technical Bulletin 8. NIRD, Reading, pp. 219–230.

Dunn, N. (1991) Organic milk – a growth market. *Dairy Industries International* 56, 14.

Esslemont, R.J. (1993) The scope for raising margins in dairy herds by improving fertility and health. *British Veterinary Journal* 149, 537–547.

Grandison, A.S., Manning, D.J. and Anderson, M. (1983) Flavour, composition and processing properties of milk from ryegrass and clover-fed cows. *Animal Production* 36, 502 (Abstract).

Hansford, P. (1992) *Introduction to Homeopathy* Helios Homeopathic Pharmacy, Kent, UK, 31 pp.

Hardy, A. and Skidmore, C.J. (1990) Kappa casein gene identification. *Holstein-Friesian Journal* 72, 471.

Holliday, R.J. (1989) White clover use on the dairy farm. *Milk Marketing Board Farm Management Services Report Number 65*. Milk Marketing Board, Thames Ditton.

Houghton, M. and Poole, A.H. (1990) Organic milk production. *Genus Information Unit. Report Number 70*. Milk Marketing Board, Thames Ditton.

Klenke, B. (1989) Investigations on cattle herds under different types of management with special reference to milk yield, health and fertility. PhD thesis, Tierarztliche Hochschule Hanover, Germany.

Lampkin, N. (1990) *Organic Farming*. Farming Press Books, Ipswich, UK, 701 pp.

Lund, P. (1991) Characterization of alternatively produced milk. *Milchwissenschaft* 46, 166–169.

Macheboeuf, D., Coulon, J.B. and D'Hour, P. (1993) Effect of breed, protein genetic variants and feeding on cows' milk coagulation properties. *Journal of Dairy Research* 60, 43–54.

Macleod, G. (1981) *The Treatment of Cattle by Homeopathy*. C.W. Daniel Company Limited, Essex, UK, 148 pp.

Macleod, G. (1983) A veterinary materia medica and clinical repertory. C.W. Daniel Company Limited, Essex, UK, 193 pp.

Mayne, C.S., Reeve, A. and Hutchinson, M. (1991) Grazing. In: Thomas, C., Reeve, A. and Fisher, G.E.J. (eds) *Milk from Grass*. pp. 53–71. Billingham Press, Cleveland, UK.

Meijs, J.A.C. (1986) Concentrate supplementation of grazing dairy cows.2. Effect of concentrate composition on herbage intake and milk production. *Grass and Forage Science* 41, 229–235.

Newton, J. (1993) *Organic Grassland*. Chalcombe Publications, Kent, UK, 128 pp.

Nix, J. and Hill, P. (1993) *Farm Management Pocketbook*, 23rd edn Wye College, Ashford, Kent, UK, 187 pp.

Phillips, C.J.C. (1993) *Cattle Behaviour*. Farming Press, Ipswich, UK, 212 pp.

Redman, M. (1992) Organic dairy costings: update. *New Farmer and Grower*. Winter Issue 1991/92, 20–21.

Reid, I.M. and Little, W. (1986) Health. In: Principles and practice of feeding dairy cows. *National Institute for Research in Dairying Technical Bulletin No 8*, pp. 231–247. NIRD, Reading, UK.

Scottish Agricultural College (SAC) and Scottish Agricultural Research Institutes (SARI) (1989) *Workshop on Priorities for Organic Research and Development*. Scottish Agricultural Colleges and Scottish Agricultural Research Institutes, Perth, Scotland, 65 pp.

Thomas, C. and Fisher, G. (1991) Forage conservation and winter feeding. In: Thomas C., Reeve, A. and Fisher, G.E.J. (eds) *Milk From Grass*. pp. 27–51. Billingham Press, Cleveland, UK.

UKROFS (1993) *The UK Register of Organic Food Standards – Standards for Organic Food Production*. UKROFS, London, UK, 88 pp.

Webster, J. (1993) *Understanding the Dairy Cow*, 2nd edn. Blackwell Scientific Publications, London, 374 pp.

Weller, R.F. (1993) Milk production during organic conversion. *Proceedings of the British Grassland Society Winter Meeting 1993*, Malvern. BGS, Reading, UK, pp. 119–120.

Weller, R.F. and Phipps, R.H.(1986) The effect of silage preference on the performance of dairy cows. *Animal Production* 42, 435 (Abstract).

Wilkins, R.J. (1993) Environmental constraints to production systems. *Proceedings of the British Grassland Society Winter Meeting 1993*, Malvern. BGS, Reading, UK, pp. 19–30.

Wilkinson, J.M. and Measures, M. (1993) Developments in feeding systems for dairy cows. *Proceedings of the 8th National Conference on Organic Farming*, Cirencester, UK, 1993.

Economic Aspects of Feeding Dairy Cows to Contain Environmental Pollution

P.B.M. Berentsen and G.W.J. Giesen

*Department of Farm Management,
Wageningen Agricultural University, Hollandseweg 1,
6706 KN Wageningen, The Netherlands*

Introduction

In intensive livestock farming regions a major cause of environmental acidification and pollution of ground and surface water is attributable to animal husbandry. In dairy farming one of the main pollutants is nitrogen (N), while phosphate (P_2O_5) is a problem on farms with a high intensity (Berentsen *et al.*, 1992; Tamminga, 1992). Dairy farming accounts for 56% of the total ammonia (NH_3) emissions in The Netherlands compared with 38% from other forms of animal husbandry (Groot Koerkamp *et al.*, 1990). Calculations based on data from Heij and Schneider (1991) indicate that NH_3 emitted in The Netherlands is responsible for 36% of the total acid deposition in The Netherlands; 44% of this total acid deposition is caused by emission abroad. A second problem related to N use is N pollution of ground water, caused by high concentrations of manure and fertilizer applied to farmland. In some areas with sandy soil, concentrations of up to $112\,mg$ of $N\,l^{-1}$ were recorded in the ground water (Goossensen and Meeuwissen, 1990). The maximum concentration allowed for drinking water in the European Union is $11.3\,mg$ of $N\,l^{-1}$ (Anonymous, 1980). Phosphate, originating from manure and fertilizer, is bound to soil particles which restricts leaching. However, if manure and fertilizer application is too great, the capacity of the soil will be exceeded in the long term and leaching of P_2O_5 will take place. Breeuwsma *et al.* (1990) estimated that 270,000 ha of grassland and maize land and 27,000 ha of arable land were saturated with P_2O_5.

In the 'manure action programme' of the Dutch government, reduction of the mineral content of fodder and improvement of the utilization of minerals in fodder are two important ways to reduce detrimental nutrient surpluses (Bloem, 1992). To reduce the mineral content of fodder, a system is needed to register the requirement and supply of minerals accurately.

This chapter quantifies possible consequences of decreasing nitrogen and phosphorus (P) input through feeding on labour income and on N and P_2O_5 losses on dairy farms situated on sandy soil. A modelling approach is used.

Feeding and the Environment

Three main environmental problems connected with dairy farming are losses of N to air and ground water and losses of P_2O_5 to ground water. To a certain extent these losses can be influenced by protein and phosphorus feeding of dairy cows.

Nitrogen losses and protein feeding

Protein fed to the dairy cow is partly used for growth and partly secreted in milk, while the remainder is excreted as organic and mineral N in manure. The concentration of organic N in the manure of dairy cattle varies only slightly with the ration (Valk *et al.*, 1990). This means that assuming a fixed milk and meat production, a change in the protein content of the ration leads to a change of the mineral N content of manure. Mineral N is mainly found in urine, whereas organic N is mainly found in faeces. When urine comes into contact with the open air, part of the mineral N is emitted as NH_3. On a dairy farm, NH_3 is emitted from the cattle housing, from the manure storage and when applying manure to the land when cows are kept indoors or from urine deposition when cows are grazing. Organic N and mineral N not emitted as NH_3 are supplied to the soil where they partly are used for growing grass and crops. The remaining N is lost through denitrification and leaching or is added to the pool of organic N in the soil. Since the N losses strongly depend on the N content of manure, the level of protein feeding influences the N losses.

For appropriate feeding of protein an adequate protein evaluation system is necessary. With concern to control N losses, the DVE (darm verteerbaar eiwit or intestine-digestible protein) system was introduced in 1991 in The Netherlands (Wever and van Vliet, 1991). The DVE system contains elements of the French PDI system (protéines digestibles dans l'intestin), the US AP system (absorbed protein) and the Scandinavian AAT-PBV system (amount of amino acids truly absorbed – protein balance in the rumen).

The DVE system consists of two parts. In the first part, requirement and supply of protein are balanced at the small intestine level. The protein supply is calculated in a similar manner to the French PDI system (Vérité *et al.*, 1979), except that feed protein values are corrected for intestinal endogenous losses.

The protein requirement for milk production (DVE milk), also determined by this system, depends totally on the amount of protein produced in milk (E) (Asijee, 1993):

$$DVE\ milk\ (g) = 1.396 \times E + 0.000195 \times E^2 (g\ day^{-1})$$

The requirement standards for maintenance, growth and gestation are taken from the US AP system and are adjusted to conditions in The Netherlands. The maintenance requirement (DVE maintenance) depends on body weight (BW, kg) (CBLF, 1991) and includes urinary endogenous N and surface protein losses (NRC, 1989), but not endogenous losses:

$$DVE\ maintenance\ (g\ day^{-1}) = (2.75 \times BW^{0.5} + 0.2 \times BW^{0.6})/0.67$$

The supply of intestine-digestible protein (DVE) is the sum of protein that is not degradable in the rumen but is digestible in the intestine plus intestine-digestible microbial protein, minus the protein needed to permit digestion, which consequently is lost in faeces and urine: the faecal metabolic protein. The protein that is not degradable in the rumen but is digestible in the intestine is calculated from the amount of crude protein in a fodder, the rumen degradability of this crude protein, and the intestine digestibility of the undegradable protein. The intestine-digestible microbial protein is calculated from the amount of rumen-degradable protein in the fodder and the energy available to form microbial protein in the rumen from this rumen-degradable protein.

The second part of the DVE system consists of a degradable protein balance (OEB, Onbestendig Eiwit Balans), which can be used to prevent unnecessary protein losses in the rumen. The rumen-degradable protein balance is taken from the Scandinavian AAT-PVB system (Madsen, 1985). The principle is quite simple. For the production of microbial protein in the rumen, rumen-degradable protein and energy are required at a certain ratio. A surplus of rumen-degradable protein leads to unnecessary protein losses, whereas a surplus of energy can give rise to digestibility problems in the long term. The OEB of a fodder is the amount of microbial protein that could be produced given the rumen-degradable protein in the fodder minus the amount of microbial protein that could be produced given the amount of usable energy in the fodder. Thus, a positive OEB indicates a protein surplus, and a negative OEB indicates an energy surplus. The OEB of a ration can be obtained by adding the OEB values of the different fodders in the ration. If the OEB of the ration is close to zero, the diet will be efficient nutritionally and environmentally.

Phosphate losses and phosphorus feeding

Like protein, P fed to the dairy cow is partly used for growth and partly secreted in the milk, while the remainder is excreted as P_2O_5 in manure. Phosphate from manure that is applied to the land can be utilized by grass or crops with the same efficiency as P_2O_5 from fertilizer (Asijee, 1993). This applies also to phosphate excreted while grazing in spite of its bad distribution. Phosphate is bound easily to soil particles and therefore it remains available for crops. If for several years more P_2O_5 is supplied with manure and fertilizer than is taken up by crops, the buffering capacity of the soil is exceeded and P_2O_5 starts to leach to the ground water. Thus P_2O_5 leaching is purely a surplus problem. Supply of more P_2O_5 than is required by crops can take place because of uncertainty of response, risk aversion, bad management or a need to get rid of a surplus of animal manure. In the first three cases, it is no use decreasing the amount of P in the diet if the use of P_2O_5 from fertilizer is increased at the same time. If uncertainty of response or the need for risk aversion is reduced or if grassland and crop management can be improved, the more economic way to decrease the supply is to cut back the use of P_2O_5 from fertilizer. Reducing P in the diet is only useful if a surplus of P_2O_5 from animal manure exists and no P_2O_5 from fertilizer is purchased. If this is the case then P_2O_5 leaching is a problem of intensity. This is confirmed by earlier calculations showing that dairy farms of an average intensity are not affected by severe legal limitations on the amount of P_2O_5 that can be used on the land (Berentsen et al., 1992). On intensive farms with a surplus of P_2O_5 from manure alone, reducing the amount of P in the diet is one of the two possible solutions to decrease P_2O_5 losses at farm level. The other solution is removal of manure from the farm, but since a P_2O_5 surplus exists at national level in The Netherlands this is not an easy solution.

Practical rationing of P is straightforward. A dairy cow requires an amount of P which depends on body weight (BW) and milk production (M) (Asijee, 1993):

$$P \ (g \ day^{-1}) = 0.042 \times BW \ (kg) + 1.5 \times M(kg)$$

This formula is based on an estimated absorption rate for dairy cows of 60% (Tamminga, 1992). This means that 60% of P supplied can be utilized by the cow. The required P has to be delivered in the ration.

Method

Analysis is based on two typical dairy farms situated on sandy soil. Both farms are characterized by a cultivated area of 24 ha and milk production per cow of $7500 \ kg \ yr^{-1}$. The first farm has an average intensity based on a quota of $12,000 \ kg \ ha^{-1}$. To analyse the effects of P feeding the second farm has a quota of $16,000 \ kg \ ha^{-1}$.

A linear programming model is used to model the dairy farms. The objective function maximizes labour income of the farm (i.e. total return to labour and management). The basic element in the model is a dairy cow, which is assumed to calve in February. This assumption is made to simplify modelling. Milk production per cow is fixed. A fixed ratio is assumed between the numbers of dairy cows and young stock. The cultivated area can be used for producing maize, fodder beet, and grass at increasing N levels (100, 200, 300, 400 and 500 kg ha^{-1}). The efficiency of N use decreases with increasing N use (Van der Meer and Van Uum Loohuyzen, 1986). In the model P_2O_5 is assumed to be utilized by grass and fodder crops with 100% efficiency which means that P_2O_5 losses can only exist as a result of a surplus of P_2O_5 from animal manure. A more extended description of the model (including values) is given in Berentsen and Giesen (1995).

The feeding part of the model is split up into four parts. The dairy cows and the young stock are fed separately, and a division is made between the summer and winter period. For this reason milk production is also divided into summer and winter production. For dairy cows, constraints are placed on the supply and requirement of energy, protein and P, on dry matter intake capacity, and on the requirement by the cows for fibre in the form of roughage. For summer and winter milk production the requirement for energy and the feed intake capacity are determined using formulas by Groen (1988). The formula Groen used for feed intake capacity is based on that of Jarrige *et al.* (1986) and ARC (1980). Given the calculated milk production per period and the lactation stage, protein and P requirements are calculated using the three equations given in the preceding sections. Constraints on fibre intake require that at least one-third of the ration dry matter be structural material (CBLF, 1991). Available fodders in the model and their characteristics are shown in Table 17.1. Prices are given only for fodders that can be purchased.

Because dietary requirements of young stock are usually less complicated, constraints are used only for energy and protein. Requirements are calculated using standards of the CBLF (1991). A fixed amount of milk powder and of starting concentrate per calf must be used. Available fodders for the summer period are grass at different N levels and standard concentrate. For the winter period grass silage at different N levels, maize silage, and standard concentrate are available.

The quantity of N in manure is calculated by subtracting N output in milk and meat from N input in feed intake. N in manure influences N lost through NH_3 volatilization, leaching, and denitrification. Other important factors that influence NH_3 emission are the method and length of storing manure and the manner in which manure is applied to the land. Here it is assumed that manure is stored partly under a slatted floor and partly in a closed manure storage for 6 months and that two-thirds of the manure is injected into the soil and one-third surface spread. This modelling interpretation of the Dutch manure legislation prescribes, among other requirements, that, in 1994, manure may only be applied on sandy soil during the growing season (February till September) and that surface spreading is allowed only during the last 2.5 months of the growing season (Bloem, 1992). Factors other than N content

Table 17.1. Different fodders with their DVE content, OEB values, P content, energy content, the yield of home grown fodders, prices of purchased fodders and the availability of fodders in summer (S) and winter (W).

Fodder	DVE content (g kg^{-1} of DM)	OEB (g kg^{-1} of DM)	P content (g kg^{-1} of DM)	Energy content[1] (MJ NEL kg^{-1} of DM)	Yield (MJ NEL ha^{-1})	Price available (Dfl[2]/MJ NEL)	Fodder availability
Grass 100 kg N ha^{-1}	93	10	4.0	6.66	50,025	–[3]	S
Grass 200 kg N ha^{-1}	98	28	4.0	6.72	63,480	–	S
Grass 300 kg N ha^{-1}	101	43	4.0	6.80	71,760	–	S
Grass 400 kg N ha^{-1}	104	60	4.0	6.87	76,590	–	S
Grass 500 kg N ha^{-1}	104	77	4.0	6.93	78,833	–	S
Grass silage 100 kg N ha^{-1}	65	20	3.8	5.80	50,025	–	W
Grass silage 200 kg N ha^{-1}	68	37	3.8	5.87	63,480	–	W
Grass silage 300 kg N ha^{-1}	71	55	3.8	5.93	71,760	–	W
Grass silage 400 kg N ha^{-1}	73	71	3.8	6.00	76,590	–	W
Grass silage 500 kg N ha^{-1}	74	84	3.8	6.07	78,830	–	W
Fodder beet	74	–51	2.3	6.21	100,740	–	W
Maize silage[4]	47	–16	2.2	6.21	93,150	0.051	S&W
Concentrate standard	100	–11	4.5	7.21	–	0.057	S&W
Concentrate low protein	89	–22	4.5	7.21	–	0.057	S
Concentrate high protein	200	139	11.0	6.90	–	0.074	W
Dried beet pulp	110	–70	0.8	7.12	–	0.058	S

Sources: Asijee (1993), CBLF (1991), Coppoolse *et al.* (1990) and Bloem and Kolkman (1992).
[1] Megajoule Net Energy for Lactation.
[2] Dfl. 1 equals about US$0.60.
[3] Not relevant.
[4] Maize silage can be grown on the farm or can be purchased.

of manure that influence leaching and denitrification are land use and the amount of N applied to grassland. Different forms of land use are available in the model, so that for example fodder crops that use nutrients more efficiently can replace less efficient crops. Lowering the N application on grassland decreases the N losses from grass production in spite of a decrease in grass production (Van der Meer and Van Uum Loohuyzen, 1986). To record N losses of the farm, the model contains an N balance.

The quantity of P_2O_5 in manure is calculated by subtraction of P output in milk and meat from P input in feed intake (in P_2O_5 equivalents). Phosphate from manure is utilized with the same 100% efficiency by grass and crops as P_2O_5 from fertilizer.

To examine the possibilities offered by the use of the DVE system the farm with an average intensity is optimized for labour income without and with restrictions on OEB. The intensive farm is optimized for labour income without and with restrictions on OEB, on P_2O_5 losses and on P_2O_5 losses and OEB at the same time, thus showing the impact of protein and P feeding.

Results of the Model

Average intensity farm

In Tables 17.2 and 17.3 two situations are compared, one where the OEB of the rations is not restricted and one where the OEB of summer and winter rations is restricted to a maximum of 0.2 kg protein day^{-1} per cow. Due to variation in dry matter intake capacities in the dairy cattle and due to variation in OEB values of fodders it is difficult to reach an OEB of zero for all cows.

In both situations the full milk quota is used, which, with the given milk production per cow, means keeping 38.4 dairy cows. Table 17.2 shows the winter and summer feed rations for the dairy cows and the land use. In Table 17.3, economic and environmental consequences are given.

Rations and land use

Given the intensity of the farm, the cheapest summer ration, which is used in the basic situation, consists mainly of grass. To encourage cows to come to the milking parlour a minimum of 1 kg concentrates is advised. The OEB shows that this ration contains a surplus of rumen-degradable protein (Table 17.2). The P surplus amounts to 15 g day^{-1} which is 26% of the P requirement. The winter ration in the basic situation consists of a minimum amount of grass silage, maize silage and two kinds of concentrates. This ration meets the requirements for energy and protein exactly. The basic situation rations strongly influence land use, consisting of growing grass and maize silage, and a required N level on grassland of 303 kg ha^{-1} which includes mineral N from

Table 17.2. Summer and winter feed rations for dairy cows and land use on a farm with an average intensity using different OEB restrictions.

	Basic situation No OEB restrictions	Maximum OEB winter and summer ration of 0.2 kg cow^{-1} day^{-1}	% change
Summer ration (kg of DM cow^{-1} day^{-1})			
Grass	17.8	15.4	−13
Concentrate standard protein	1	0	−100
Dried beet pulp	0	3.3	–
OEB of this ration (kg protein day^{-1})	0.76	0.2	−74
P surplus of this ration (g day^{-1})	15	4	−73
Winter ration (kg of DM cow^{-1} day^{-1})			
Grass silage	2	2	0
Maize silage	6.4	6.7	5
Concentrate standard	7.6	7.2	−5
Concentrate high protein	0.5	0.6	20
OEB of this ration (kg protein day^{-1})	0	0	0
P surplus of this ration (g day^{-1})	6	7	17
Grassland (ha)	19.2	18.8	−2
N applied to grassland (kg ha^{-1})	303	222	−27
Maize silage (ha)	4.8	5.2	8
Maize silage purchased (ha)	0.6	0.3	−50

Table 17.3. Economic and environmental effects of using different OEB restrictions on a farm with an average intensity.

	Basic situation No OEB restrictions	Maximum OEB winter and summer ration of 0.2 kg cow^{-1} day^{-1}	% change
Economic results (Dfl)			
Gross returns	256,049	256,049	0
Variable costs			
Purchased concentrate	25,070	31,929	27
Purchased roughage	2326	1142	−51
Purchased fertilizer	5876	4231	−28
Growing roughage costs (excl. fertilizer)	21,921	22,190	1
Other	33,842	33,842	0
Fixed costs	128,371	128,371	0
Total costs (excl. labour)	217,406	221,705	2
Labour income	38,643	34,344	−11
Nitrogen balance (kg ha^{-1})			
N input			
Purchased concentrate	63.0	72.8	16
Purchased roughage	4.5	2.2	−51
Purchased fertilizer	205.3	143.8	−30
Deposition	49.0	49.0	0
N output (milk and meat)	70.4	70.4	0
N losses	251.4	197.4	−21
of which NH$_3$ emission	39.2	35.4	−10
P$_2$O$_5$ losses (kg ha^{-1})	2.5	2.3	−8

manure. Fodder beet is not grown on the farm because it is quite expensive to grow and harvest and requires additional investment in machinery for feeding. To fulfil the roughage requirement a small amount of maize silage has to be bought.

Tightening the OEB surplus to 0.2 kg day^{-1} per cow drastically changes the summer ration (Table 17.2). The amount of grass in the ration decreases by 13% but more importantly the N use on grassland decreases by 27%. The N level of grassland has a great influence on the OEB of grass (see Table 17.1). Furthermore, dried beet pulp is used in the ration since it has a high negative OEB. This results in a summer ration with the maximum OEB allowed of 0.2 kg. The small changes in the winter ration arise from the decreased N use on grassland. The decrease in protein from silage grass is compensated for by an increase in high protein concentrate.

Economic and environmental aspects

Table 17.3 shows returns and costs resulting in an assessed labour income of the farm (i.e. total return to labour and management) in the basic situation of Dfl. 38,643 (US$23,185). In the OEB restricted situation, the changes in rations cause an increase in the cost of purchased concentrates (including dried beet pulp) by 27%, which is partly compensated for by lower costs of purchased roughage and fertilizer. The decrease in fertilizer costs is caused by the lower N level of grassland. As a result labour income of the farm decreases by Dfl. 4300 (US$2580, 11%).

As could be expected, feeding protein according to the OEB standards decreases the N input at farm level. Although input of N through concentrates increases, input through fertilizer decreases much more, resulting in a decrease of N losses by 54 kg ha^{-1}. This decrease is mainly the result of a decrease of the N applied to grassland which leads also to a reduction of the amount of protein fed. Grass production at higher N levels is less efficient in N use. Phosphate losses are small and, since P$_2$O$_5$ fertilizer is purchased in both situations, it makes no sense to decrease the P surplus in the rations.

Sensitivity of labour income of the farm and N losses to decreasing OEB requirements

To get an impression of the effect on N losses and labour income of the farm when the OEB requirements are tightened, the upper OEB limit for the summer as well as for the winter ration was decreased stepwise from 0.8 to 0 kg day^{-1} per cow in steps of 0.1 kg. Figure 17.1 shows the consequences for N losses and labour income. At an OEB limit of 0.8 kg day^{-1} the basic situation prevails. Labour income then decreases more than proportionally with a decreasing upper OEB limit. This occurs because the cheapest means of meeting the requirements are used first. The N losses decrease sharply until an OEB limit of 0.4 kg day^{-1} is reached. Beyond this point further decrease is small. In the first part the decrease of OEB input in summer is reached mainly

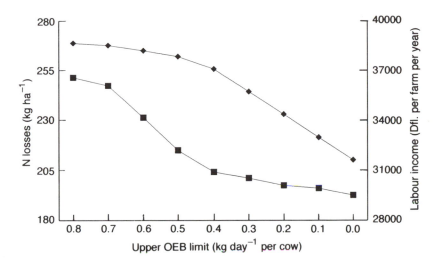

Fig. 17.1 Labour income (◆) and N losses (■) at different upper limits for rumen-degradable protein balance (OEB).

through a decrease of the N applied to grassland. To maintain total grass production the area of grassland is extended and the area of maize is decreased. The decrease in maize silage production is compensated for by increased purchase of maize silage. A decrease of the N application to grassland has a great influence on N losses because not only does N input decrease but also the efficiency of the use of N by grass increases (Van der Meer and Van Uum Loohuyzen, 1986). Beyond an OEB of 0.4 kg day^{-1}, N application is not changed further and OEB input in summer is now decreased by replacing grass in the summer ration by dried beet pulp. Thus less grass is produced, production of maize silage increases again and purchase of silage maize decreases. At farm level N input from purchased maize silage is replaced by lower N input in the form of dried beet pulp. However, this difference in N input is relatively small.

The high intensity farm

Tables 17.4 and 17.5 show the results of a farm with a high stocking rate using different restrictions on maximum OEB and P_2O_5 losses. As previously explained, if a P_2O_5 surplus exists and no P_2O_5 fertilizer is purchased, the surplus can only be decreased by decreasing the P input of the cows. A P_2O_5 loss of 5 kg ha^{-1} is considered the maximum acceptable loss by the Dutch government (Anonymous, 1993) and is used here as a constraint.

In all situations the milk quota is fully used which means that the farm

Table 17.4. Summer and winter feed ration for dairy cows and land use on a farm with a high intensity using different restrictions on OEB and on P_2O_5 losses.

	Basic situation no additional restrictions	Maximum OEB winter and summer ration of 0.2 kg cow^{-1} day^{-1}	Maximum P_2O_5 losses of 5 kg ha^{-1}	Maximum P_2O_5 losses of 5 kg ha^{-1}, maximum OEB of 0.2 kg cow^{-1} day^{-1}
Summer ration (kg of DM cow^{-1} day^{-1})				
Grass	17.6	14.9	14.1	12.6
Concentrate standard protein	1	0	0	0
Concentrate low protein	0	0	0	1.2
Dried beet pulp	0	3.8	4.3	4.9
OEB of this ration (kg protein cow^{-1} day^{-1})	1.05	0.20	0.55	0.20
P surplus of this (g cow^{-1} day^{-1})	14	3	0	0
Winter ration (kg of DM cow^{-1} day^{-1})				
Grass silage	2	2	6.1	4.4
Maize silage	6.4	6.7	4.6	3.7
Fodder beet	0	0	0	3.4
Concentrate standard protein	7.6	7.2	5.9	5.0
Concentrate high protein	0.4	0.6	0	0
OEB of this ration (kg protein cow^{-1} day^{-1})	0.02	0	0.39	0.02
P surplus of this ratio (g cow^{-1} day^{-1})	5	7	4	0
Grassland (ha)	24	24	24	21.6
N level grassland (kg ha^{-1})	400	237	423	338
Fodder beet (ha)	0	0	0	2.4
Maize silage purchased	7.2	7.3	5.7	5.0

has 51.2 cows. Table 17.4 shows the winter and summer feed rations for the dairy cows and land use. In Table 17.5, economic and environmental consequences are presented.

Rations and land use
Summer and winter rations in the basic situation resemble the rations on the average farm (Table 17.2). The small differences are caused by the higher N application to grassland. Grass grown at a higher N level contains more energy and protein per kg dry matter and has a higher OEB (Table 17.1). On the intensive farm, grass is grown at an N level of $400 \, kg \, ha^{-1}$ to ensure enough grass production given the greater number of cows. For the same reason the total area is used as grassland. To fulfil the feeding requirements, 7.2 ha of maize silage is purchased.

A maximum OEB of $0.2 \, kg \, day^{-1}$ changes the summer ration in the same way as it did on the average farm. Dried beet pulp takes the place of grass and grass is grown at a lower N level which decreases the OEB of grass. The small changes in the winter ration follow from the decreased N application to grassland. It should be noticed here that the decrease in N use as well as in the OEB of the summer ration is much bigger on the intensive farm than on the average farm.

On the intensive farm a P_2O_5 surplus of $20 \, kg \, ha^{-1}$ exists when no P_2O_5 fertilizer is purchased (Table 17.5). A maximum P_2O_5 loss of $5 \, kg \, ha^{-1}$ can be reached by changing the rations (Table 17.4). In the summer ration, dried beet pulp, which has a very low P content (Table 17.1), takes the place of grass, thereby decreasing the P surplus of the summer ration to zero. In the winter ration, part of the concentrate is replaced by grass silage which has a lower P content. To fulfil the DVE requirements, grass is grown at a higher N level and in addition, part of the maize silage is replaced by grass. As a result the amount of maize silage purchased decreases.

When restrictions are placed on the OEB of the rations and on the P_2O_5 surplus at the same time, summer and winter rations both change land use. To get a low OEB in the summer, dried beet pulp enters the ration and the N level of grassland is decreased. The latter, however, results also in a lower DVE content of grass silage and this affects the DVE supply in the winter ration. In the winter ration, a fodder with a negative OEB, a low P content and a high DVE content is needed to solve the problem. Fodder beet, which has these qualities (Table 17.1), is therefore used in the winter ration. 2.4 hectares of fodder beet are grown on the farm. As a result there is not enough grass for the summer ration and concentrate is used instead. In this situation, the P surplus of both the summer and winter ration is decreased to zero, thereby reaching a P_2O_5 loss of $5 \, kg \, ha^{-1}$. It is necessary to decrease the P surplus in the rations to zero because fodder beet requires less P_2O_5 for growing than grass.

Comparing the rations in the different situations, it is obvious that in the

Table 17.5. Economic and environmental effects of using different restrictions on OEB and P_2O_5 losses on a farm with a high intensity.

	Basic situation, no additional restrictions	Maximum OEB winter and summer ration of 0.2 kg cow^{-1} day^{-1}	Maximum P_2O_5 losses of 5 kg ha^{-1}	Maximum P_2O_5 losses of 5 kg ha^{-1}, maximum OEB of 0.2 kg cow^{-1} day^{-1}
Economic results (DFl)				
Gross returns	341,399	341,399	341,399	341,399
Variable costs:				
Purchased concentrate	33,408	44,408	39,479	42,766
Purchased roughage	28,274	28,664	22,333	19,607
Purchased fertilizer	8489	4306	9033	6612
Growing roughage costs (excl. fertilizer)	13,504	11,854	19,619	24,598
Other	44,272	44,272	44,272	44,272
Fixed costs	145,367	145,367	145,367	147,957
Total costs (excl. labour)	273,314	278,871	280,103	285,811
Labour income	68,086	62,528	61,296	55,588

Nitrogen balance (kg ha^{-1})

N input:				
Purchased concentrate	83.7	100.0	80.3	84.9
Purchased roughage	54.5	55.2	43.0	37.8
Purchased fertilizer	307.6	156.0	327.3	239.6
Deposition	49.0	49.0	49.0	49.0
N output (milk and meat)	93.9	93.9	93.9	93.9
N losses	400.9	266.4	405.7	317.4
of which NH$_3$ emission	55.2	47.3	59.0	48.9

Phosphate balance (kg ha^{-1})

P$_2$O$_5$ input:				
Purchased concentrate	40.9	40.3	29.9	31.7
Purchased roughage	18.9	19.1	14.9	13.1
P$_2$O$_5$ output (milk and meat)	39.8	39.8	39.8	39.8
P$_2$O$_5$ losses	20.0	19.6	5.0	5.0

summer ration dried beet pulp serves both the objectives of decreasing the OEB as well as decreasing the P_2O_5 losses. In the winter ration, fodder beet can do this but because growing and feeding fodder beet is quite expensive, they are used only if no other solution is available.

Economic and environmental aspects

Table 17.5 shows returns and costs resulting in a labour income of the intensive farm in the basic situation of Dfl. 68,086 (US$40,850). When this is compared with the income on the average farm (Table 17.2), it appears that the additional quota of 92,000 kg on the intensive farm results in an income that is Dfl. 29,443 (US$17,665) greater, i.e. an extra income of Dfl. 0.32 (US$ 0.19) kg^{-1} additional quota (milk price in The Netherlands is about Dfl. 0.80 (US$0.48) kg^{-1}).

Comparing returns, costs and labour income in the situation with a maximum OEB of 0.2 kg day^{-1} with those in the basic situation, cost of concentrates (including beet pulp) are higher while costs of fertilizer are lower. The fall in grass production is compensated for by extra concentrate purchase. Costs of purchased roughage remain almost the same. Costs of growing grass decrease as grassland used at a lower N level needs to be renewed less often. The changing costs result in a labour income which is Dfl. 5558 (US$3335) lower than in the basic situation. Compared with the average farm, most changes are greater due to the higher N level on grassland in the basic situation.

The restriction on P_2O_5 losses increases the costs of concentrate. Due to the increased amount of concentrate fed and to the slightly greater grass production (at an increased N level) the costs of purchased roughage decrease substantially. The increased N level increases the costs of fertilizer. The costs of growing roughage are increased by the increased N level and by the greater amount of grass that is ensiled. As a result, labour income is Dfl. 6790 (US$4047) less than in the basic situation.

The combination of a restriction on OEB and on P_2O_5 losses increases costs of concentrates while costs of purchased roughage decrease by the same amount. Due to the decreased N applied to grassland and the lower nutrient requirement of fodder beet, the fertilizer costs decrease. The high costs of growing and harvesting fodder beet increase the costs of growing roughage, and the additional investment in machinery for feeding increases fixed costs. As a result, labour income of the farm is decreased by Dfl. 12,498 (US$7499).

The nutrient balances show a loss of N of 400.9 kg ha^{-1} and a loss of P_2O_5 of 20 kg ha^{-1} in the basic situation. An upper OEB limit of 0.2 kg day^{-1}, increases N input through feed by 17.1 kg ha^{-1}. The N input through fertilizer decreases by 152 kg ha^{-1}, total N losses decrease by 135 kg ha^{-1} while P_2O_5 losses are hardly affected. Compared with the average intensity farm, the decrease in N loss is almost three times as high on the intensive farm. This is mainly caused by the N application to grassland in the basic situation which is 97 kg ha^{-1} greater on the intensive farm. Obviously a higher N level on grassland (with a lower N efficiency) offers more possibilities to reduce N losses.

A restriction on P_2O_5 losses has only a small influence on N input and N losses. Nitrogen input in purchased roughage is decreased but fertilizer input is increased by almost the same amount. Phosphate input is influenced by replacement of concentrate with a high P content by beet pulp with a low P content. Phosphate input through purchased roughage decreases with the amount of maize silage purchased, as a result the required P_2O_5 losses of 5 kg ha^{-1} are realized.

Restricting both the OEB and P_2O_5 losses decreases the N input through purchased roughage and through fertilizer. The result is a decrease of the N losses by 83.5 kg ha^{-1}. This decrease is far smaller than in the situation with only a restriction on OEB. In the latter situation the OEB limit is mainly achieved by decreasing the N application to grassland followed by changes of the ration. With both restrictions the OEB limit is met by a greater change of the rations and by a smaller change of the N application to grassland. A decrease of the N applied to grassland has a greater influence on N losses than a change in the ration. The difference in P_2O_5 losses is easier to explain. The input from concentrate as well as purchased roughage decrease in about the same way as in the situation with a restriction only on P_2O_5 losses. Consequently P_2O_5 losses of 5 kg ha^{-1} are reached.

Discussion

Labour income is maximized given the restrictions imposed by the feeding standards among other things. One could argue that, in reality, farmers strive to feed according to standards but due to various reasons (i.e. imperfect information, risk aversion, etc.) often do not succeed, making the absolute value of the results questionable. In this chapter, however, the focus is more on differences between situations than on the absolute results. If the difference between reality and model calculations does not differ between situations, then the calculated changes between situations give a good estimation of what could be achieved in reality.

In the calculations the rations of young stock are kept constant. For the OEB of the winter ration as well as for P of the winter and summer rations this is sensible because the OEB of the winter ration is approximately zero and the P surpluses of the summer and winter ration are small. The OEB of the summer ration, which consists only of grass, is high. This offers further possibilities for reducing the N losses. In practice, however, young stock usually graze all summer and are not fed supplements because it would involve too much labour. If young stock could be fed according to the OEB requirements, N losses would be reduced further.

To simplify modelling, all dairy cows are assumed to calve in February, which means that about 45% of the milk is produced in the winter period, and the rest is produced in summer. If the cows should calve earlier, the amount of milk produced in winter would increase, therefore the feed intake in winter would increase. From the results it appears that the summer ration of the dairy cows in the basic situation offers more possibilities to reduce N losses and P_2O_5 losses than the winter ration (OEB and P surplus are higher in summer). A shift of the feed intake from summer to winter would therefore reduce the possibilities to decrease these losses by different feeding.

The OEB values of different fodders given in Table 17.1 are averages. In practice the OEB for grass at a given N level can differ greatly, but the DVE value remains almost constant (Wever and Van Vliet, 1991). For concentrates, so far there are no standard values for OEB so, in practice these values too can differ from the values given in Table 17.1. The OEB of dried beet pulp depends heavily on the sugar content of the beet pulp. Since the sugar content may vary also the OEB may differ from the value given in Table 17.1. An optimal use of the protein evaluation system requires adequate information about the protein contents of fodders that are available for farmers.

With restrictions on P_2O_5 losses, in some cases the P surplus of the ration is zero which means that cows are fed exactly according to their requirement. One could argue that this is risky because due to differences in feed intake caused by age and lactation stage some cows could receive too little P while others get too much. However, when a cow gets a surplus of P, part of this surplus is stored in the skeleton of the animal which functions as a buffer for P (NARC, 1990). When P supply is inadequate, P can be mobilized from the skeleton which means that periods with a shortage of P supply through feed can be overcome. In addition, P requirements for dairy cows seem to be inadequately established as feeding standards for P differ widely among countries (Tamminga, 1992). Since The Netherlands use an intermediate standard this could mean that a safety margin has been included.

The imposition of an OEB of 0.2 kg day^{-1} decreases labour income per kilogram reduction of N losses by between Dfl 1.72 on the intensive farm to Dfl 3.32 on the average farm. Compared with other measures to reduce N losses on dairy farms (Leneman *et al.*, 1992) feeding according to these OEB standards is very cost-effective. For instance, measures to reduce N losses that require an investment in machinery or in adaptation of buildings have 5–25 times higher costs per kilogram reduction of N losses.

Decreasing P_2O_5 losses through a decreased P input by feed leads to an income loss of Dfl. 20 per kg reduction of P_2O_5 losses. A possible future alternative could be removal of the manure surplus to a manure processing plant with estimated costs of Dfl. 30 (18 US$) per tonne of manure (Luesink and Van der Veen, 1989). These costs must be corrected for extra purchase of N which is removed in the manure (about Dfl. 3 per tonne) and for avoided cost of applying manure to the land (about Dfl. 6.50 per tonne). As a tonne of manure from dairy cows contains about 1.8 kg P_2O_5, removal and processing of surplus manure seems to be a cheaper solution. However, manure processing is still in the experimental stage and the uncertain export possibilities of the product make the costs of removal and processing highly uncertain.

Conclusions

Tight OEB restrictions lead to considerable changes, particularly in the summer feed ration. The land use becomes more extensive (N application to grassland is lowered) and the amount of concentrate fed (especially dried beet pulp) greatly increases. This results in sharp decreases of both labour income and N losses. In particular the N application to grassland strongly influences N losses. The animal density on the farm also influences the N losses and the possibilities to reduce N losses by using tight OEB restrictions. The model calculations support use of the protein evaluation system to reduce N losses cost effectively.

Reduction of P input through feed to reduce P_2O_5 losses is only relevant for farms with a high stocking rate. Lower P_2O_5 losses can be reached by replacing fodders with a high P content like grass by fodders with a low P content like dried beet pulp. The result is a sharp decrease of income and a decrease of P_2O_5 losses.

A substantial decrease of N losses and P_2O_5 losses at the same time through feeding is only possible by feeding fodder beet in winter and dried beet pulp in summer. Both measures result in a significant decrease of income. As a result, farmers will have to be persuaded by financial stimuli or legal regulations to feed according to strict OEB requirements and with a low P surplus.

References

Agricultural Research Council (ARC) (1980) *The Nutrient Requirements of Ruminant Livestock.* Commonwealth Agricultural Bureaux and Agricultural Research Council, Farnham Royal, England.

Anonymous (1980) Council directive on the quality of water for human consumption. *Official Journal of the European Economic Community* 23, 229, 11–29.

Anonymous (1993) *Notitie Mest-en Ammoniakbeleid 3e Fase.* Dutch Parliament 1992–1993, 19 882, nr 34, The Hague, Netherlands (in Dutch).

Asijee, K. (ed.) (1993) Handboek voor de Rundveehouderij. Ministry of Agriculture, Nature Management and Fisheries, Lelystad, Netherlands (in Dutch).

Berentsen, P.B.M. and Giesen, G.W.J. (1995) An environmental-economic model at farm level to analyse institutional and technical change in dairy farming. *Agricultural Systems* 49, 153–175.

Berentsen, P.B.M., Giesen, G.W.J. and Verduyn, S.C. (1992) Manure legislation effects on income and on N, P and K losses in dairy farming. *Livestock Production Science* 31, 43–56.

Bloem, L. (ed.) (1992) *Hoofdlijnen van Beleid en Regelgeving; Mest-en Ammoniak-maatregelen.* Ministry of Agriculture, Nature Management and Fisheries, Ede, Netherlands (in Dutch).

Bloem, L. and Kolkman, G. (eds) (1992) *Kwantitatieve Informatie Veehouderij 1992–1993.* Ministry of Agriculture, Nature Management and Fisheries, Ede, Netherlands (in Dutch).

Breeuwsma, A., Reijerink, J.G.A. and Schoumans, O.F. (1990) *Fosfaatverzadigde Gronden in het Oostelijk, Centraal en Zuidelijk Zandgebied.* Report 68. Staring Centre,

Wageningen (in Dutch).

Central Bureau for Livestock Feeding (CBLF) (1991) *Voedernormen Landbouwhuisdieren en Voederwaarde Veevoeders*. Lelystad, Netherlands (in Dutch).

Coppoolse, J., Vuuren, A.M. van, Huisman, J., Janssen, W.M.M.A., Jongbloed, A.W., Lenis, N.P. and Simons, P.C.M. (1990) *De Uitscheiding van Stikstof, Fosfor en Kalium door Landbouwhuisdieren, Nu en Morgen*. Research concerning manure and ammonia problems in animal husbandry, Report no. 5, Wageningen, Netherlands (in Dutch).

Goossensen, F.R. and Meeuwissen, P.C. (eds) (1990) *Advies van de Commissie Stikstof*. Ministry of Agriculture, Nature Management and Fisheries, Ede, Netherlands (in Dutch).

Groen, A.F. (1988) Derivation of economic values in cattle breeding. A model at farm level. *Agricultural Systems* 27, 195–213.

Groot Koerkamp, P.W.G., Verdoes, N. and Monteny, G.J. (1990) *Naar Stallen met een Berkte Ammoniakuitstoot; Deelrapport Bronnen, Processen en Factoren*. Ministry of Agriculture, Nature Management and Fisheries, Wageningen, Netherlands (in Dutch).

Heij, G.H. and Schneider, T. (eds) (1991) *Dutch Priority Programme on Acidification*. National Institute of Public Health and Environmental Protection, Bilthoven, Netherlands.

Jarrige, R., Demarquilly, C., Dulphy, J.P., Hoden, A., Robelin, J., Beranger, C., Geay, Y., Journet, M., Malterre, C., Micol, D. and Petit, M. (1986) The INRA 'Fill Unit' system for predicting the voluntary intake of forage based diets in ruminants: a review. *Journal of Animal Science* 63, 1737–1758.

Leneman, H., Giesen, G.W.J. and Berentsen, P.B.M. (1992) *Costs of Reduction of Nitrogen and Phosphate Losses on Farms*. Research Report. Department of Farm Management, Agricultural University Wageningen, Netherlands (in Dutch).

Luesink, H.H and Van der Veen, M.Q. (1989) *Twee Modellen voor de economische Evaluatie van de Mestproblematiek*. LEI research report 47, Agricultural Economic Institute, The Hague, Netherlands (in Dutch).

Madsen, J. (1985) The basis for the proposed nordic protein evaluation system for ruminants. The AAT-PBV System. *Acta Agricultura Scandinavica*, Supplementum 25.

National Agricultural Research Council (NARC) (1990) *Handleiding Mineralenonderzoek bij Rundvee in de Praktijk*, 4th edn. NARC, The Hague, Netherlands (in Dutch).

National Research Council (NRC) (1989) *Nutrient Requirements of Dairy Cattle*, 6th revised edn. National Academic Press, Washington, DC.

Tamminga, S. (1992) Nutrition management of dairy cows as a contribution to pollution control. *Journal of Dairy Science* 75, 345–357.

Van der Meer, H.G. and Van Uum Loohuyzen, M.G. (1986) The relationships between inputs and outputs of nitrogen in intensive grassland systems. In: Van der Meer, H.G., Ryden, J.C. and Ennik, G.C. (eds) *Nitrogen Fluxes in Intensive Grassland Systems*. Developments in Plant and Soil Sciences, volume 23, 1–18.

Valk, H., Klein Poelhuis, H.W. and Wentink, H.J. (1990) *Snijmais of Krachtvoerbijvoeding naast Gras in het Rantsoen voor hoogproductief Melkvee*. Research Institute for Livestock Feeding, report nr. 213, Lelystad, Netherlands (in Dutch).

Verite, R., Journet, M. and Jarrige, R. (1979) A new system for the protein feeding of ruminants; the PDI system. *Livestock Production Science* 6, 349–367.

Wever, C. and Van Vliet, J. (1991) *Het Nieuwe Eiwitwaarderingssysteem voor Herkauwers*. Ministry of Agriculture, Nature Management and Fisheries, Ede, Netherlands (in Dutch).

Ethical Impacts of Biotechnology in Dairying | 18

T.B. Mepham

Centre for Applied Bioethics, Faculty of Agricultural and Food Sciences, University of Nottingham, Sutton Bonington Campus, Loughborough LE12 5RD, UK

Introduction

If science is concerned with acquiring knowledge of how things are, ethics is concerned with how we should respond to that knowledge. In the present context, our concerns form part of the field of bioethics, which may be defined for these purposes as: 'the systematic consideration of biotechnological knowledge and technique in the light of certain standards and values'. This definition gives bioethics a very broad remit: certainly much broader than the common perception that it is about objections to biotechnological practices which offend 'decency'.

Biotechnology impacts on dairying in an extraordinarily wide range of ways. In relation to breeding, it is represented by techniques such as artificial insemination (AI), semen preservation, semen sexing, synchronization of oestrus, multiple ovulation and embryo transfer (MOET), embryo sexing, *in vitro* maturation (IVM) and fertilization (IVF) of oocytes, cloning, nuclear transfer and gene transfer. In relation to nutrition, biotechnology has led, for example, to production of inoculants and enzymes for improving the nutritive value of silage, and the use of probiotics to enhance fibre breakdown and protein flow to the abomasum; while research is in progress to produce transgenic rumen bacteria to improve digestive capabilities. In terms of productivity promotion, biotechnology has achieved a high profile through the employment of recombinant bovine somatotrophin (rbST), which when injected into dairy cows can increase milk yield substantially. Moreover, the efficiency of traditional genetic selection programmes for milk yield has been enhanced by the inclusion in the selection index of specific biochemical markers. Biotechnology is also having an increasingly important role in animal health control by its use in production of vaccines, hormones and therapeutic proteins; and in the development of monoclonal antibodies and DNA probes used as diagnostic tools. Finally, a new style of 'dairying' employs genetically engineered animals to secrete in their milk novel pharmaceuticals and nutrients, or enhanced or diminished amounts of normal constituents.

Any analysis of the complex range of biotechnologies available or planned for use in dairying that is confined to a single chapter must necessarily be partial and superficial; and what follows is merely a preliminary step in a much larger programme. The principal aim is to propose a methodology for bioethical analysis and demonstrate its applicability to test cases.

It is important to establish at the outset that the aim is not prescriptive: readers will not find a list of which biotechnologies are 'ethically acceptable' and which not. Rather, the object is to clarify the issues by performing analyses that will facilitate resolution of ethical dilemmas by rational means. Thus, it is claimed that performing ethical evaluations of biotechnologies involves a three-stage process. At the first stage, it is necessary to establish both the relevant *scientific facts* and the appropriate *ethical criteria*. At the second stage, these two inputs are combined to provide an *ethical analysis*. At the third stage, the ethical analysis is interpreted in the light of certain attitudes and ideological viewpoints to produce, typically, a range of *ethical evaluations*.

An important characteristic of the approach to be adopted is that it relates to *social ethics*, which has a major bearing on the formulation of public policy. The highly pluralistic and multicultural character of modern society demands no less, for as Rawls (1993) has pointed out: 'For political purposes, a plurality of reasonable yet incompatible comprehensive doctrines is the normal result of the exercise of human reason within the framework of the free institutions of a democratic regime'. Clearly, in formulating public policy relating to the implementation of laws, codes of practice and financial subsidies, economics and public opinion (expressed most tangibly via the market) are of major importance, but often decisions are also crucially influenced by ethical concerns.

Because rationality is a necessary, but not a sufficient, element of ethical evaluation, an important question concerns how far an ethical analysis can proceed before it has to submit to the differing 'reasonable yet incompatible comprehensive doctrines' (one might call them 'world views') which characterize democratic societies. The author proposes that ethical analysis *per se* is impartial, so that in principle it should be possible to reach universal agreement at this stage even if evaluations subsequently diverge as different elements in the analysis are accorded different values. This, at any rate, is the project in hand: to perform ethical analyses of several biotechnologies which impact on modern dairying with a view to facilitating rational ethical evaluation and hence sound decision-making on their appropriate application.

Ethical Principles

Rational solutions to ethical dilemmas depend on appeal to ethical theories. In Western societies these are generally represented as belonging to one of two principal ethical traditions, namely consequentialism, of which the most prominent version is utilitarianism, and deontology, of which that articulated by Kant is exemplary. Consequentialist theories adjudge right and wrong actions in terms of their outcomes and, in the case of utilitarianism, the yardstick is net benefit. Deontological theories, on the other hand, assert the primacy of rights and duties irrespective of consequences: according to Kant one should 'act only on the maxim that you can at the same time will that it be a universal law', the so-called 'categorical imperative'.

In practice, it is almost impossible to sustain a personal ethical position based exclusively on either of these two theories. For example, the right of parents to indulge in a favourite pastime might conflict with duties towards their children, so that a utilitarian criterion of benefit (to the children), constraining the time spent on the leisure pursuit, would generally be considered ethically applicable. Not infrequently there is a tension between utilitarian and Kantian motivations and individuals seek to achieve a 'reflective equilibrium' by matching ethical theory with life experience. In the context of social and political life, an important recent ethical theory, advanced by Rawls (1973), interprets justice as 'fairness' and regards a fair system as one to which different parties would agree in advance without knowing how they, personally, would benefit.

Writing on medical ethics, Beauchamp and Childress (1989) suggested that the main tenets of consequentialist and deontological theories were encompassed by four principles and that a sound basis for patient care was to observe *prima facie* duties to respect these principles, namely nonmaleficence, beneficence, autonomy and justice. The former two principles relate to cost and benefit and thus correspond to the types of factor which would be important in a utilitarian ethical analysis; whereas the principles of autonomy and justice correspond more closely to Kantian criteria, although they may also form part of, so-called, rule-utilitarian theory. In more familiar terminology, the four principles equate approximately to harmlessness, kindness, freedom and fairness, respectively.

As a development of the approach of Beauchamp and Childress (1989), the author (Mepham, 1993, 1994, 1995) has applied the four principles separately to the interests of different groups affected by several biotechnologies. (Independently, a somewhat similar evaluative framework has been suggested by Brom and Schroten, 1993.) The same approach is pursued in this chapter, but in addition an attempt is made to compare the ethical impacts of the different biotechnologies by employing a diagrammatic summary. Moreover, since respect for beneficence and nonmaleficence are closely related, these two principles have been combined as *wellbeing*. Thus, the now three principles may be seen as representing three important strands of contemporary ethical theory: respect for wellbeing encompassing utilitarian criteria; for autonomy, Kantian criteria; and for justice (as fairness), Rawlsian criteria.

While it is clear that the mere application of ethical principles cannot, and perhaps should not, prescribe policy, the approach may nevertheless provide a framework for policy decisions as well as a means of making the reasoning employed by decision-makers more transparent to the public.

An Ethical Matrix

A crucial aspect of bioethical analysis pertaining to public policy is the need to consider the impacts of implementation of biotechnologies from the perspectives of different affected groups. While it would be possible to identify countless distinct interest groups, for simplicity, only four are considered here: animals; producers (usually dairy farmers); consumers of dairy products and meat produced within the industry; and biota, defined as 'the animal and plant life of a region' (*Longman's Dictionary of the English Language*, 1984). The perspective adopted is Eurocentric or, at least, Western and, inevitably, this limits the validity of the exercise. However, some general comments about the impacts of biotechnology in dairying on a global scale are made at the end of the chapter.

Table 18.1 suggests how the three ethical principles, i.e. (respect for) wellbeing, autonomy and justice, might be interpreted for the four different interest groups. Some comments are necessary on each of the resulting 12 principles in the table, which collectively form an *ethical matrix* for biotechnology evaluation.

Dairy animals
In modern industrialized society there has been a progressive awareness that animals have moral significance and that to cause them distress, show lack of respect for their intrinsic nature (termed *telos* by Aristotle), or use them as mere instruments of human desires, requires intellectual justification. Most people admit that use of animals, e.g. for food or in research, demands that a balance be struck between the importance of the human objective and the extent to which there is diminished respect for animals' wellbeing. Wellbeing (**DW**) and autonomy (**DA**) are protected in the UK by the 'five freedoms' defined by FAWC (the Farm Animal Welfare Council of the Ministry of Agriculture, Fisheries and Food), which form the basis of a code of practice for dairy farmers (MAFF, 1983). Other countries have similar provisions.

The notion of 'respect for telos' (**DJ**) requires more discussion than is possible here. Rollin (1989) has defined it as: 'the unique, evolutionarily determined, genetically encoded, environmentally shaped set of needs and interests which characterize the animal in question, (see also Verhoog, 1992). Suffice it to say that those who show respect for animal telos extend Kant's categorical imperative to nonhuman animals, considering it a duty that they regard animals, like people, as ends in themselves. Analyses of impacts in this category refer only to animals treated with the biotechnology in question.

Produceers
An important ethical aspect of any technology is the extent to which it affects the income and working conditions of the primary users (**PW**). Moreover, in view of the many objections which have been voiced against modern bio-technologies, producers should be free to choose whether or not to adopt them (**PA**). Respect for the principle of justice suggests that there should be fair treatment of dairy farmers, both in relation to the law and in commercial practice (**PJ**). Ethical analyses in this category include all dairy farmers, whether or not they adopt the technologies discussed. In one case discussed below the producers are pharmaceutical companies.

Consumers
In industrialized societies, in which food is produced, processed and marketed by largely anonymous, mass distribution systems, there is a clear need to ensure the supply of safe food to consumers (**CW**) by implementation of laws and codes of practice. Moreover, the diversity of opinion in society, e.g. with respect to attitudes to the treatment of farm animals, implies that customer freedom of choice needs to be ensured by adequate public information on new biotechnologies and by appropriate labelling of food (**CA**). The fact that food is a basic human need demands that its universal affordability be recognized as a fundamental right of all people (Byron, 1988) (**CJ**), although a right regrettably denied many people in the modern world. The analyses in this category refer to all consumers of dairy products, whether or not they choose to buy products of biotechnology, or are aware of such developments.

Biota
Inclusion of 'biota' as an entity worthy of respect in ethical terms may strike some readers as unjustifiable anthropomorphism. Yet consideration of this interest group would seem to be crucial in both utilitarian terms for humanity (in view of the essential role of wildlife populations in maintaining ecological stability) and for many people who consider that all life forms possess an intrinsic value worthy of moral respect. (Such concerns are more usually represented in terms of 'the environment' or 'the biosphere'.) Even from an exclusively human perspective, protecting (**BW**) and ensuring sustainability (**BJ**) of the biota, and maintaining biodiversity (**BA**), would appear to be essential for the wellbeing of current and future human populations. For some environmental ethicists, the moral imperatives are less anthropocentric but no less valid, and legitimize reference to the *biotic community* (Rolston, 1993).

Table 18.1. Definition of the ethical matrix. For details see the text. In subsequent analyses the 12 principles are referred to by their initials, i.e. **DW**, **BJ**, etc.

	Wellbeing	Autonomy	Justice
Dairy animals	**DW** Freedom from pain and stress	**DA** Behavioural freedom	**DJ** Respect for animal 'telos'
Producers (dairy farmers)	**PW** Adequate income and working conditions	**PA** Freedom to adopt or not adopt the biotechnology	**PJ** Fair treatment in law and trade
Consumers of dairy products	**CW** Availability of safe food	**CA** Respect for consumer choice	**CJ** Universal affordability of food
Biota	**BW** Protection of the biota	**BA** Maintenance of biodiversity	**BJ** Sustainability of biotic populations

In making ethical analyses based on this approach several points need to be borne in mind.

- The principles defined are *prima facie* duties: there will clearly be occasions when a trade-off has to be made in observing respect for one at the expense of another. It is indeed doubtful whether *any* technology could produce positive responses for all twelve principles.

- In the analyses to be described, impacts are compared with the situation existing without the technology. But in many cases the existing situation is far from ethically acceptable so that the mere absence of a negative impact on an ethical principle would hardly represent a satisfactory outcome.

- Any attempt to arrive at a summary position, e.g. to the effect that a particular biotechnology will adversely affect respect for **CW** (with reference to Table 18.1) is subject to limitations imposed by the extent of available knowledge. One is on slightly firmer ground in referring to *risk* – and this is the strategy adopted here.

- For reasons of space, the analysis omits factors that might be crucial in a comprehensive ethical analysis, e.g. impacts on less-developed countries. Brief reference to such issues is made at the end of the chapter.

- At its most basic level, the matrix defines a strategy for ethical enquiry, an alternative to those which conform to the norms of 'value-free' science, or which are aimed at commercial exploitability.

Table 18.2. An ethical matrix. Claimed ethical impacts of biotechnologies in dairying. Each cell of the matrix shows risks to the respect for ethical principles which the author claims are presented by the identified biotechnologies. Symbols: ○ indicates respect for the principle; ● indicates infringement of the principle; * indicates a negligible impact. Where ◑ is shown the biotechnology is considered to have both positive and negative ethical impacts. No grading of the impacts has been attempted. For fuller details and explanation of **DW**, **DA**, etc. see text and Table 18.1.

	Dairy cows			Producers			Consumers			Biota		
	DW	DA	DJ	PW	PA	PJ	CW	CA	CJ	BW	BA	BJ
AI/MOET	◑	●	●	◑	◑	*	*	●	○	●	●	●
Transgenesis hLF	●	●	●	○	○	*	◑	◑	◑	●	●	●
bST	●	●	●	◑	◑	●	●	●	*	●	●	●
Vaccine	◑	○	◑	◑	◑	*	○	*	*	○	○	○

In the following sections, the author examines the ethical impacts of a number of representative dairy biotechnologies, using the ethical matrix described above. Some of these are established biotechnologies, others prospective. Claimed summary 'ethical risk assessments' for these biotechnologies are shown in Table 18.2.

Reproductive Biotechnologies

Such technologies are wide ranging and are of ethical concern not only in themselves but because their adoption in agriculture often presages moves to apply them in human medicine.

In general, low conception rates and high rates of embryo mortality represent substantial problems to farmers wishing to maximize milk yield per cow, both by ensuring the birth of calves on an annual cycle and by sustaining a progressive increase in the genetic potential of the dairy cows. AI has been the single most important technique for genetic enhancement of milk yield, enabling spermatozoa from one highly selected bull to inseminate thousands of females (up to 50,000) per annum and facilitating the distribution of valuable genes, even to small farmers (Hafez, 1993). MOET has the advantage of promoting genetic improvements through both sire and dam; and the combination of AI and MOET is the means by which the newer genetic and reproductive technologies, such as embryo manipulation, will be propagated. The animal biotechnologist's agenda consists of producing, to order, animals of desired sex and productive potential, resistant to diseases and/or capable of secreting valuable pharmaceuticals in their milk (e.g. see Wall and Seidel, 1992; Robinson and McEvoy, 1993).

In the cases of AI and MOET the assessments are so similar that they have been combined in Table 18.2.

Artificial insemination

Animals

DW is both positively and negatively affected by AI. Natural service can be stressful and painful when there is a mismatch between the two sexes, particularly, for example, when a beef bull is used as a 'sweeper'. The poor detection of oestrus in heifers means that they are most likely to be served naturally and hence suffer vaginal and cervical bruising during copulation; and dystocia, or caesarian section, with the birth of overweight calves. Bulls on farms are often poorly managed, partly because their danger to humans discourages close attention to their welfare. AI may thus be thought to improve respect for **DW**. However, because semen of a single bull is used to inseminate many cows, genetic defects may spread rapidly through a wide population of cattle, and inappropriate use of semen can also produce oversize calves at the time of parturition. Moreover, the increased milk yields of cattle which result from 'genetic improvement' also increase the incidence of production related diseases, such as ketosis, mastitis and lameness; and the latter two may be exacerbated when modern, larger, cows occupy undersize cubicles, which force them to lie in slurry. Thus, from the animals' perspective it is doubtful whether on balance welfare is improved by AI.

Whether or not utilitarian criteria favour AI, the technique clearly fails to respect **DA**. Bulls and cows are denied normal sexual behavioural expression and some would consider the repetitive teasing of bulls prior to ejaculation and the practice of electroejaculation (Hafez, 1993) are serious violations of animal telos (**DJ**).

Producers

Dairy farmers benefit by access to a large variety of genes, which results in a higher profitability of their herds, and by the ability to avoid problems with keeping a bull (**PW, PA**). Nonadopters may, however, experience financial difficulty (**PW**) in maintaining their traditional farming practices (**PA**). Such difficulties are unlikely to be restricted to nonadopters of technology because, as the 'efficiency' of production increases, profit margins decline leaving many farmers victims of the 'technological treadmill'. It is difficult to decide whether this represents unfairness since many of these changes have been in progress for some time and a degree of accommodation has been achieved: impacts on fairness (**PJ**) are thus recorded as neutral.

Consumers

In itself, AI would appear to have had no significant effects on the safety of milk and meat produced in the dairy herd (**CW**). Impacts on affordability are difficult to separate from general economic trends but, in view of the substantial reduction over the last ten years in the average time taken to earn the price of a pint of milk (Milk Marketing Board, 1992), it seems likely that use of AI has increased respect for **CJ**. However, respect for consumer choice (**CA**) has hardly been observed, since it is almost impossible for a consumer to choose to purchase milk not produced by artificially inseminated cows.

Biota
Adverse impacts of AI on biota are indirect but no less important for that reason. For example in the UK, the high productivity of dairy herds depends *inter alia* on substantial use of concentrate feeds and silage, and high levels of artificial fertilizer application to promote forage growth (**BJ**). Economic factors have resulted in a reduction in herd numbers but in increasing herd size. This intensification of the dairy industry, in no small measure made possible by AI, has had significant detrimental impacts on biota (**BW, BA**).

Multiple ovulation and embryo transfer

Animals
There are no intrinsic benefits of MOET to either the donor or recipient cows and according to the chairman of the UK Royal College of Veterinary Surgeons' Embryo Transfer Working Party: '. . . the risks of pain, suffering and long-term harmful effects are that much greater in embryo transfer than in AI' (Lucke, 1991).

Superovulation involves injection of gonadotrophins, principally FSH or PMSG, to increase ovulation rate (up to 20), followed some days later by injection of prostaglandin to cause luteolysis. However, responses are very variable both in terms of ovulation rate and the yield of viable embryos. As noted by Armstrong (1993): '. . . the practice of superovulation is still a primitive art form that has not advanced appreciably since it was first performed over 60 years ago'. As cows may receive many injections, their blood hormone concentrations are substantially altered. Frequently, valuable cows are held back from normal breeding and submitted to repeated superovulation. After AI, embryos are flushed from the donor cow with the aid of a three-way flexible catheter incorporating a plastic balloon, inflation of which seals off the uterine horn into which the catheter is introduced. As the procedure takes place a week after oestrus, it is often difficult to penetrate the cervix, particularly in heifers, and there are risks of causing bleeding and uterine rupture. After flushing, the donor cow is administered prostaglandin (which may have adverse effects) to ensure the return of cyclicity and prevent embryos left in the uterus from causing a multiple pregnancy.

Embryo transfer into the recipient cow (usually hormonally synchronized by prostaglandin injection) is generally nonsurgical, though exceptionally, e.g. for 'valuable' embryos, surgical transfer via a flank incision is practised. For nonsurgical transfer, embryos are placed in an AI straw, which is loaded into an embryo transfer gun. Because the gun has to pass further into the uterus than for AI, great care is needed to avoid traumatizing the uterus and introducing infection (Sreenan and Diskin, 1992). Since such skills can only be acquired with much practice, it is likely that many cows experience significant degrees of stress. In the UK, under The Veterinary Surgery (Epidural Anaesthesia) Order 1992, epidural anaesthesia is now required for both nonsurgical embryo collection and transfer, although there is concern in the veterinary profession that the Order permits 'suitably-trained' non-veterinarians to administer anaesthetic and collect embryos (e.g. Coulthard, 1992). In the USA, 'Embryos are usually transferred surgically' (Hafez, 1993).

However, the most significant welfare question relates to the mismatch betwen the embryo and recipient cow. 'Major problems arise when embryos, obtained from large dairy breeds and particularly from large double-muscled beef breeds, are placed in recipients which are unlikely to be able to give birth to them and therefore require surgery to deliver the fetus' (Murray and Ward, 1993). Because breeders are loath to risk infertility in good breeding cows should surgical complications arise, use of 'expendable' maiden heifers has been commonplace. The consequences are illustrated by the report of a Simmental MOET study, in which 23% of heifer recipients experienced dystocia and 10% required caesarian section (Anon, 1992), compared with a rate of 1% in normally-bred cattle. When MOET is used to induce twin-calving the incidence of retained afterbirth can be quadrupled and calf mortality increased by 60% (Sreenan and Diskin, 1992).

MOET is still very inefficient: the median number of calves produced by superovulaton is two and the modal number is zero (Hafez, 1993). There is little doubt that MOET involves considerable risks to obligations to respect **DW** and **DA**.

Producers, consumers and biota
Essentially the same comments apply for MOET as for AI with respect to **PW**, **PA**, **PJ**, **CW**, **CA**, **CJ**, **BW**, **BA** and **BJ** (see above). Additionally, because of its abortifacient properties, administering prostaglandin can present risks to the welfare of pregnant women and their unborn children (**PW**).

Transgenesis

Transgenic cattle producing human lactoferrin

Although transgenesis has been largely applied to laboratory animals for medical research, a number of transgenic farm animals have been produced. The most notable instance in cattle is the production, in The Netherlands, of cows that are designed to secrete human lactoferrin (hLF) in their milk. This provides a good example for ethical analysis because it seems likely to presage a whole range of transgenic animals secreting pharmaceuticals and, so-called, nutraceuticals.

The Dutch Government has adopted a 'no – unless' policy with respect to animal transgenesis, i.e. there is a presumption that use of transgenesis in animals will be illegal unless a strong case can be made for its employment. In the latter case, the onus is on the proposer to establish the need for use of such animals. The so-called 'Herman the bull' project was approved on the grounds that hLF would be used in human and veterinary medicine as a 'possible treatment for sepsis' (Postma, 1992). But the subsequent disclosure that the company involved, Gene Pharming Europe, had a contract with a baby food manufacturer (B.V. Verenigde Bedrijven Nutricia) indicated that a quite different, and potentially more lucrative, market was being sought. The fact that hLF, a constituent of human breastmilk, is absent from cows' milk diminishes the suitability of baby foods manufactured from the latter as breastmilk substitutes. Thus, pharmaceutical companies envisage considerable financial profit in manufacturing a baby food which could be claimed to be closer in composition to human milk (Dutch Society for the Protection of Animals, 1994).

Officially, in The Netherlands, the objective of adding hLF to baby foods has been suspended. But for the purposes of this analysis it is still considered a likely objective for transgenic technology because, even if this particular proposal is not renewed, there is no reason to believe that other companies will be deterred by the far less restrictive legislation that applies outside The Netherlands.

Animals

Transgenesis in cattle generally involves microinjection of several hundred copies of the cloned prospective transgene into a pronucleus (made visible by high speed centrifugation) of an early embryo recovered from a superovulated cow. The efficiency of the process is very low, since even under the best conditions only 25% of transferred ova survive to term and less than 1% of the injected eggs develop to become transgenic offspring (Wilmut *et al.*, 1990). In the 'Herman project', oocytes were collected by aspiration of follicles from ovaries obtained from a slaughterhouse (Krimpenfort *et al.*, 1991). Following maturation (IVM), the oocytes were fertilized (IVF), microinjected and transferred to recipients as 9-day embryos.

Certain applications of animal transgenesis have a poor welfare record. According to Pinkert *et al.* (1990), pioneers of transgenesis in farm animals: 'Essentially all transgenic GH (i.e. with additional growth hormone genes) pigs and sheep produced to date have had serious physiological as well as anatomical problems . . . including altered endocrine profiles and metabolism, insufficient thermoregulatory capacity, joint pathology (lameness and arthritis), low libido, infertility and an increased susceptibilty to pneumonia'. Thus, it is clear that animal transgenesis can seriously undermine welfare.

However, welfare problems are claimed to be absent in transgenic sheep in which expression of the transgene is tissue specific, e.g. when the human α-1 antitrypsin (hα_1-AT) gene was fused to the functional ovine β-lactoglobulin gene promoter (Wright *et al.*, 1991). One sheep secreted up to $35\,\mathrm{g\,l^{-1}}$ hα_1-AT in its milk. On this evidence, transgenic cattle secreting hLF might be thought unlikely to suffer diminished welfare. What should not be overlooked is that the large numbers of fetuses and liveborn animals which do not express the transgene may suffer decreased welfare. Moreover, embryo culture often produces overweight offspring, which results in dystocia or the need for caesarian delivery (Heap and Moor, 1995).

Welfare problems may also arise in the longer term. Transgene incorporation into the genome is a random process and multiple copies of transgenes are usually found as tandem repeats at a single insertion site. The phenomenon of pleiotropy means that animals may suffer quite unpredictable effects and these may also occur as a result of mutations in endogenous genes which result from inaccurate repair of chromosomes broken in the microinjection process (Wilmut, 1995). Even if no adverse effects are apparent in the first generation of transgenic animals there is no reason to assume they will not occur in subsequent generations. Several expert bodies in the UK (such as FAWC and the British Veterinary Association) have expressed concern over the inadequate protection afforded to transgenic animals by current legislation (see Mepham, 1994). Thus, while in certain respects obligations to respect **DW** may be unaffected, in others they are infringed. Clearly, because the animal's telos is altered, respect for **DJ** is not observed; while the close confinement of such valuable animals will almost inevitably infringe respect for **DA**.

Adverse ethical impacts on animals are accentuated in this case by the existence of other means of hLF production. For example, Ward *et al.* (1992) reported production of hLF in recombinant *Aspergillus oryzae*, a filamentous fungus. According to an ethical committee set up by the Senate of the Dutch Parliament, this system, e.g. as developed by the US company, Aggennix, provides a promising alternative to animal transgenesis: clinical trials are planned for 1995 (Provisional Committee, 1993).

Producers

Valuable products like hLF will not be produced on traditional dairy farms: as noted, the producer in this case is a commercial company, Gene Pharming Europe. Were this technology to be applied to infant formula (artificial baby food), it seems likely that there would be substantial returns on investment (**PW**) since global sales of infant formula currently exceed US$7 billion p.a. and are increasing at 15% p.a. (see Mepham, 1995). Companies' freedom to adopt the technology (cf. **PA** and **PJ**) seems assured, even though restrictions may apply in some countries, such as The Netherlands.

Consumers

In this case, assuming that hLF was used for baby food manufacture, the consumers would be the infant population fed infant formula in place of breastmilk, although the interests of parents (or other carers) also need to be included. It has been argued (Mepham, 1995) that the superiority of breastmilk for human babies, in terms of nutritional, immunological, prophylactic and other properties, implies that use of infant formula in cases other than medical need severely disadvantages the welfare (**CW**) of many children. This is particularly true for most infants in less developed countries and in lower income sectors of developed countries. In that infants are deprived of breastmilk, neither autonomy (**CA**) nor justice (**CJ**) are respected.

While the inclusion of hLF may be deemed to improve the suitability of infant formulae, the superiority of breastmilk is attributable to a wide range of chemical and immunological properties which it would be impossible to simulate in any commercial product. To imply otherwise fails to respect the need for 'disclosure of information' which is a crucial aspect of customer autonomy (**CA**) – in this case, referring to the infant's carer. Moreover, the superiority of breastmilk, for both infant and mother, is partly attributable to its means of delivery – breastfeeding, which it is quite impossible to simulate. For the small number of infants who, for medical reasons, cannot be breastfed, hLF-formula might enhance **CW**, **CA** and **CJ** – but this would depend on factors such as the affordability of the infant formula, appropriately hygienic preparation of the feed, etc., failure to observe which might well offset perceived advantages (Mepham, 1995).

Biota

Technologies depending on animal production systems, with their low efficiency of energy conversion, which replace more sustainable systems (in this case breastfeeding), represent unnecessary infringements of the ethical obligation to respect the wellbeing (**BW**), diversity (**BA**) and sustainability (**BJ**) of the biota. For a fuller discussion of this point, see Mepham (1995).

Productivity Promotion

Recombinant bovine somatotrophin

Injected into dairy cows at fortnightly intervals, rbST can substantially increase milk yield. Whilst this might be seen as beneficial to producers in that yield per cow is increased (and thus also economic efficiency, by dilution of maintenance costs), several ethical concerns have been expressed, relating to animal welfare, consumer safety and socioeconomic impacts.

Animals

When Monsanto's bST product, Posilac, became available commercially in the USA in 1994, a full list of side-effects was made public for the first time. The manufacturers warn users that administration of Posilac may result in the following:

> reduced pregnancy rates; increases in cystic ovaries and disorders of the uterus; small decreases in gestation length and birth weight of calves; increased twinning rates; higher incidence of retained placenta; increased risk of clinical and subclinical mastitis; increased frequency of use of medication in cows for mastitis and other health problems; increased body temperature; increased digestive disorders such as indigestion, bloat and diarrhea; increased numbers of cows experiencing periods of 'off-feed'; increased numbers of enlarged hocks and lesions (e.g. lacerations, enlargements, calluses) of the knee; disorders of the foot region; reductions in hemoglobin and hematocrit values; injection site swellings, which may remain permanent.

The list is a remarkable admission of the range of risks of diseases (many of them painful) to which dairy cows are subjected by treatment with bST.

A more revealing, because quantified, demonstration of the reduced welfare associated with bST was provided by Cole *et al.* (1992) of the Monsanto Company. In a study conducted over two lactations, there were eight deaths among 64 bST-treated cows compared with none in a group of 21 control cows. Of the cows that died, four were mastitis cases (one cow receiving the commercial dose and three cows higher dose levels), two pneumonia cases (one receiving the commercial dose and one a higher dose), one abomasal displacement case (receiving a higher dose) and one case of Johne's disease (receiving the commercial dose).

The impact of bST on the incidence of mastitis is a matter of major concern, since this often produces clinically recognizable signs of mammary inflammation and pain, and may produce toxins which cause generalized sickness. An independent analysis of the results of a large number of studies (Willeberg, 1993) shows that use of bST leads to an increase in clinical mastitis of 15–40%. In a situation where bST was universally adopted, e.g. in Europe, 20–30% of all cases of clinical mastitis occurring after day 60 of lactation would be attributable to bST.

While it is sometimes argued that morbidity associated with bST use is a consequence of high yield rather than bST itself (an argument of dubious validity and relevance), it is clear that this cannot be applied to lesions produced by injections. In a study reported by Pell *et al.* (1992), of the Monsanto Company, cows received subcutaneous injections of bST, or injectate vehicle only (controls), at 14-day intervals. Ten days after the injection, severe swellings (>16 cm long, >2 cm high, or both, or other complications) were experienced by 10.1% of treated cows but by no control cows; moderate swellings (10–16 cm long, 1–2 cm high, or both) by 49.9% of treated cows and by 0.6% of control cows; and minimal swellings (<10 cm long, 0–1 cm high or both) by 33.5% of treated cows and by 9.7% of control cows. Almost 90% of control cows showed no swellings, indicating that it was the injectate rather than the injection procedure that caused most of the problems.

There is little doubt that respect for **DW**, **DA** and **DJ** are all infringed by use of bST.

Producers

Essentially the same types of impact as those noted for AI are to be anticipated. Some dairy farmers will benefit financially (**PW**) but, in view of increased cow morbidity and culling rates and the increased cost of veterinary care and medicines (the likely scale of which some farmers may be unaware), respect for **PA** and **PJ** may well be jeopardized. According to Persley, Biotechnology Project Manager of the World Bank; '... use of ... BST in dairy production is expected to accentuate the competitive advantage of large-scale, high-production dairy enterprises to the disadvantage of family-scale operations' (Persley, 1990).

Consumers

There is strong public resistance to bST, reflected in an overwhelming rejection of this biotechnology in responses from consumer groups to the UK Ministry of Agriculture, Fisheries and Food's public consultation exercise in July, 1994 (e.g. see Coghlan, 1994). Claims that it is impossible to identify rbST-milk and therefore label it (Bauman, 1992) are contradicted by reports such as that of Erhard *et al.* (1994), which described a monoclonal antibody method for identifying rbST in milk. Without labelling, as is currently the situation in the USA, consumer choice (**CA**) is undermined.

There are also risks to consumers (**CW**) from the elevated concentrations of IGF-1 in milk of bST-treated cows, since studies in rats demonstrate that IGF-1, at the concentrations at which it occurs in bovine milk, stimulates intestinal cell proliferation (see Mepham, 1992; Mepham *et al.*, 1994a,b). Public opinion surveys show that if bST is licensed for use in Europe, a substantial number of people will reduce or totally abandon consumption of milk and dairy products (Commission of the European Community, 1993). Because milk provides several important nutrients, most notably calcium, reduced consumption is likely to have serious public health consequences (**CW**, **CJ**), e.g in exacerbating the already rapidly increasing rates of osteoporosis (Mepham, 1992).

Biota

It has been claimed that bST use will reduce environmental pollution because the milk required will be produced by less cows than would be the case without bST use, hence reducing production of methane, manure and urine (Bauman, 1992). However, this assertion is highly questionable, certainly in relation to effects on regional biota. In the first place, pollutants such as manure and silage effluents are serious in terms of their local impacts (Archer and Nicholson, 1993), which are likely to be exacerbated by the intensification of dairying which bST use encourages, and to have negative effects on **BW** and **BA**. Secondly, since the galactopoietic effect of bST is dependent on high feed energy intake, adverse effects on the environment will be accentuated by increased dependence on fossil fuels, artificial fertilizers, farm and industrial machinery and transportation. Such consequences represent an infringement of respect for **BJ**.

Animal Health

Mastitis vaccine

This final example refers to a prospective development in veterinary medicine. In Britain, there are approximately 40 cases of clinical mastitis per 100 cows per annum, each case costing the farmer between £40 and £186. The availability of new vaccines, based on antibody responses to specific antigenic epitopes, would thus be a valuable element of mastitis control programmes. While already available in some countries, such vaccines are not yet in use in the UK. The complexity of mastitis means that the first generation of products will be specific to a single type of mastitis only, e.g. coliform infections. According to results of trials in the USA, the incidence of clinical coliform mastitis can be reduced by 80% (Anonymous, 1994). Immunity will probably persist for at least a single lactation period, requiring booster vaccinations during dry periods (Hillerton, 1994).

Animals
Mastitis is often a serious condition, producing mammary inflammation, pain and generalized sickness. Vaccine use would thus appear to have positive ethical impacts in all respects (**DW, DA, DJ**) by reducing suffering, allowing normal behavioural expression and respecting animals' telos. However, the vaccination process is not without side-effects, since in cases of intramammary presentation it can, itself, cause mastitis. The most likely method of administration will involve 2–3 subcutaneous inoculations, at 2–3 week intervals, but these may induce adverse site reactions. Moreover, since 50% of cows never suffer from mastitis, any adverse responses will have been unnecessary from those animals' perspective (**DW, DJ**) (Hillerton, 1994).

It must also be admitted that vaccination, as a largely untried system in this context, may not prove to be a sustainable approach if, for example, cows become vaccine-dependent and/or more virulent microorganisms appear.

Producers

Those adopting (**PA**) the vaccine will benefit (**PW**) from increased productivity and reduced losses. Given the Eurocentric basis of this assessment, the impacts of failure to adopt could hardly be considered unfair (**PJ**).

Consumers

In reducing the risks of zoonotic disease and of exposure to antibiotics otherwise used in mastitis treatment, respect is shown for **CW**, but impacts on **CA** and **CJ** appear negligible.

Biota

Reduced use of medicines in vaccinated animals will diminish environmental pollution and hence promote biotic conservation (**BW**) and biodiversity (**BA**), while the resulting improved energetic efficiency of animal production will enhance biotic sustainability (**BJ**).

Ethical Evaluation

The analyses summarized in Table 18.2 suggest that ethical impacts of biotechnology are context-specific. Thus, vaccination for mastitis prevention respects many of the identified ethical principles and infringes few, whereas there are numerous infringements with respect to animals, consumers and biota for the four other biotechnologies discussed. For these four technologies, profitability and productivity are often achieved at a cost to animal welfare, environmental preservation and, in some cases, human welfare.

Ethical evaluations of biotechnologies require that such analyses be subjected to two separate processes. Firstly, some form of grading is required to assess the ethical impacts with more precision. For example, in the case of bST, both the scale of the animal welfare problems caused and the alleged benefits due to reduced methane production need to be assessed. Such questions are the subject of ongoing enquiry, which in the former case (animal welfare) has received some, but not enough, attention, but in the latter case (methane production) hardly any at all.

Secondly, ethical evaluation involves screening the (graded) analyses according to 'world view'. In effect, this amounts to weighing the relative importance of the different cells of the matrix. For example, in the case of AI/MOET, the financial benefits accruing to the dairy industry might be deemed to outweigh claimed infringements to the wellbeing and behavioural freedom of cows and negative impacts on biotic conservation, sustainability and diversity.

The crux of the ethical dilemma is 'do the benefits outweigh the costs?'. However, it is stressed that posing the question in that form does not imply that a utilitarian cost/benefit analysis is all that is required: for most people, the consideration of ethical principles needs to extend beyond effects on wellbeing to impacts on freedom and justice.

But even were the dilemma to be couched in solely utilitarian terms, the sustainability of modern dairy practice as a whole would need to feature prominently in the equation; and it is by no means obvious that the current emphasis on increased production is a wise strategy, let alone an ethical one. Yet, trendsetting herds such as that at the National Agricultural Centre, Stoneleigh, UK, with a yield average of $6200 \, l \, annum^{-1}$ in 1994, are targetted on an average of $8000 \, l \, annum^{-1}$ (Hope, 1994). There are strong arguments for adopting an entirely different approach, in which dairying plays a vital element in a sustainable organic system of agriculture (Phillips and Sorensen, 1993).

The Global Dimension

The analyses performed have been limited perforce to those applied or in prospect in western societies. But ethics is not partisan: it abjures parochialism and self-interest. We need to face, however inadequately, the ethical impacts of western biotechnology on less-developed countries.

Technology transfer to less-developed countries is subject to many constraints of infrastructural, environmental, financial, social and cultural nature (Aboul-Ela and Abdel-Aziz, 1989). For example, feed resources for cattle are frequently the key determinants of yield, so that technologies aimed at genetic improvement or hormonally-induced galactopoiesis are generally ineffective (Mepham, 1991). Appropriate application of biotechnology thus requires attention to: technical and economic feasibility; compatibility with existing farming systems and infrastructures; safety in use; and social acceptability (Aboul-Ela and Abdel-Aziz, 1989). While it is notoriously difficult to generalize about the less-developed world as a whole, it must be admitted that few of the bio-technologies considered here show prospect of having a significant positive impact on sustainable food security without the implementation of fundamental political change (Mepham, 1987). Such change needs to be founded on rational ethical evaluation, and the ethical matrix described in this chapter is proposed as a modest step in promoting discussion to that end.

As stated in the opening paragraph, ethics is about how we should respond to our knowledge of 'how things are'. This includes not only knowledge of the latest scientific discovery but also of the persistent remorseless facts of poverty, hunger and deprivation experienced by many of the world's population. For some ethicists the implication is plain: we need to work for the evolution of a global morality: 'Shared morality is the stuff out of which communities are made. A global community needs a shared morality to develop both world government and world peace' (MacNiven, 1993).

Clearly, such an objective is fraught with difficulties. But those who consider it misconceived because they fear that its achievement would induce a dull uniformity or, worse, moral totalitarianism, might find reassurance in the analogy of language. Without rules, the rules of grammar, language would be meaningless: with grammar, we have a seemingly boundless richness of expression. Similarly, human society needs to work out an acceptable ethical code for the application of biotechnology if its viability is to be assured.

Acknowledgements

The author gratefully acknowledges advice received from Dr Michael Linskens, of the Dutch Society for the Protection of Animals, on recent developments in the 'Herman' project; from Dr Eric Hillerton, of the BBSRC Institute for Animal Health, on mastitis vaccines; and from Mr Colin Moore and Ms Kate Millar, both of the Centre for Applied Bioethics, for comments on the draft version of this chapter.

References

Aboul-Ela, M.B. and Abdel-Aziz, A.S. (1989) The incorporation of new techniques into cattle production in developing countries. In: Phillips, C.J.C. (ed.) *New Techniques in Cattle Production*. Butterworths, London, pp. 214–220.

Anonymous (1992) Developments in embryo transfer. *Veterinary Record* 130, 502.

Anonymous (1994) British Mastitis Conference. *Dairy Farmer* November, 74–76.

Archer, J.P. and Nicholson, R.J. (1993) Liquid wastes from farm animal enterprises. In: Phillips, C.J.C. and Piggins, D. (eds) *Farm Animals and the Environment*. CAB International, Wallingford, pp. 325–343.

Armstrong, D.T. (1993) Recent advances in superovulation in cattle. *Theriogenology* 39, 7–24.

Bauman, D.F. (1922) Bovine somatotropin: review of an emerging animal technology. *Journal of Dairy Science* 75, 3432–3451.

Beauchamp, T.L. and Childress, J.F. (1989) *Principles of Biomedical Ethics*, 3rd edn. Oxford University Press, Oxford.

Brom, F.W.A. and Schroten, E. (1993) Ethical questions around animal biotechnology. The Dutch approach. *Livestock Production Science* 36, 99–107.

Byron, W.J. (1988) On the protection and promotion of the right to food. In: Le May, B.J.W. (ed.) *Science, Ethics and Food*. Smithsonian Institution Press, Washington, pp. 14–30.

Coghlan, A. (1994) Keep milk hormone ban, say farmers. *New Scientist* Aug 13, 8.

Cole, W.J., Eppard, P.J., Boysen, B.G., Madsen, K.I., Sorbet, R.H., Miller, M.A., Hintz, R.L., White, T.C., Ribelin, W.E., Hammond, B.G., Collier, R.J. and Lanza, G.M. (1992) Response of dairy cows to high doses of sustained release bovine somatotropin administration, during two lactations. 2. Health and reproduction. *Journal of Dairy Science* 75, 111–123.

Commission of the European Communities (1993) Concerning bovine somatotrophin (B.S.T.) COM (93) 331 Final.

Coulthard, K.L. (1992) Embryo transfer legislation. *Veterinary Record* 130, 278.

Dutch Society for the Protection of Animals (1994) *Herman: the First Genetically Engineered Bull in the Netherlands*. DSPA, The Hague, The Netherlands.

Erhard, M.H., Kellner, J., Schmidhuber, S., Schams, D. and Losch, U. (1994) Identification of antigenic differences of recombinant and pituitary bovine growth hormone using monoclonal antibodies. *Journal of Immunoassay* 15, 1–19.

Hafez, E.S.E. (1993) Artificial insemination [pp. 424–439] and assisted reproductive technology: ovulation manipulation, in vitro fertilization/embryo transfer (IVF/ET) [pp. 461–502] In: Hafez, E.S.E. (ed.) *Reproduction in Farm Animals*. Lea and Febiger, Philadelphia.

Heap, R.B. and Moor, R.M. (1995) Reproductive technology in farm animals. In: Mepham, T.B., Tucker, G.A. and Wiseman, J. (eds) *Issues in Agricultural Bioethics*. University Press, Nottingham, pp. 247–268.

Hillerton, J.E. (1994) Vaccines – have they a future? *Proceeding of the British Mastitis Conference*. Genus/Institute for Animal Health/Ciba Agriculture, Stoneleigh, pp. 9–15.

Hope, H. (1994) NAC dairy unit: 'has ability to improve'. *Farmers Weekly* Sept 16, S21–S22.

Krimpenfort, P., Rademaker, A., Eyeston, W., Schans, A., Broek, S., Kooiman, P., Kootwijk, E., Platenburg, G., Pieper, F., Striker, R. and Boer, H. (1991) Generation of transgenic dairy cattle using 'in vitro' embryo production. *Biotechnology* 9, 844–847.

Lucke, J.N. (1991) Embryo transfer: putting welfare first. *Veterinary Record* 129, 474.

MacNiven, D. (1993) *Creative Morality*. Routledge, London and New York, pp. 167–186.

MAFF (1983) *Codes of Recommendations for the Welfare of Cattle*. Ministry of Agriculture, Fisheries and Food, London.

Mepham, T.B. (1987) Changing prospects and perspectives in dairy research. *Outlook on Agriculture* 16, 182–187.

Mepham, T.B. (1991) Control of milk production. In: Hunter, A.G. (ed.) *Biotechnology in Livestock in Developing Countries*. Centre for Tropical Veterinary Medicine, Edinburgh, pp. 186–200.

Mepham, T.B. (1992) Public health implications of bovine somatotrophin use in dairying: discussion paper. *Journal of the Royal Society of Medicine* 85, 736–739.

Mepham, T.B. (1993) Approaches to the ethical evaluation of animal biotechnologies. *Animal Production* 57, 353–359.

Mepham, T.B. (1994) Transgenesis in farm animals: ethical implications for public policy. *Politics and the Life Sciences* 13, 195–203.

Mepham, T.B. (1995) Bioethical issues in the marketing of infant foods. In: Mepham, T.B., Tucker, G.A. and Wiseman, J. (eds) *Issues in Agricultural Bioethics*. University Press, Nottingham, pp. 73–89.

Mepham, T.B., Schofield, P.N., Zumkeller, W. and Cotterill, A.M. (1994a) Safety of milk from cows treated with bovine somatotrophin. *Lancet* 344, 197–198.

Mepham. T.B., Schofield, P.N., Zumkeller, W. and Cotterill, A.M. (1994b) Safety of milk from cows treated with bovine somatotrophin. *Lancet* 344, 1445–1446.

Milk Marketing Board (1992) *Dairy Facts and Figures*. MMB, Thames Ditton, p. 187.

Murray, R.D. and Ward, W.R. (1993) Welfare implications of modern artificial breeding techniques for dairy cattle and sheep. *Veterinary Record* 133, 283–285.

Pell, A.N., Tsang, D.S., Howlett, B.A., Huyler, M.T., Meserole, V.K., Samuels, W.A., Hartnell, G.F. and Hintz, R.L. (1992) Effects of prolonged release formulation of sometribove (n-methionyl bovine somatotropin) on Jersey cows. *Journal of Dairy Science* 75, 3416–3431.

Persley, G.J. (1990) *Beyond Mendel's Garden: Biotechnology in the Service of World Agriculture*. CAB International, Wallingford, p. 30.

Phillips, C.J.C. and Sorensen, J.T. (1993) Sustainability in cattle production systems. *Journal of Agricultural and Environmental Ethics* 6, 61–73.

Pinkert, C.H., Dyer, T.J., Kooyman, D.L., and Kiehm, D.J. (1990) Characterization of transgenic livestock production. *Domestic Animal Endocrinology* 7, 1–18.

Postma, O. (1992) Transgenics and animal ethics. *Biotechnology* 40, 864–865.

Provisional Committee (1993) The assessment of possible alternatives in the context of the ethical evaluation of gene transfer in animals. Senate of the Netherlands 'Provisional Committee on Ethical Evaluation of Genetic Modification of Animals' (Chairman: Professor E. Schroten), The Hague.

Rawls, J. (1973) *A Theory of Justice*. Oxford University Press, Oxford.

Rawls, J. (1993) *Political Liberalism*. Columbia University Press, New York.

Robinson, J.J. and McEvoy, T.G. (1993) Biotechnology – the possibilities. *Animal Production* 57, 335–352.

Rollin, E.B. (1989) *The Unheeded Cry: Animal Consciousness, Animal Pain and Science*. Oxford University Press, Oxford.

Rolston, H. (1993) Environmental ethics: values in and duties to the natural world. In: Winkler, E.R. and Coombs, J.R. (eds) *Applied Ethics*. Blackwell Scientific Publications, Oxford, pp. 271–292.

Sreenan, J. and Diskin, M. (1992) *Breeding the Dairy Herd*. Teagasc, Dublin, pp. 76–85.

Verhoog, H. (1992) The concept of intrinsic value and transgenic animals. *Journal of Agricultural and Environmental Ethics* 5, 147–160.

Wall, R.J. and Seidel, G.F. (1992) Transgenic farm animals: a critical analysis. *Theriogenology* 38, 337–357.

Ward, P.P., Lo, J.-Y., Duke, M., May, G.S., Headon, D.R. and Connedy, O.M. (1992) Production of biologically active recombinant human lactoferrin in *Aspergillus oryzae*. *Biotechnology* 10, 784–789.

Willeberg, P. (1993) Bovine somatotropin and clinical mastitis: epidemiological assessment of the welfare risk. *Livestock Production Science* 36, 55–66.

Wilmut, I. (1995) Modification of farm animals by genetic engineering and immunomodulation. In: Mepham, T.B., Tucker, G.A. and Wiseman, J. (eds) *Issues in Agricultural Bioethics*. University Press, Nottingham, pp. 229–246.

Wilmut, I., Archibald, A.L., Harris, S., McClenaghan, M., Simons, J.P., Whitelaw, C.B.A. and Clark, A.J. (1990) Methods of gene transfer and their potential use to modify milk composition. *Theriogenology* 33, 113–123.

Wright, G., Carver, A., Cottam, D., Reeves, D., Scott, A., Simon, P., Wilmut, I., Garner, I. and Colman, A. (1991) High level expression of active human α-antitrypsin in the milk of transgenic sheep. *Biotechnology* 9, 830–834.

Index

Note: page numbers in *italics* refer to figures and tables